Lecture Notes in Control and Information Sciences 397

Editors: M. Thoma, F. Allgöwer, M. Morari

Hao Yang, Bin Jiang, and Vincent Cocquempot

Fault Tolerant Control Design for Hybrid Systems

Springer

Authors

Dr. Hao Yang

Nanjing University of
Aeronautics and Astronautics
29 YuDao Street
Nanjing, 210016
China
E-mail: younghao82@yahoo.com.cn

Prof. Vincent Cocquempot

Université des Sciences et
Technologies de Lille
59655 Villeneuve d'Ascq cedex
France
E-mail: vincent.cocquempot@univ-lille1.fr

Prof. Bin Jiang

Nanjing University of
Aeronautics and Astronautics
29 YuDao Street
Nanjing, 210016
China
E-mail: binjiang@nuaa.edu.cn

ISBN 978-3-642-10680-4 e-ISBN 978-3-642-10681-1

DOI 10.1007/978-3-642-10681-1

Lecture Notes in Control and Information Sciences ISSN 0170-8643

Library of Congress Control Number: 2009942891

Typeset & Cover Design: Scientific Publishing Services Pvt. Ltd., Chennai, India.

Printed in acid-free paper

5 4 3 2 1 0

springer.com

This work is dedicated to our parents and

Dongqing

——————— *Hao Yang*

Wen and Xinhao

——————— *Bin Jiang*

Delphine, Clément, Bastien, Nathan and Anaïs

——— *Vincent Cocquempot*

Preface

Hybrid systems (HS) are dynamical systems that involve the interaction of continuous and discrete dynamics. The study of HS is motivated by the fundamentally hybrid nature of many real life applications. Over the last decade, significant progress has taken place in modeling and simulation, verification, stability and controller synthesis for HS.

Faults in automated processes often cause undesired reactions and shut-down of a controlled plant, and the consequences could be damage to technical parts of the plant or to its environment. *Fault diagnosis* (FD) and *fault tolerant control* (FTC) are highly required for safety purpose, and aim at guaranteeing certain system performances and/or properties to be maintained in spite of faults. In the past more than 30 years, fruitful theoretical results on FD and FTC have been reported for various linear and nonlinear systems with many successful engineering applications in practical systems.

FD problem for HS has attracted some attentions. However, to the best of the authors' knowledge, until now, the FTC issue for HS has not yet been intensively studied. FTC method for HS deserves further investigations due to its academic meaning as well as practical one.

1. Motivation from academic research
   It is well known that the stability and some specifications of HS can be achieved under quite rigorous conditions. Most of existing results are devoted to off-line analysis and design, such that the HS works well as what it is expected. However, faults may abruptly change system behavior, FTC strategies must be applied on-line, not only to keep the stability but also to maintain some specifications of the HS in presence of faults. This results in a great theoretical challenge, since many classical FTC methods for non-hybrid systems can not be easily extended to HS. FTC theory for HS needs to be developed.

2. Motivation from practical applications
   Many practical systems have to be modeled by hybrid models, e.g. chemical processes, switched RLC circuits, intelligent transportation systems, etc. The safety and reliability of these systems are needed, and FTC techniques for HS are highly required.

The HS considered in this book consists of a series of continuous modes and a switching logic. Switching from one mode to another is due to a switching law generated from the switching logic. Faulty behaviors of HS are investigated systematically. Two main kinds of faults are considered: *Continuous faults* that affect continuous modes; *Discrete faults* that affect the desired switching. In these two faulty cases, the FTC design has two main objectives as follows:

1) maintain the continuous performances including various stabilities (e.g. Lyapunov stability, asymptotical stability and input-to-state stability) of the origin and the output tracking/regulation behaviors along the trajectories of HS.

2) maintain the discrete specifications that have to be followed by HS, e.g. a desired switching sequence.

For HS with various switching, e.g., time-dependent switching, state-dependent switching, impulsive switching and stochastic switching, a set of FTC methods based on continuous system theories are proposed to maintain the continuous performance. Two natural ideas are considered: One way is first to follow the general FTC idea for non-hybrid systems, i.e., design FTC law in each faulty mode such that its origin is stable (Lyapunov stable, asymptotical stable, input-to-state stable) or the output regulation problem is solvable, and second to apply the standard stability results of HS. Another way is to research directly the stability of HS without reconfiguring the controller in each unstable mode. It will be shown that FTC of HS can be achieved through the balance among different modes i.e., the negative effects resulting from unstable faulty modes are compensated for by that of stable modes. This provides us a new clue to design FTC for HS.

For HS with certain discrete specifications, i.e., it has to follow some specifications imposed on the discrete part of the system, a discrete fault would violate these specifications. As for such fault, one natural idea is to reconfigure the discrete part after faults occur to maintain the specification, which can be achieved from discrete event system (DES) point of view. Two major DES models, i.e. *finite state machine* and *Petri net* are utilized. However, compared with pure discrete event systems, continuous system behaviors must be taken into account in HS. A set of novel schemes are derived.

Some novel supervisory FTC techniques are also developed based on HS methods to improve non-hybrid (linear and nonlinear) system performances during FTC period. Unlike the multiple model FDI /FTC method or supervisory control technique, the proposed supervision schemes do not need a series of models or filters to work concurrently with the plant in order to identify the current situation, but only rely on a simple switching scheme among candidate controllers.

The materials in the monograph have explicit and broad practical backgrounds. Many examples are taken to illustrate the theoretical results, e.g. Circuit systems;

DC motors; CPU process; Manufacturing system; Intelligent transportation systems and electric automated vehicles, etc.

This book intends to provide the readers a good understanding on how to achieve FTC goal of HS. The book can be used as a reference for the academic research on FTC and HS or used in Ph.D. study of control theory and engineering. The knowledge background for this monograph would be some undergraduate and graduate courses on FD and FTC theory, linear sytem theory, nonlinear system theory, HS theory and DES theory.

There are totally seven chapters in this book. Chapter 1 introduces some background knowledge on HS and FTC design, and also describes the main work of the book. Chapters 2-4 provide new theoretical developments for the analysis and design of FTC for HS with various switching properties, which are based on continuous system theories and FTC goals aim at maintaining the continuous performance. Chapter 5 considers the HS with discrete specifications, which need to be maintained in spite of faults, FTC issue is addressed from DES point of view. In Chapter 6, some new supervisory FTC results based on HS methods are reported. A four-wheel-steering and four-wheel-driving electric vehicle in LAGIS laboratory is particularly focused on whose actuator faults are analyzed systematically and the hybrid fault tolerant tracking control approach is applied. At last, in Chapter 7, the perspectives of FTC for HS are predicated.

The authors are grateful to Prof. Marcel Staroswiecki, of laboratory SATIE from Ecole Normale Supérieur de Cachan, France; Prof. Zhenyu Yang, of Technical University of Denmark, Denmark; Prof. Peng Shi, of University of Glamorgan, UK and Prof. Gang Tao, of University of Virginia, USA, who have greatly aided this research. Finally, we would like to acknowledge the support of research grants, including National Natural Science Foundation of China (60874051), International campus on safety and intermodality in transportation (CISIT), and Graduate innovation research funding of Jiangsu Province (CX07B-112z).

Nanjing, China
September 2009

Hao Yang
Bin Jiang
Vincent Cocquempot

Contents

Symbols and Acronyms

$\Re$	the field of real numbers
$\Re^r$	the r-dimensional real vector space
$\lvert\cdot\rvert$	the Euclidean norm
$\mathscr{L}_1$	$a(t) \in \mathscr{L}_1$ if $\int_0^\infty \lvert a(t)\rvert dt < \infty$
$\mathscr{C}^k$	the set of k times continuously differentiable functions
$\lVert\cdot\rVert_{[a,b]}$	the supremum norm of a signal on the time interval $[a,b]$
Class $\mathscr{K}$	a class of strictly increasing and continuous functions $[0,\infty) \to [0,\infty)$ which are zero at zero
Class $\mathscr{K}_\infty$	the subset of $\mathscr{K}$ consisting of all those functions that are unbounded
Class $\mathscr{K}\mathscr{L}$	$\beta : [0,\infty) \times [0,\infty) \to [0,\infty)$ belongs to class $\mathscr{K}\mathscr{L}$ if $\beta(\cdot,t)$ is of class $\mathscr{K}$ for each fixed $t \geq 0$ and $\beta(s,t) \to 0$ as $t \to \infty$ for each fixed $s \geq 0$
$\lambda_{\max}(\cdot)$	the maximal eigenvalue
$\lambda_{\min}(\cdot)$	the minimal eigenvalue
$L_g h$	the Lie derivative of h along a vector field g
$[f,g]$	the Lie bracket of the vector fields f and g
$(\cdot)^{(i)}$	the ith time derivative of the function
t^-	the left limit time instant of t
$(\cdot)^\top$	the transposition
$\forall$	for all
$\exists$	there exists
$\subset$	subset of
$\in$	belongs to
$\triangleq$	define
$\cap$	intersect
$\cup$	union
Σ	sum
Π	product

FDI	fault detection and isolation
FD	fault diagnosis
FTC	fault tolerant control
HS	hybrid systems
HIS	hybrid impulsive systems
PNs	Petri nets
HPNs	hybrid Petri nets
SDP	stochastic diffusion process
FTRP	fault tolerant regulation problem
OFTRP	overall fault tolerant regulation problem
ADT	average dwell time
sADT	stochastic average dwell time
ISS	input-to-state stable
ISpS	input-to-state practically stable
SPR	strictly positive real
MLFs	multiple Lyapunov functions
CLF	control Lyapunov function
DES	discrete event systems
QA	qualitative abstraction
ITFMS	ideal tolerable fault marked sequence
ATFMS	acceptable tolerable fault marked sequence
GMEC	general mutual exclusion constraints
PA	progressive accommodation
LQ	linear quadratic
ITS	intelligent transportation systems
EV	electric vehicle
4WS4WD	four-wheel-steering and four-wheel-driving

Chapter 1
Fault Tolerant Control and Hybrid Systems

Both research areas of fault tolerant control (FTC) and hybrid systems (HS) have been developed separately for several decades, and fruitful results appeared respectively. However, until now, the FTC problem of HS has not yet attracted enough attention, and needs to be investigated due to its academic meaning as well as practical one. Many modern complex systems have to be modeled by HS and their safety and reliability are quite important. This naturally motivates to study FTC for HS, which is the topic of this book. In this chapter, we shall describe the relations between HS and FTC, and present some examples of HS as well as their fault behaviors. Based on these examples, we formulate the problems to be solved in this book.

1.1 Background

1.1.1 Hybrid Systems

HS are dynamical systems that often consist of continuous time (CT) and/or discrete time (DT) processes interfaced with some logical or decision-making (LDM) process. The continuous/ discrete time (C/DT) component may consist of differential/difference equations or continuous/discrete time state models. The LDM component might be a finite automaton or a more general discrete event system. The C/DT processes affect the state transitions of the LDM, and the LDM processes affect the dynamic motions of the C/DT processes [22, 80]. The study of HS is motivated by the fundamentally hybrid nature of many real life systems, e.g., circuit systems, flight management system, process control and intelligent transportation systems. Over the last decade, significant progress has taken place in modeling and simulation [80], verification [122], [46], stability [22], [47] and controller synthesis [110], [123] for HS.

The HS considered in this book can be illustrated using Fig.1.1, which consists of a series of continuous/discrete time modes (N maybe a finite or infinite number) and a switching logic. These modes are switched among each other according to a switching law generated from the switching logic. The framework in Fig.1.1 is general and covers several different kinds of HS that have different switching properties

H. Yang et al.: Fault Tolerant Control Design for Hybrid Systems, LNCIS 397, pp. 1–9.
springerlink.com

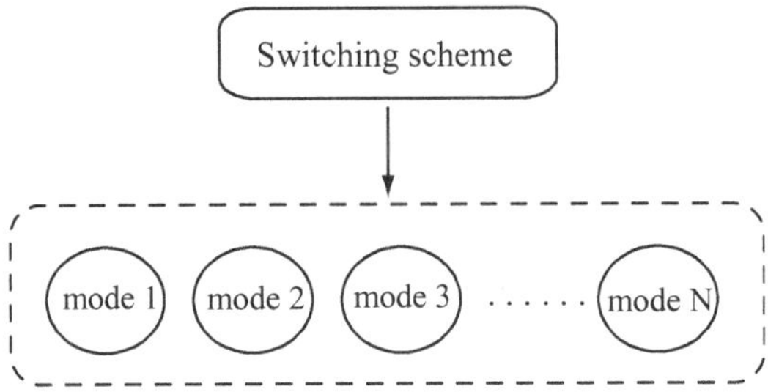

Fig. 1.1 The HS model

or performance requirements. Let us take three interesting examples which will be discussed in details in the following chapters.

Example 1.1: [138] A simplified CPU processing control system is shown in Fig.1.2. The key control problem is to deal with the trade-off between the high-speed computing and the physical constraints. The CPU needs to operate at high clock frequency (voltage) to realize high-speed computing, while a high clock frequency spends much energy and raises the CPU temperature, which often leads to hardware trouble.

The system is naturally modeled as a HS with two modes.

Mode 1 (busy mode): the amount of CPU tasks is large while CPU temperature is not too high.

Mode 2 (usual mode): the amount of CPU tasks is not large and more energy is used for decreasing the temperature.

A state dependent switching law could be designed i.e., switching occurs when the temperature or the amount of CPU reaches some given values.

Example 1.2: [140] A hose insertion task shown in Fig. 1.3 is a typical example of manipulation of deformable objects. The fingertip of the robot arm inserts a deformable hose on the plug. The motion of the hose and the fingertip are restricted in $x_1 - x_2$ plane. The completed work is to insert the hose onto the plug. Such task

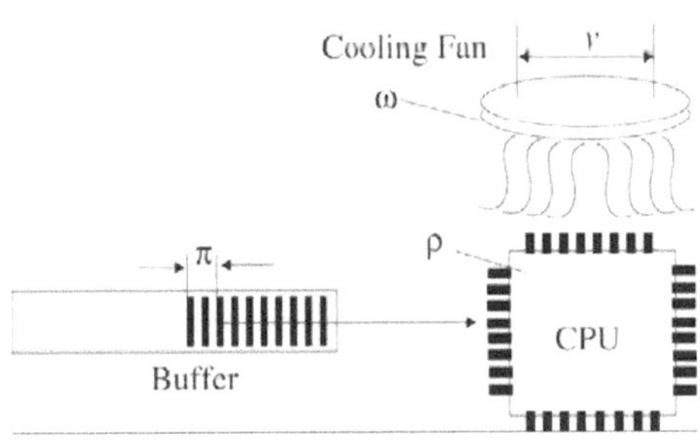

Fig. 1.2 The CPU model

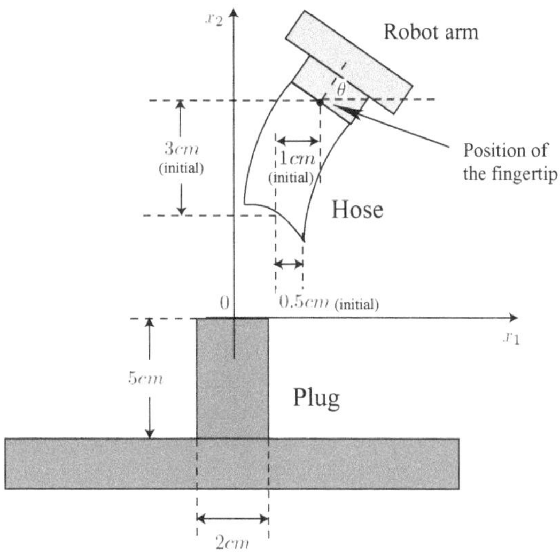

Fig. 1.3 Hose insertion task

can be modeled as a HS according to different contact configurations between the hose and the plug when the fingertip is at different positions.

Example 1.3: [139] Consider a traffic flow control problem in intelligent transportation systems at the terminator of the bridge, where six roads are interconnected with the bridge, as in Fig. 1.4. The roads r_1^{out}, r_2^{out} and r_3^{out} are the output roads to which the autonomous vehicles (AVs) go from bridge, whereas the roads r_1^{in}, r_2^{in} and r_3^{in} are the input roads from which AVs go to the bridge. There is a supervisor consisting of a series of internal logic lights (similar to traffic lights for man-driven cars) for input roads, such that the traffic flows from each input roads get into the bridge with the prescribed sequence. The overall system is also a hybrid system involves the interaction of continuous (AVs flows) and discrete dynamics (traffic lights).

It can be seen from the above examples that the structure of HS is very special and complex, the analysis of fundamental properties of HS is also difficult and quite different from that of the non-hybrid systems. This is because both continuous and discrete dynamics and their relations have to be fully taken in account. The general models of HS include hybrid automata [80], hybrid inclusions [42], and switched systems [48]. These modes capture both continuous and discrete dynamics of HS, under which some properties of HS can be analyzed systematically.

The stability and performance of HS are related to many factors including the initial states, the decreasing rate of Lyapunov function of each continuous mode, the frequency of switching, the switching sequence, etc. Two basic stability methods can be applied: multiple Lyapunov functions (MLFs) technique [22, 13, 156] and dwell-time scheme [72, 125, 129]. MLFs method claims that the stability of HS can be achieved if the value of each mode's Lyapunov function 1) does not increase

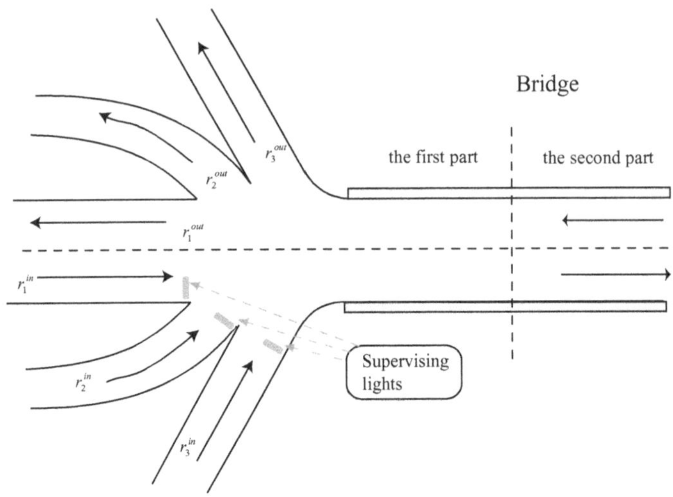

Fig. 1.4 One terminator of the bridge

when the mode works and 2) is non-increasing over the consecutive time sequence when the corresponding mode is just switched on. The dwell-time scheme introduces a minimum time interval called "dwell time", and claims that the HS is stable if the interval between any two consecutive switching instants is not smaller than the "dwell time". The above two methods have been extended to HS with various switching properties and stability requirements.

The control design of HS consists of continuous controller design in each continuous mode and switching scheme design (e.g., switching instants design, switching sequence design, this is also called "discrete controller design" [122]). The former one is similar to that for non-hybrid systems. Each continuous mode can often be stabilized by its corresponding continuous controller. However, the individual stability of each mode is not enough to make HS stable. The latter one is special and more required for HS, which plays an important role to stabilize the HS globally.

It will be shown throughout the book that the above two stability methods are the basis of FTC design for HS, and both controller design clues will be followed. The reader is refer to several excellent survey papers [22, 42] for the analysis of other interesting properties of HS which is not closely related to the topic of the book and thus is omitted here.

1.1.2 *Fault Tolerant Control*

Faults in automated processes will often cause undesired reactions and shut-down of a controlled plant, and the consequences could be damages to technical parts of the plant or to its environment, so fault diagnosis (FD) and FTC are highly required for modern complex control systems. FD is concerned with detecting, isolating and estimating the faults [18, 23, 36, 95], while Fault Tolerant Control (FTC) aims at guaranteeing the system goal to be achieved in spite of faults [10, 54, 154].

In the past 30 years, fruitful results have been obtained in the area of FTC. Generally speaking, FTC can be categorized into two main classes: passive and active. Passive FTC is designed with the consideration of a set of presumed failure modes. The resulting control system performance tends to be conservative. It also has the limitation to deal with unanticipated faults. In contrast, Active FTC reacts to the occurrence of system faults on-line in real-time in an attempt to maintain the overall system stability and performance. Two main potential advantages of Active FTC are 1) the ability to deal with previously unknown faults with explicit FD and controller reconfiguration, and 2) the possibility to achieve the optimal performance. The reader are referred to [154] for more detailed development and bibliography.

1.2 FTC Problems of HS

Although the FD problem for HS has been addressed in some literatures recently using Petri net technique in [155], bond graphs method in [90], observer techniques in [126], and parity space method in [20], etc., until now, few results have been reported about FTC for HS.

It is well known that the stability of HS is achieved under quite rigorous conditions as stated previously. Most of existing results are devoted to off-line analysis and design, such that the HS works well as what is expected. However, faults may abruptly change system behavior, FTC strategy must be applied on-line to keep the system performance including stability of the HS in spite of faults. This prevents many classic FTC methods for non-hybrid systems from being applied to HS.

Two main kinds of faults have been defined for HS [20] with respect to the process (C/DT or LDM) that is affected by : One is a *continuous fault* that affects each continuous system mode, which corrupts the continuous state behavior of the related mode. Recall example 1.1, if there exists a fault in voltage input channel or clock frequency input channel, the system behavior may become unexpected in busy mode or usual mode. Another one is a *discrete fault* that affects the switching sequence. In example 1.2, if there exists an abrupt change of the fingertip's position due to physical faults of the robot arm, the prescribed motion sequence may be changed. In example 1.3, the discrete faults represent the unexpected behaviors of traffic lights, whereas the continuous faults describe the abnormal situations of AVs flows.

Now we define a general HS model with faults.

Definition 1.1. *A* hybrid automaton with fault *is a collection*

$$\mathscr{H} = (Q, X, U, V, \mathscr{F}, Y, F, Init, Inv, E, G, R) \tag{1.1}$$

where

- $Q = \{1, 2, \ldots, N\}$ *is the finite set of discrete states;*
- X *is the set of continuous states;*
- U *defines the set of continuous inputs;*
- V *defines the set of discrete inputs;*

- $F = F_c \cup F_d$ *denotes the set of faults, with* F_c *and* F_d *respectively, continuous and discrete.*
- $\mathscr{F}: Q \times X \times U \times F_c \to X$ *represents the set of vector fields for each mode;*
- Y *is the set of continuous outputs;*
- $Init \subseteq Q \times X$ *is the set of initial states;*
- $Inv: Q \to 2^X$ *assigns to each mode an invariant set;*
- $E: V \times F_d \to Q \times Q$ *is the set of discrete transitions between modes;*
- $G: E \times F_d \to 2^X$ *defines a guard set related to each* $(i,i') \in E$, *where the system can be switched from mode* i *to* i'.
- $R: Q \times Q \times X \to X$ *is the set of reset maps.*

The above model is an extension of usual hybrid automaton as in e.g., [80] and [46] to the faulty cases. This model is also more general than that in [155] where only the parameter faulty cases are considered.

It can be seen from model (1.1) that

- Continuous faults F_c corrupt the equality constraints of the related mode. Such kind of faults are similar to that considered in non-hybrid systems.
- discrete faults F_d affect the mode transitions by changing discrete transition set E or the guard sets G. Both the switching instants and switching sequences may be changed unexpectedly. Such faults are special for HS.

The FTC objective for HS is concerned with the system requirement, i.e., to guarantee the system goal to be achieved in spite of continuous and discrete faults. In this book, two main system requirements are considered:

- Continuous performance goal, e.g., the origin of the HS is stable (Lyapunov stable, asymptotical stable, input-to-state stable) and the output regulation/tracking problem is solvable.
- Discrete specification goal, i.e., the HS has to satisfy some constraints on discrete modes, e.g., the switching sequence.

To investigate continuous performance goal, a class of HS (1.1) named switched systems are considered which take the form

$$\begin{aligned} \dot{x} &= g_\sigma(x, u_\sigma, f_\sigma) \\ y &= h_\sigma(x) \end{aligned} \tag{1.2}$$

where $x \in X$, $u_\sigma \in U$, $y \in Y$, $f_\sigma \in F_c$. $\sigma(t) : [t_0, \infty) \to Q$ denotes the *switching function*, which is assumed to be a piecewise constant function continuous from the right. The *dwell period* of a mode represents the time period during which this mode is activated. The switched system model (1.2) emphasizes the vector fields $\mathscr{F}$ in (1.1), and captures the behavior of continuous dynamics using ordinary differential equations. The affect of the switching on each continuous mode is also clearly represented. Such model allows us to analyze FTC problems using continuous system theories, and to extend the existing FTC techniques of non-hybrid systems to the hybrid cases.

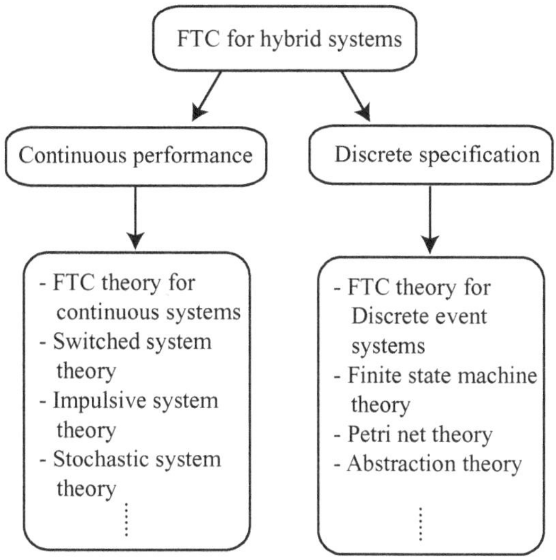

Fig. 1.5 The FTC clue for HS

Four kinds of switchings are considered:

- *Time-dependent switching*. Such switching occurs at a certain time instant. These switching instants can be prescribed *a priori* and fixed, or designed arbitrarily by engineers. The continuous states x are continuous at switching instants.
- *State-dependent switching*. Such switching occurs whenever the states reach some given surfaces or satisfy an inequality. x are also continuous at switching instants.
- *Impulsive switching*. Under such switching, x abruptly change due to the impulse effect at each switching instant.
- *Stochastic switching*. Such switching is governed by some random processes, i.e. Markov process.

The above various switchings are related with the guard set G, the discrete transitions set E and the reset maps set R in (1.1), which determines switching properties of system (1.2). As for above different HS, the continuous performance can be investigated using various continuous system theories as shown in Fig 1.5. Some existing FTC results for non-hybrid systems could be potentially applied and combined with the stability conditions of HS. The main idea is to design the FTC law in each faulty mode and develop an appropriate switching scheme such that the continuous performance goal is maintained.

As for the discrete specification goal, one natural idea is to reconfigure the discrete part of the HS after faults occur to maintain such specification. The continuous system theories are limited in this case. However, the discrete-event system (DES) supervisory control theories can be applied as also indicated in Fig. 1.5.

A well known DES model named finite state machines will be utilized to abstract the discrete part of (1.1) as

$$(Q, E, T_d, Q_{d0}, Q_{dm})$$

where T_d denotes the activated discrete transition, $Q_{d0} = \bigcup_{\forall(x,q)\in Init} q$. $Q_{dm} \subseteq Q$ is the set of marked states. Such mode captures the behavior of discrete dynamics. The affect of the switching sequences is particularly emphasized. DES supervisory control theory [101] can be developed to reconfigure the switching sequence after faults occur, which, together with some criteria imposed on continuous dynamics of HS, achieves the discrete specification goal.

Another important HS model named Hybrid Petri net (HPN) originating from the DES model Petri net (PN) are also considered. HPN inherits all the advantages of the PN and effectively captures behaviors including concurrency, synchronization and conflicts, which often appear in complex systems, e.g., the traffic flow control problem in example 1.3. A HPN structure is the 5-tuple

$$(P, T, Pre, Post, h)$$

where P is a set of places, T is a set of transitions; The set of places P (resp. transitions T) is split into two subsets: discrete places (resp. discrete transitions) and continuous places (resp. continuous transitions). *Pre* and *Post* assign the weights between transitions and places. More detailed formulations will be given in Section 5.2. HPN is closely connected with hybrid automaton (1.1), a hybrid automaton can be constructed associated with a given HPN as reported in [109]. Different control schemes can be designed for continuous part and discrete part of HPNs respectively such that the desired discrete specifications are maintained.

One of the motivations of HS research arises from the hybrid control problem. HS may present different control configurations. Commutation from one configuration to another one is described using discrete event system model as claimed in [117]. Thus the controlled system becomes hybrid due to the switching control. Some novel supervisory FTC techniques are also developed based on HS methods to improve non-hybrid (linear and nonlinear) performance during FTC period. The hybrid automaton model (1.1) can be applied after a minor modification, where each mode denotes respectively faulty or healthy situations of the system. All the switching among modes can be controlled by the user. The discrete fault disappears.

1.3 The Structure of the Book

The rest of this book is organized as follows: Chapters 2-3 provide new theoretical developments of FTC analysis and design for HS with time-dependent and state-dependent switchings respectively. Chapter 4 discusses the HS with impulsive and stochastic switchings based on some results in Chapter 2. These new approaches are based on continuous system theories and FTC goals aim at maintaining the continuous performance. The switched system model (1.2) is utilized in Chapters 2, 4

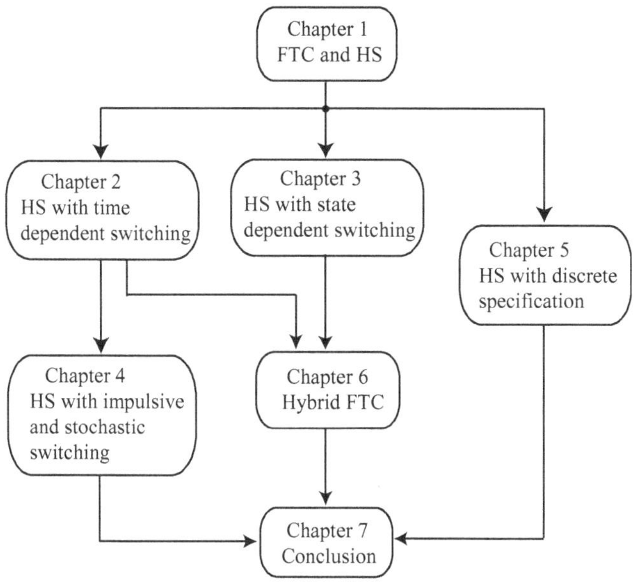

Fig. 1.6 The chapter relations

and Section 3.3. Chapter 5 considers the HS with discrete specifications, FTC issue is addressed from DES point of view and the discrete specification goal is emphasized. HPNs model is applied in Section 5.2. The Hybrid automaton model (1.1) is considered in Sections 3.2 and 5.1. As an important related issue of HS, supervisory control problems are addressed in Chapter 6, some new supervisory FTC results are reported based on HS approaches developed in Chapters 2-3. A four-wheel-steering and four-wheel-driving electric vehicle in LAGIS laboratory is particularly focused on whose actuator faults are analyzed systematically and the hybrid fault tolerant tracking control approach is applied. In the final Chapter, several future research directions are predicated related to FTC of HS.

Fig.1.6 shows the relations among chapters. One can follow the arrowhead sequence to read the book. the reader who is interested in continuous system FTC theories can read Chapters 2, 3 and 4. The reader who focuses on supervisory control can read Chapters 2, 3 and 6. Chapter 5 is independent from Chapters 2-4 and 6, the reader who cares about DES only can read Chapter 5 directly.

Chapter 2
Hybrid Systems with Time-Dependent Switching

This chapter considers a broad class of HS whose switchings are activated according to time functions, i.e., a switching occurs at a certain time instant. These switching instants can be prescribed *a priori* and fixed, or designed arbitrarily by engineers. The motivation of researching HS appears from many practical systems e.g., circuit system, and also the switching control ideas. In this chapter, several FTC methods are presented for such HS. Two natural ideas follow: One way is to design FTC law in each faulty mode such that it is stable (Lyapunov stable, asymptotical stable or input-to-state stable) or the output regulation problem of each mode is solvable, then apply the standard stability results on HS (see sections 2.1-2.3). Another way is to research directly the stability of HS without reconfiguring the controller in each unstable mode (see sections 2.4-2.5). These two ideas will be developed in this chapter. The switching control techniques as developed in Chapter 6 also have their roots in this chapter.

2.1 Output-Input Stability Technique

In this section, we apply the output-input stability concept proposed in [70, 71] to the FTC design of HS with continuous faults.

The concept of *output−input stability* (OIS) [70, 71] is a robust variant of the minimum-phase property for general smooth nonlinear control systems. Its definition requires the state and the input of the system to be bounded by a suitable function of the output and derivatives of the output. Our objective is to provide a fault tolerant strategy for a class of hybrid nonlinear systems, in which each mode is output−input stable in the healthy situation and without full state measurements. The main ideas are that:

1 An observer-based FTC method is proposed for each output−input stable mode to make each mode asymptotically stable whenever faults occur during its dwell period;
2 A set of switching laws based on this FTC method are designed to guarantee the asymptotic stability of the overall HS.

H. Yang et al.: Fault Tolerant Control Design for Hybrid Systems, LNCIS 397, pp. 11–58.
springerlink.com

To make this section more readable, we first discuss the FTC for nonlinear systems in the following two subsections 2.1.1 and 2.1.2, then extend the obtained results to hybrid case in subsection 2.1.3.

2.1.1 State Feedback Control for Nonlinear System

Consider the following affine nonlinear system with faults

$$\begin{aligned} \dot{x} &= f(x) + G(x)u + E(x)f_a \\ y &= h(x) \end{aligned} \tag{2.1}$$

where $x \in \Re^n$ is the non measured state, $u \in \Re^m$ is the input, $y \in \Re^p$ is the output, and only the case $m \leq p$ is considered. Functions $f(\cdot)$, $G(\cdot)$, $E(\cdot)$ and $h(\cdot)$ are smooth, and it is assumed that $u \in \mathscr{C}^k$, the set of k times continuously differentiable functions $u : [0;\infty) \to \Re^m$, with $k \geq 1$. For all $u \in \mathscr{C}^k$, derivatives $\dot{y}, \ddot{y}, \ldots, y^{(k+1)}$ are assumed to exist and to be continuous.

The fault effect is modelled by a "fault pattern", described by the distribution matrix $E(x)$ and a "fault parameter" $f_a \in \Re^d$, which can be time varying, and is supposed to be norm bounded, i.e., $\exists f_1 : |f_a| < f_1$. The fault pattern describes the family of faults that are investigated [152], as identified e.g. through standard methods like failure modes and effect analysis (FMEA) [10]. The fault parameter describes the size of the fault, and its time evolution. It is assumed that the distribution matrix $E(x)$ satisfies the so-called matching condition

$$E(x) = G(x) \cdot W(x) \tag{2.2}$$

i.e. it can be factorized as (2.2) for some $m \times d$ continuous matrix $W(x)$. The interpretation of the matching condition is that the effect of faults can be described by a deviation of the control signal. This model covers actuator faults and a large number of system faults.

Definition 2.1. *[70] System (2.1) with $f_a = \mathbf{0}$ is called* output-input stable *if there exist a positive integer N, a function β of class $\mathscr{KL}$, and a function γ of class $\mathscr{K}_\infty$ such that for every initial state $x(0)$ and every input $u \in \mathscr{C}^{N-1}$ its solution $x(t)$ satisfies*

$$\left| \begin{pmatrix} x(t) \\ u(t) \end{pmatrix} \right| \leq \beta(|x(0)|, t) + \gamma\left(\left\| \underline{y}_N \right\|_{[0,t]} \right) \tag{2.3}$$

for all t, where $\underline{y}_k \triangleq (y^\top, \dot{y}^\top, \ldots, y^{(k)\top})^\top$.

Note that (2.3) implies

$$|x(t)| \leq \beta(|x(0)|, t) + \gamma\left(\left\| \underline{y}_N \right\|_{[0,t]} \right) \tag{2.4}$$

According to [70], the system is said to be *weakly uniformly 0-detectable of order* N if inequality (2.4) holds, or just *weakly uniformly 0-detectable* when an order is not specified.

The weak uniform 0-detectability is independent on any input, which implies that even when the faulty system is not output-input stable any more, it is still weakly uniformly 0-detectable if faults satisfy the matching condition (2.2). This property is very useful for FTC.

The following structure algorithm will be helpful to construct the feedback controller later. Due to the structure of the fault distribution matrix (2.2), the term $G(x)u + E(x)f_a$ is written as $G(x)\bar{u}$ where $\bar{u} = u + W(x)f_a$.

Algorithm 2.1. *nonlinear structure algorithm*

Step 1: *Define* $\tilde{h}_1(x) \triangleq L_f h(x)$, $\tilde{J}_1(x) \triangleq L_G h(x)$. *Differentiating* y *with respect to time along the trajectories of (2.1) gives*

$$\dot{y} = \tilde{h}_1(x) + \tilde{J}_1(x)\bar{u} \tag{2.5}$$

Assume that matrix $\tilde{J}_1(x)$ *has constant rank* r_1 *and a fixed set of* r_1 *rows that are linearly independent for all x, these rows are taken as the first* r_1 *rows of* $\tilde{J}_1(x)$.

Denote $\check{h}_1(x)$ *and* $\hat{h}_1(x)$ *as respectively the first* r_1 *and the last* $p - r_1$ *components of* $\tilde{h}_1(x)$, *then Eq.(2.5) is divided into two parts as*

$$\dot{y}_{1\ldots r_1} = \check{h}_1(x) + J_1(x)\bar{u}$$

and

$$\dot{y}_{r_1+1\ldots p} = \hat{h}_1(x) + \hat{J}_1(x)\bar{u} \tag{2.6}$$

where $(\cdot)_{1\ldots k}$ *denotes the first k elements of the signal.* $J_1(x)$ *is a matrix of full row rank, and* $\hat{J}_1(x) = f_1(x)J_1(x)$ *for some* $(p - r_1) \times r_1$ *matrix* $f_1(x)$.

Define $\bar{h}_1(x, \dot{y}_{1\ldots r_1}) \triangleq \hat{h}_1(x) + f_1(x)(\dot{y}_{1\ldots r_1} - \check{h}_1(x))$. *Eq.(2.6) can be rewritten as*

$$\dot{y}_{r_1+1\ldots p} = \bar{h}_1(x, \dot{y}_{1\ldots r_1}) \tag{2.7}$$

Step 2: *Similar to Step 1, define*

$$\tilde{h}_2(x, \dot{y}_{1\ldots r_1}, \ddot{y}_{1\ldots r_1}) \triangleq L_f \bar{h}_1(x) + \sum_{i=1}^{r_1} \frac{\partial \bar{h}_1}{\partial \dot{y}_i}(x, \dot{y}_{1\ldots r_1})\ddot{y}_i$$

$$\tilde{J}_2(x, \dot{y}_{1\ldots r_1}) \triangleq L_G \bar{h}_1(x)$$

Differentiating (2.7) leads to

$$\ddot{y}_{r_1+1\ldots p} = \tilde{h}_2(x, \dot{y}_{1\ldots r_1}, \ddot{y}_{1\ldots r_1}) + \tilde{J}_2(x, \dot{y}_{1\ldots r_1})\bar{u} \tag{2.8}$$

The termination condition of the structure algorithm at Step 2, denoted as C 1, is as follows:

C 1: The matrix $\begin{bmatrix} J_1(x) \\ \tilde{J}_2(x,\dot{y}_{1\ldots r_1}) \end{bmatrix}$ *is continuous and has constant rank m and there is a fixed set of* $m-r_1$ *rows of* $\tilde{J}_2(x,\dot{y}_{1\ldots r_1})$ *which together with the rows of* $J_1(x)$ *form a linearly independent set for all x and* $\dot{y}_{1\ldots r_1}$*. These rows are taken as the first* $m-r_1$ *rows of* $\tilde{J}_2(x,\dot{y}_{1\ldots r_1})$*.*

Denote $\check{h}_2(x)$ and $\hat{h}_2(x)$ as respectively the first $m-r_1$ and the last $p-m$ components of $\tilde{h}_2(x)$. Under C 1, since $m \leq p$, Eq.(2.8) can be written similarly to Step 1 as

$$\ddot{y}_{r_1+1\ldots m} = \check{h}_2(x,\dot{y}_{1\ldots r_1},\ddot{y}_{1\ldots r_1}) + J_2(x,\dot{y}_{1\ldots r_1})\bar{u}$$

and

$$\ddot{y}_{m+1\ldots p} = \hat{h}_2(x,\dot{y}_{1\ldots r_1},\ddot{y}_{1\ldots r_1}) + \hat{J}_2(x,\dot{y}_{1\ldots r_1})\bar{u} \tag{2.9}$$

The following Lemma is a special case of Theorem 1 in [71], therefore its proof is omitted. It gives a necessary and sufficient OIS condition.

Lemma 2.1. *Under the termination condition C 1, the system (2.1) with* $f_a = \mathbf{0}$ *is output-input stable if and only if it is weakly uniformly 0-detectable.*

Based on Algorithm 2.1, a state feedback controller is now designed for the healthy system, $m = p$ is considered, the extension to $m \leq p$ is straightforward. Two assumptions are imposed.

Assumption 2.1. *The vector* $\dot{y}_{r_1+1,\ldots,m}$ *is not affected directly by input signals, which results, for an output-input stable system (2.1) with* $f_a = \mathbf{0}$*, in the fact that* $f_1(x) = \mathbf{0}$*.*

Assumption 2.2. *Let* $\chi \in \Re^{2m-r_1} \triangleq (y_{1\ldots r_1}^{\top}, y_{r_1+1\ldots m}^{\top}, \dot{y}_{r_1+1\ldots m}^{\top})^{\top}$*. When* $f_a = \mathbf{0}$*, there exists an invertible map* $T : \Re^n \to \Re^{2m-r_1}$*, such that* $\chi = T(x)$*.*

Since $m = p$, Eq.(2.9) is removed. Under C 1 and assumptions 2.1-2.2, the algorithm 2.1 leads to

$$\begin{bmatrix} \dot{y}_{1\ldots r_1} \\ \ddot{y}_{r_1+1\ldots m} \end{bmatrix} = \begin{bmatrix} \check{h}_1(x) \\ \check{h}_2(x) \end{bmatrix} + \begin{bmatrix} J_1(x) \\ J_2(x) \end{bmatrix} \bar{u} \tag{2.10}$$

where $\check{h}_2 = \tilde{h}_2$, $J_2 = \tilde{J}_2$.

The state feedback control design consists of the following three steps:

Step 1: Choose a Hurwitz matrix A_{10}, which gives $y_{1\ldots r_1} = A_{10}y_{1\ldots r_1}$ provided that $J_1(x)\bar{u} = \vartheta_1(x)$ with

$$\vartheta_1(x) \triangleq A_{10}y_{1\ldots r_1} - \check{h}_1(x)$$

Step 2: Choose two $(m-r_1)\times(m-r_1)$ matrices A_{21} and A_{20} such that

$$\ddot{y}_{r_1+1\ldots m} = A_{21}\dot{y}_{r_1+1\ldots m} + A_{20}y_{r_1+1\ldots m}$$

The matrix $\begin{bmatrix} \mathbf{0} & I_{(m-r_1)\times(m-r_1)} \\ A_{20} & A_{21} \end{bmatrix}$ is Hurwitz provided that $J_2(x)\bar{u} = \vartheta_2(x)$ and

$$\vartheta_2(x) \triangleq A_{21}\dot{y}_{r_1+1\ldots m} + A_{20}y_{r_1+1\ldots m} - \check{h}_2(x)$$

Step 3: Design the state feedback controller $u_n(x)$ as

$$u_n(x) \triangleq \begin{bmatrix} J_1(x) \\ J_2(x) \end{bmatrix}^{-1} \begin{bmatrix} \vartheta_1(x) \\ \vartheta_2(x) \end{bmatrix} \tag{2.11}$$

Define

$$h_\chi(x) \triangleq \begin{bmatrix} \check{h}_1(x) \\ \mathbf{0} \\ \check{h}_2(x) \end{bmatrix}, \quad J_\chi(x) \triangleq \begin{bmatrix} J_1(x) \\ \mathbf{0} \\ J_2(x) \end{bmatrix}$$

$$\bar{A} \triangleq \begin{bmatrix} A_{10} & \mathbf{0} & \mathbf{0} \\ \mathbf{0} & \mathbf{0} & I_{(m-r_1)\times(m-r_1)} \\ \mathbf{0} & A_{20} & A_{21} \end{bmatrix}$$

Then under the control $u_n(x)$, the system (2.10) is augmented as

$$\dot{\chi} = h_\chi(x) + J_\chi(x)u_n = \bar{A}\chi \tag{2.12}$$

Therefore, $u_n(x)$ in (2.11) asymptotically stabilizes system (2.12) if A_{10}, A_{20}, and A_{21} are chosen such that $\bar{A}$ is Hurwitz. An "optimized" choice of $\bar{A}$ can be refered to [61]. The weak uniform 0-detectability implies that the closed-loop system is stabilized.

2.1.2 Observer-Based FTC for Nonlinear System

Now we provide an observer-based method to stabilize system (2.1) under both healthy and faulty conditions.

The FD scheme in [56] is first applied to provide rapid and accurate estimation of states and faults. Denote $\hat{x}$ and $\hat{f}_a$ as the estimates states and faults respectively. Using the differential geometry theory, we can obtain (see [56] for details) a global diffeomorphism $z = N(x)$ with $N(0) = 0$ and $z \in \Re^n$ that satisfies

$$|\tilde{z}| \le \mu(\lambda^*)|\tilde{z}(0)|\exp(-\lambda^* t) \tag{2.13}$$

where $\tilde{z} \triangleq z - \hat{z}$, $\lambda^* > 0$, $\mu(\lambda^*) > 0$ is polynomial in λ^*. We can also get from [56] that $f_a(t) - \hat{f}_a(t) \to 0$ when $z(t) - \hat{z}(t) = 0$. This means that rapid and accurate fault estimates can always be obtained when faults occur.

The following two lemmas provide the control strategy for the healthy case and faulty case respectively.

Lemma 2.2. *Suppose that the output-input stable system (2.1) with $f_a = \mathbf{0}$ and $m = p$ satisfies C 1 and assumptions 2.1-2.2. Given an initial $x(0)$, there exists a constant $\varepsilon_1 > 0$ such that if $|\bar{\tilde{z}}(0)| \le \varepsilon_1$, then the control $u(\hat{x}) = u_n(\hat{x})$ makes the origin of the closed-loop system asymptotically stable.*

Proof: In the healthy case, system (2.12) controlled by $u_n(\hat{x})$ is rewritten as

$$\dot{\chi} = \bar{A}\chi + J_\chi(x)(u(\hat{x}) - u(x)) \tag{2.14}$$

Let P be the symmetric positive definite solution of the Lyapunov equation $\bar{A}^\top P + P\bar{A} = -Q$ with a given matrix $Q > 0$. Consider the Lyapunov function $V = \chi^\top P\chi$, its time derivative with respect to (2.14) is

$$\begin{aligned} \dot{V} &= -\chi^\top Q\chi + 2\chi^\top PJ_\chi(x)(u(\hat{x}) - u(x)) \\ &\leq -\lambda_{\min}(Q)|\chi|^2 + 2|\chi| \cdot |P| \cdot |J_\chi(x)| \cdot |u(\hat{x}) - u(x)| \end{aligned} \tag{2.15}$$

Consider the given initial $x(0)$, and define $\Omega \triangleq \{\chi : V(\chi) \leq \chi(0)^\top P\chi(0)\}$, which are the level sets of V with respect to χ (see Chapter 4 in [62]).

Note that $|u(\hat{x}) - u(x)|$ is continuous within the region Ω, and vanishes when $\hat{x} - x = \mathbf{0}$, i.e., $\tilde{z} = \mathbf{0}$. There exists two constants $\bar{\varepsilon}_1 > 0$ and $\kappa_1 > 0$, such that $|\bar{\tilde{z}}_2| \leq \bar{\varepsilon}_1 \Longrightarrow |u(\hat{x}) - u(x)| \leq \kappa_1|\tilde{z}|$. From inequality (2.15) it follows

$$\begin{aligned} \dot{V} &\leq -\lambda_{\min}(Q)|\chi|^2 + 2\kappa_1|\chi| \cdot |P| \cdot |\tilde{z}|\sqrt{(\lambda_{\max}(J_\chi^\top(x)J_\chi(x)))_{(\chi\in\Omega)}} \\ &\leq -(1-r)\lambda_{\min}(Q)|\chi|^2 \end{aligned} \tag{2.16}$$

$$\forall|\chi| \geq \sqrt{\frac{2\kappa_1|P| \cdot |\bar{\tilde{z}}_2|\sqrt{(\lambda_{\max}(J_\chi^\top(x)J_\chi(x)))_{(\chi\in\Omega)}}}{r\lambda_{\min}(Q)}} \triangleq \bar{\gamma}(|\bar{\tilde{z}}_2|), 0 < r \leq 1 \tag{2.17}$$

where $\bar{\gamma}(\cdot)$ is a class $\mathscr{K}$ function. There exists a constant $\bar{\varepsilon}_2$ such that $|\tilde{z}| \leq \bar{\varepsilon}_2$ satisfies (2.17). Based on [62], the choice of $|\tilde{z}(0)| \leq \varepsilon_1$ where $\varepsilon_1 = \min(\bar{\varepsilon}_1, \bar{\varepsilon}_2)$, clearly results in χ being input-to-state stable with respect to $\tilde{z}$. Note that $\lim_{t\to\infty}\tilde{z}(t) = 0$. Hence the origin of the system (2.14) is asymptotically stable. On the other hand, the map $T(x)$ is invertible and not affected by the observer, and system (2.1) is weakly uniformly 0-detectable, which leads to the asymptotic stability of the origin of the system. □

Lemma 2.3. *Consider the output-input stable system (2.1) with $f_a = \mathbf{0}$ and $m = p$ satisfying C 1 and assumptions 2.1-2.2. Let a fault occur at $t = 0$. Given an initial $x(0)$, there exists a constant $\varepsilon_2 > 0$ such that for all $|\bar{\tilde{z}}_2(0)| \leq \varepsilon_2$, the control $u(\hat{x}) = u_n(\hat{x}) - W(\hat{x})\hat{f}_a$ makes the origin of the closed-loop faulty system (2.1) asymptotically stable.*

Proof: In the faulty case, the system (2.10) controlled by $u_n(\hat{x}) - W(\hat{x})\hat{f}_a$ is rewritten as

$$\dot{\chi} = \bar{A}\chi + J_\chi(x)\Big(u_n(\hat{x}) - u_n(x)\Big) + J_\chi(x)W(\hat{x})(f_a - \hat{f}_a) + J_\chi(x)\Big(W(x) - W(\hat{x})\Big)f_a \tag{2.18}$$

The time derivative of V along (2.18) is

$$\begin{aligned} \dot{V} = -\chi^\top Q\chi + 2\chi^\top PJ_\chi(x)\Big[\Big(u_n(\hat{x}) - u_n(x)\Big) \\ +W(x)(f_a - \hat{f}_a) + \Big(W(x) - W(\hat{x})\Big)f_a\Big] \end{aligned} \tag{2.19}$$

There exist two constants $\bar{\varepsilon}_3 > 0$ and $\kappa_2 > 0$, such that $|\tilde{z}| \leq \bar{\varepsilon}_3 \Longrightarrow |N(t) - \hat{N}(t)| \leq \kappa_2 |\tilde{z}|$ within Ω. Similarly, there exist two constants $\bar{\varepsilon}_4 > 0$ and $\kappa_3 > 0$, such that $|\tilde{z}| \leq \bar{\varepsilon}_4 \Longrightarrow |W(x) - W(\hat{x})| \leq \kappa_3 |\tilde{\bar{z}}_2|$. Following (2.19), appropriate selection of $\hat{\bar{z}}_2$ leads to

$$\dot{V} \leq -\lambda_{\min}(Q)|\chi|^2 + \Xi \tag{2.20}$$

$$\begin{aligned} \Xi \triangleq\; & 2|\chi| \cdot |P| \cdot |\tilde{z}| \cdot \sqrt{(\lambda_{\max}(J_\chi^\top J_\chi))_{(\chi \in \Omega)}} \\ & \cdot \left[\kappa_1 + \kappa_2 \sqrt{((\lambda_{\max}(\eta^\top \eta) \cdot \lambda_{\max}(W^\top W))_{(\chi \in \Omega)}} + \kappa_3 f_1\right] \end{aligned} \tag{2.21}$$

where η is defined as in [56]. Given a physical bound of control signals and f_1, the value of $\lambda_{\max}[\eta^\top \eta]$ within Ω can be estimated. As in Lemma 2.2, there exists a constant $\varepsilon_2 > 0$ such that $|\tilde{z}(0)| \leq \varepsilon_2$ makes the origin of system (2.14) asymptotically stable. On the other hand, from the structure of faults in (2.2) and Assumption 2.2, $T(x)$ exists and is still invertible, the faulty system (2.1) is still weakly uniformly 0-detectable, which leads to the asymptotic stability of the origin of the closed-loop system. □

The following theorem provides a reconfiguration strategy based on the previous analysis.

Theorem 2.1. *Assume the output-input stable system (2.1) with $f_a = \mathbf{0}$ and $m = p$ satisfies C 1, assumptions 2.1-2.3. Faults are assumed to occur at $t = t_f$. Given a $x(0)$, there exists a constant $\omega = \min(\varepsilon_1, \varepsilon_2)$ such that for all $|\tilde{z}(0)| \leq \omega$, the following control*

$$u_s(\hat{x}, t_{fd}) \triangleq \begin{cases} u_n(\hat{x}), & t \in [0, t_{fd}) \\ u_n(\hat{x}) - W(\hat{x})\hat{f}_a, & t \in [t_{fd}, \infty) \end{cases} \tag{2.22}$$

makes the origin of the closed-loop system asymptotically stable, where t_{fd} is the time instant when the fault has been estimated.

Proof: From Lemma 2.2, under the control $u_n(\hat{x})$ with the initial $|\tilde{z}(0)| \leq \omega$, one has $\dot{V} < 0, \forall t \in [0, t_f)$, and $\chi(t_f) \in \bar{\Omega}$, where $\bar{\Omega} \subset \Omega$. Eq.(2.13) implies $|\tilde{z}(t_f)| \leq |\tilde{z}(0)|$. On the other hand, the fault can be detected at $t_{fd} = t_f$ if $|\tilde{z}(0)| \leq \omega$ (see [56] and [142]), which means the faults are detected rapidly. Therefore, after $t = t_{fd}$, inequality (2.20) holds under the control $u_n(\hat{x}) - W(\hat{x})\hat{f}_a$. The result of Lemma 2.3 is then applied to complete the proof. □

Remark 2.1. *Theorem 2.1 provides a flexible control architecture which guarantees that $\dot{V} < 0$ $\forall t \in [0, \infty)$ whenever the faults occur, this property is very suitable for HS [142]. The proposed strategy treats the healthy system and the faulty system with different controllers, which leads to good system performance in the sense of FTC.*

Example 2.1: [142] A DC motor example is employed to illustrate a potential application field of this approach. A series DC motor is a DC motor where the field circuit is connected in series with the armature circuit [19]. Under the hypothesis that there is no magnetic saturation, the modified model of this system is expressed as follows:

$$\begin{bmatrix}\dot{x}_1\\ \dot{x}_2\\ \dot{x}_3\end{bmatrix} = \begin{bmatrix}-k_1x_1x_2 - \frac{R}{L}x_1 + u_1 + Lf_a\\ -k_2x_2 + \frac{k_1}{JL}x_1^2 - \frac{x_3}{J}\\ u_2 + 2k_1x_1f_a\end{bmatrix} \tag{2.23}$$
$$\begin{bmatrix}y_1\\ y_2\end{bmatrix} = \begin{bmatrix}x_1\\ x_2\end{bmatrix}$$

where $x_1 = \phi_f$ denotes the flux, $x_2 = \omega_f$ denotes the speed, $x_3 = T_L$ denotes time varying load torque, u_1 and u_2 are the voltage inputs. the speed and the flux are measured.

Let us first consider the healthy case ($f_a = 0$). Since $x_1 = y_1$, $x_2 = y_2$, and $|x_3| = J|\dot{y}_2^2 + k_2y_2 - \frac{k_1}{JL}y_1^2| \leq J|\dot{y}_2|^2 + Jk_2|y_2| + \frac{k_1}{L}|y_1|^2$, it is seen that the healthy system is weakly uniformly 0-detectable of order 1. The output derivatives are

$$\begin{bmatrix}\dot{y}_1\\ \dot{y}_2\end{bmatrix} = \begin{bmatrix}-k_1x_1x_2 - \frac{R}{L}x_1\\ -k_2x_2 + \frac{k_1}{JL}x_1^2 - \frac{x_3}{J}\end{bmatrix} + \begin{bmatrix}1 & 0\\ 0 & 0\end{bmatrix}\begin{bmatrix}u_1\\ u_2\end{bmatrix}$$

so $r_1 = 1$, differentiating the equality of $\dot{y}_2$ leads to

$$\ddot{y}_2 = k_2^2x_2 - \frac{k_1k_2}{JL}x_1^2 + \frac{k_2}{J}x_3 - \frac{2k_1^2}{JL}x_1^2x_2 - \frac{2k_1R}{JL^2}x_1^2 + \left[\ \frac{2k_1}{JL}x_1\ \ -\frac{1}{J}\ \right]\begin{bmatrix}u_1\\ u_2\end{bmatrix}$$

The matrix $\begin{bmatrix}1 & 0\\ \frac{2k_1}{JL}x_1 & -\frac{1}{J}\end{bmatrix}$ is always nonsingular. The map $T : x \rightarrow \chi$ is also invertible and not affected by the observer. C 1 and assumptions 2.1-2.2 are satisfied. From (2.11), u_n can be designed as

$$u_n = \begin{bmatrix}k_1x_1x_2 + (\frac{R}{L} - 1)x_1\\ (Jk_2^2 - \frac{k_1}{JL})x_1^2 - (\frac{2k_1}{L} + \frac{k_1k_2}{L})x_1^2 + (k_2 + \frac{1}{J})x_3\end{bmatrix}$$

which makes $\bar{A}$ Hurwitz.

Now consider the faulty case. It is clear that $W(x) = (L,\ 2k_1x_1)^\top$, f_a is an actuator fault that affects both control channels. The invertible transformation $z_1 = x_2$, $z_2 = -\frac{x_3}{J} + \frac{k_1}{JL}x_1^2$, $z_3 = x_1$ puts system (2.23) into the form

$$\begin{bmatrix}\dot{z}_1\\ \dot{z}_2\end{bmatrix} = \begin{bmatrix}z_2 - k_2y_1\\ -2\frac{k_1}{JL}y_2(k_1y_1y_2 + \frac{R}{L}y_2 - u_1) - \frac{u_2}{J}\end{bmatrix} \tag{2.24}$$
$$y_2 = z_1 \tag{2.25}$$
$$\dot{z}_3 = -k_1y_1y_2 - \frac{R}{L}y_1 + u_1 + Lf_a \tag{2.26}$$
$$y_1 = z_3$$

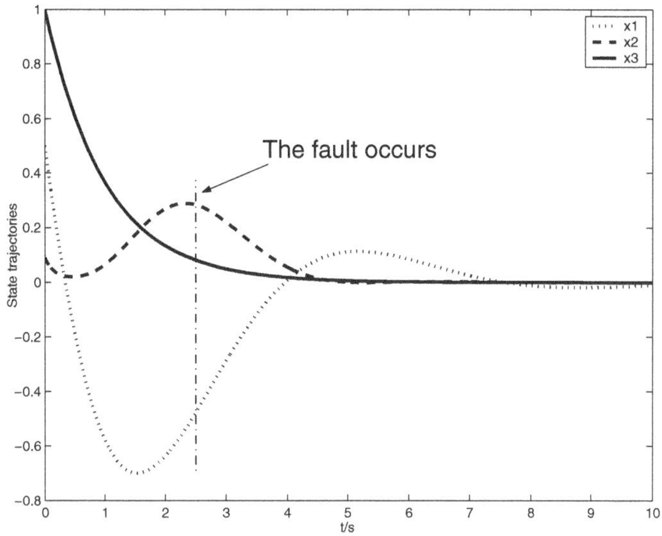

Fig. 2.1 State trajectories

Eq.(2.26) does not involve the estimation of z_1 and z_2, which implies that fault estimates are obtained without any estimation error. So the fault can be detected and compensated immediately after the fault occurs. Under control $u_n(x)$, one has $\dot{\chi} = \bar{A}\chi$, where $\chi = (y_1, y_2, \dot{y}_2)^\top$.

In the simulation, the parameters are [19]: $R = 0.0247, L = 0.06, J = 30.1, k_1 = 0.04329$, $k_2 = 0.0033$. The initial $x(0) = (0.5, 0.1, 1)^\top$. $\hat{x}(3) = 0.85$. The fault is considered as

$$f_a = \begin{cases} 0, & 0s \leq t < 2.5s \\ 0.5 + 0.2\sin(5t), & 2.5s \leq t < 10s \end{cases} \tag{2.27}$$

Fig.2.1 shows state trajectories, the origin of the closed-loop system is asymptotically stable in spite of faults.

2.1.3 FTC for Hybrid Systems

The above FTC solution is now extended to a class of switched systems taking the form

$$\begin{aligned} \dot{x} &= f_\sigma(x) + G_\sigma(x)u_\sigma + E_\sigma(x)f_{a\sigma} \\ y &= h(x) \end{aligned} \tag{2.28}$$

where each mode satisfies all the conditions in Theorem 2.1. $\sigma(t) : [t_0, \infty) \to Q = \{1, 2, \ldots, N\}$ is a switching signal, which is assumed to be a piecewise constant function continuous from the right.

The switching property is considered as in [29]: (a) the switching sequence is fixed, (b) there is a series of dwell periods Δt_{kj} for mode k when it is activated for the jth time and mode k switches to mode $(k+1)$ for the jth time at $t = t_{kj}$ when $\Delta t_{(k+1)j}$ is elapsed, (c) the states do not jump at the switching instants.

The observer-based method in Section 2.1.2 is modified for the HS as follows:

- The observer and the fault estimates scheme are switched according to the current mode at each switching time.
- The initial states of the current observer are chosen as the final states of the previous observer. The fault estimates are set to zero at each switching instant.

We also need to impose a condition on the above switching law such that the weak uniform 0-detectability of the overall HS can be guaranteed.

Assumption 2.3. *$\Delta t_{kj}(k = 1,2,\ldots,N)$ are large enough such that for any $s \in \Re+$, we have*

$$\beta_{k+1}(2\beta_k(2s,\Delta t_{kj}),\Delta t_{(k+1)j}) \leq \bar{\lambda} s < s \quad \forall k \in Q \tag{2.29}$$

where $0 < \bar{\lambda} < 1$ and $\beta_k(k \in Q)$ satisfies (2.4) for mode k.

Lemma 2.4. *Consider the HS (2.28) satisfying Assumption 2.3 in the healthy case. Then, the overall HS is weakly uniformly 0-detectable.*

Proof: Lemma 2.4 is an extension of Theorem 1 in [129] to the weak uniform 0-detectability case, its proof is omitted. □

Let V_k, $u_{sk}(\hat{x},t_{fdk})$, ω_k be respectively V, $u_s(\hat{x},t_{fd})$, ω for mode k. The FTC problem for the system (2.28) with unfixed dwell periods and fixed dwell periods will be discussed respectively.

Theorem 2.2. *Under Assumption 2.3, consider the HS (2.28) under a family of control laws $u_k(\hat{x},t_{fdk})$. There exists a constant ω_k such that $|\bar{\tilde{z}}_2(0)| \leq \omega_k$ with a given $x(0)$. If, at any time instant $\bar{t}$, the following conditions hold:*

$$|\tilde{z}(\bar{t})| \leq \omega_{k+1} \tag{2.30}$$

$$V_{k+1}(\chi(\bar{t})) < V_{k+1}(\chi(t_{(k+1)(j-1)})), \quad j > 0 \tag{2.31}$$

then, choosing $\Delta t_{kj} \geq \bar{t} - t_{kj}$, which satisfies (2.29), and setting $\sigma(t) = k+1$ at $t = t_{kj} + \Delta t_{kj}$ guarantee that the origin of the overall HS is asymptotically stable.

Proof: If the initial $|\tilde{z}(0)| \leq \omega_k$ for some $k \in Q$, it follows from Theorem 2.1 that $\dot{V}_k < 0$ as long as mode k remains active. If at some time instant $\bar{t}$ one has $|\tilde{z}(\bar{t})| \leq \omega_{k+1}$, and $\sigma(t) = k+1$ is set on, then for all $t \in [\bar{t}, t_{kj} + \Delta t_{kj})$, $\dot{V}_{k+1} < 0$ as long as $\sigma(t) = k+1$. It is concluded that if the kth mode is activated only when $|\tilde{z}(t)| \leq \omega_k$, then

$$\dot{V}_\sigma(t) < 0, \quad \forall \sigma(t) = k \tag{2.32}$$

Moreover, from (2.31), for any admissible switching time t_{kj} one has

$$V_{k+1}(\chi(t_{(k+1)j})) < V_{k+1}(\chi(t_{(k+1)(j-1)})) \tag{2.33}$$

Since the kth faulty mode is still weakly uniformly 0-detectable, and T always exists, the Multiple Lyapunov function method [22] can be applied to conclude that the origin of the hybrid system is Lyapunov stable. On the other hand, for each switching time t_{kj}, $j = 1,2,\ldots$ such that $\sigma(t_{kj}^+) = k$, the sequence $V_{\sigma(t_{kj})}$ is decreasing and positive, and therefore has a limit $\zeta \geq 0$. One has

$$\lim_{j\to\infty}\left[V_{k+1}(\chi(t_{(k+1)(j+1)})) - V_{k+1}(\chi(t_{(k+1)j}))\right] = \zeta - \zeta = 0$$

Note that there exists a class $\mathscr{K}$ function α such that

$$\begin{aligned} 0 &= \lim_{j\to\infty}\left[V_{k+1}(\chi(t_{(k+1)(j+1)})) - V_{k+1}(\chi(t_{(k+1)j}))\right] \\ &\leq \lim_{j\to\infty}[-\alpha(\|\chi(t_{(k+1)j}\|)] \leq 0 \end{aligned} \tag{2.34}$$

Inequality (2.34) together with Lemma 2.4 implies that $x(t)$ converges to the origin, which combined with Lyapunov stability, leads to the asymptotic stability of the origin of the HS. This completes the proof. □

Remark 2.2. *Inequality (2.31) is used only when the target $k+1^{th}$ mode has been previously activated. Actually, when only a finite number of switchings is considered over the infinite time-interval, Inequality (2.31) can be relaxed to allow for finite increases in V_{k+1}, (see [28] and [29] for some analysis). In this case, inequality (2.30) alone is sufficient to enforce the asymptotic stability of the origin.*

Many real systems work under a series of prescribed dwell periods, i.e., Δt_{kj} is fixed. In this case, the goal of FTC must be achieved before each switching time whenever the faults occur. This is possible because the decay rate of V_k can be estimated. We have the following corollary.

Corollary 2.1. *Consider the HS (2.28) under a family of control laws $u_k(\hat{x}, t_{fdk})$ with fixed Δt_{kj} $k \in Q$ which satisfies (2.29). If each faulty mode satisfies (iv), T exists and is still invertible, and there exists a constant ω_k such that $|\bar{\bar{z}}_2(0)| \leq \omega_k$, then the origin of the overall hybrid system is asymptotically stable.*

Proof: It is clear from (2.13) that appropriate selection of λ makes (2.30) hold at a given $t_{(k+1)j}$. On the other hand, inequality (2.20) in Lemma 2.3 leads to

$$\dot{V}_k \leq -\lambda_{\min}(Q_k)|\chi|^2 + \Xi_k \leq -\iota_k V_k + \Xi_k, \quad \iota_k \triangleq \frac{\lambda_{\min}(Q_k)}{\lambda_{\max}(P_k)} \tag{2.35}$$

Note that Ξ_k is bounded within a known region and converges to zero, so the trajectory of V_k can be estimated by (2.35). The results of Theorem 2.2 can be applied to guarantee the asymptotic stability of the origin of the HS. □

2.2 Overall Fault Tolerant Regulation

This section extends the classical output regulation theories to hybrid nonlinear systems and analyzes its fault tolerance in the presence of continuous faults modeled by the exosignals.

2.2.1 *Fault Tolerant Regulation for Nonlinear Systems*

The considered system takes the general nonlinear form

$$\dot{x}(t) = G(x(t), u(t), f(t)) \tag{2.36}$$
$$y(t) = H(x(t), f(t)) \tag{2.37}$$
$$\dot{f}(t) = S(f(t)) \ \forall t \geq t_f, \text{ with } f(t) = 0 \ \forall t \in [0, t_f) \tag{2.38}$$
$$e(t) = y(t) - y_r(x(t)) \tag{2.39}$$

with measurable state $x \in \Re^n$, input $u \in \Re^p$, output $y \in \Re^m$. The regulated error e denotes the output tracking error between y and the continuous reference signal $y_r(x) : \Re^n \to \Re^m$. The vector fields G, H are assumed to be smooth and known.

Once the fault occurs, the fault signal $f \in \mathscr{F} \subset \Re^q$ is generated by the *neurally stable* exosystem (2.38), i.e., $\partial S(0)/\partial f$ has all its eigenvalues on the imaginary axis, which means that f is always bounded. The function S is also assumed to be smooth and known. Such model effectively describes process, actuator and sensor faults.

The following assumption is a basic requirement for the state feedback output regulation design [55].

Assumption 2.4. *There exist some $u = \alpha(x, f)$ with $f = 0$ such that $x = 0$ of healthy system (2.36) $\dot{x} = G(x, \alpha(x, 0), 0)$ is asymptotically stable.*

Definition 2.2. Fault tolerant regulation problem *(FTRP) for system (2.36)-(2.39) is to find a FTC law $u = \alpha(x, f)$ such that $\forall x(0) \in \mathscr{X}$ with $\mathscr{X} \subset \Re^n$ a neighborhood of 0 and $\forall f \in \mathscr{F}$, the trajectory of the closed-loop system (2.36) $\dot{x} = G(x, \alpha(x, f), f)$ is bounded $\forall t \geq 0$ and $\lim_{t\to\infty} e(t) = 0$.*

Theorem 2.3. *Suppose that the fault f can be detected/approximated accurately, and there exists a $u = \alpha(x, f)$ satisfying Assumption 2.4. The FTRP for system (2.36)-(2.39) is solvable if and only if there exists a $\mathscr{C}^k$ mapping $x = \pi(f)$ with $\pi(0) = 0$ defined for $(x, f) \in \mathscr{X} \times \mathscr{F}$ satisfying*

$$\frac{\partial \pi}{\partial f} S(f) = G(\pi(f), \alpha(\pi(f), f), f) \tag{2.40}$$
$$0 = H(\pi(f), f) - y_r(\pi(f)) \tag{2.41}$$

Proof: The proof follows the same way as that of Theorem 8.3.2 in [55], which is thus omitted. □

Remark 2.3. *It can be seen that FTRP is similar to the general output regulation problem with disturbances. Theorem 2.3 provides necessary and sufficient conditions to solve FTRP in the classical faulty case. The existence and the design of $\pi(f)$ and $\alpha(x, f)$ have been deeply investigated in many literatures, e.g. [55], [52], which are not focused on here.*

2.2.2 *Overall Fault Tolerant Regulation*

Now we consider the hybrid case. The system is

$$\dot{x}(t) = G_{\sigma(t)}(x(t), u_{\sigma(t)}(t), f_{\sigma(t)}(t)) \tag{2.42}$$

$$y(t) = H(x(t), f_{\sigma(t)}(t)) \tag{2.43}$$

$$\dot{f}_{\sigma(t)}(t) = S_{\sigma(t)}(f_{\sigma(t)}(t)) \ \forall t \geq t_f, \text{ with } f_{\sigma(t)}(t) = 0 \ \forall t \in [0, t_f) \tag{2.44}$$

where $\sigma(t) : [0, \infty) \to Q$ also denotes a piecewise constant switching function.

Assumption 2.5. *There exists a family of controllers $u_i = \alpha_i(x, f_i)$ for $i \in Q$ solving the FTRP for system (2.39) and (2.42)-(2.44) with $\sigma(t) = i$.*

Assumption 2.5 means that the FTRP of each mode is solvable individually. The following definition is an extension of FTRP to the successional faulty case.

Definition 2.3. Overall fault tolerant regulation problem *(OFTRP) for system (2.39) and (2.42)-(2.44) is to find a switching scheme among $u_i = \alpha_i(x, f_i)$, $i \in Q$ such that $\forall x(0) \in \mathscr{X}$ and $\forall f_i \in \mathscr{F}$, the trajectory of the closed-loop system (2.42) is bounded $\forall t \geq 0$ and $\lim_{t\to\infty} e(t) = 0$.*

Before solving the OFTRP, we give an important concept as follows

Definition 2.4. *[49]: Let $N_\sigma(T, t)$ denote the number of switchings of σ over the interval (t, T), if there exists a positive number τ_a such that*

$$N_\sigma(T, t) \leq N_0 + \frac{T - t}{\tau_a}, \quad \forall T \geq t \geq 0 \tag{2.45}$$

where $N_0 > 0$ denotes the chattering bound, then the positive constant τ_a is called average dwell time *(ADT) of σ over (t, T).*

Definition 2.4 means that there may exist some switchings separated by less than τ_a, but the average dwell period among switchings of modes is not less than τ_a.

The following theorem establishes the sufficient conditions to solve OFTRP.

Theorem 2.4. *Consider a system (2.39) and (2.42)-(2.44) satisfying Assumption 2.5. Suppose that each fault can be diagnosed without delay, and each FTC law u_i is applied once a fault f_i occurs. The OFTRP is solvable if*

C1) $\tau_a > \frac{\ln B}{a}$, where $B \triangleq \max_{i\in Q} B_i$, $a \triangleq \min_{i\in Q} a_i$.

and either C2) or C3) holds for $k = 1, 2, \ldots$

C2) $\pi_{\sigma(t_{k-1})}(f_{\sigma(t_{k-1})}(t_k)) = \pi_{\sigma(t_k)}(f_{\sigma(t_k)}(t_k))$.

C3) $-(a - \frac{\ln B}{\tau_a})(t - t_k) + \ln k < -a^* t$, for $t \geq t_k$ and $a^* > 0$.

Remark 2.4. *Before proving Theorem 2.4, we provide some insight into the conditions C1)-C3): C1) requires that the switching of modes is slow averagely, i.e., the frequency of switching is not too much. C2) imposes a condition on the mapping π_i and the fault value f_i. It can be seen that if there is a common mapping $x = \pi(f_i)$ for all modes, and $f_{\sigma(t_{k-1})}(t_k)) = 0$, then C2) holds. Generally, C2) is hard to satisfy even in the linear case [76]. In the absence of C2), C3) requires that the dwell period of each mode is long enough. C3) can be verified by checking whether $\ln k + (a - \frac{\ln B}{\tau_a})t_k < (a - \frac{\ln B}{\tau_a} - a^*)t$ holds or not for $t \in [t_k, t_{k+1})$.*

Proof of Theorem 2.4: Since mode $\sigma(t_k)$ in the time interval $[t_k, t_{k+1})$ is controlled by $u_{\sigma(t_k)}$, thus its FTRP is solved from Assumption 2.5. According to Theorem 8.3.2 in [55], a center manifold $x = \pi_{\sigma(t_k)}(f_{\sigma(t_k)})$ of mode $\sigma(t_k)$ is locally attractive, i.e.,

$$|x(t) - \pi_{\sigma(t_k)}(f_{\sigma(t_k)}(t))| \leq Be^{-a(t-t_k)}|x(t_k) - \pi_{\sigma(t_k)}(f_{\sigma(t_k)}(t_k))|, \ \ t_k \leq t < t_{k+1} \quad (2.46)$$

Similarly, in $[t_{k-1}, t_k)$ one has

$$|x(t_k^-) - \pi_{\sigma(t_{k-1})}(f_{\sigma(t_{k-1})}(t_k^-))| \leq Be^{-a(t_k^- - t_{k-1})}|x(t_{k-1}) - \pi_{\sigma(t_{k-1})}(f_{\sigma(t_{k-1})}(t_{k-1}))| \quad (2.47)$$

Combining (2.46) with (2.47) yields

$$\begin{aligned}|x(t) - \pi_{\sigma(t_k)}(f_{\sigma(t_k)}(t))| &\leq Be^{-a(t-t_k)}\Big|x(t_k) - \pi_{\sigma(t_{k-1})}(f_{\sigma(t_{k-1})}(t_k)) \\ &\quad + \pi_{\sigma(t_{k-1})}(f_{\sigma(t_{k-1})}(t_k)) - \pi_{\sigma(t_k)}(f_{\sigma(t_k)}(t_k))\Big| \\ &\leq B^2 e^{-a(t-t_{k-1})}|x(t_{k-1}) - \pi_{\sigma(t_{k-1})}(f_{\sigma(t_{k-1})}(t_{k-1}))| \\ &\quad + Be^{-a(t-t_k)}|\pi_{\sigma(t_{k-1})}(f_{\sigma(t_{k-1})}(t_k)) - \pi_{\sigma(t_k)}(f_{\sigma(t_k)}(t_k))| \end{aligned} \quad (2.48)$$

By induction, we obtain

$$\begin{aligned}|x(t) - \pi_{\sigma(t_k)}(f_{\sigma(t_k)}(t))| &\leq B^{k+1}e^{-at}|x(0) - \pi_{\sigma(0)}(f_{\sigma(0)}(0))| \\ &\quad + \sum_{s=1}^{k}\Big(B^s e^{-a(t-t_{k-s+1})}|\pi_{\sigma(t_{k-s})}(f_{\sigma(t_{k-s})}(t_{k-s+1})) \\ &\quad - \pi_{\sigma(t_{k-s+1})}(f_{\sigma(t_{k-s+1})}(t_{k-s+1}))|\Big) \end{aligned} \quad (2.49)$$

From C1), we can pick $\lambda = a - \frac{\ln B}{\tau_a}$, we have $\tau_a = \frac{\ln B}{(a-\lambda)}$. Based on (2.45), we have

$$B^{k+1}e^{-at} \leq B^{N_0+1}e^{\frac{t}{\tau_a}\ln B - at} < B^{N_0+1}e^{-\lambda t} \quad (2.50)$$

If C2) holds, each term of the sum in (2.49) is zero. Substituting (2.50) into (2.49), we further have

$$|x(t) - \pi_{\sigma(t_k)}(f_{\sigma(t_k)}(t))| \leq B^{N_0+1}e^{-\lambda t}|x(0) \quad \pi_{\sigma(0)}(f_{\sigma(0)}(0))| \quad (2.51)$$

Inequality (2.51) means that $x - \pi_{\sigma(t_k)}(f_{\sigma(t_k)})$ still converges to zero $\forall t \geq t_k$, $\forall x(0) \in \mathscr{X}$ and $\forall f_i \in \mathscr{F}$. By continuity of H and y_r in each $[t_{k-1}, t_k)$, it follows that $\lim_{t\to 0} e(t) = 0$.

If C2) does not hold, one has from C1) and (2.45) that

$$\begin{aligned} B^s e^{-a(t-t_{k-s+1})} &\leq B^{N_0 + \frac{t - t_{k-s+1}}{\tau_a}} e^{-a(t-t_{k-s+1})} \\ &\leq B^{N_0} e^{\frac{t-t_{k-s+1}}{\tau_a}\ln B - a(t-t_{k-s+1})} \\ &\leq B^{N_0} e^{-\lambda(t-t_{k-s+1})} \end{aligned} \quad (2.52)$$

Since each f_i is bounded due to the neurally stable exosystems, there exists a constant $\xi > 0$ such that $\forall k = 1,2,...,$ and $1 \leq s \leq k$

$$\left|\pi_{\sigma(t_{k-s})}(f_{\sigma(t_{k-s})}(t_{k-s+1})) - \pi_{\sigma(t_{k-s+1})}(f_{\sigma(t_{k-s+1})}(t_{k-s+1}))\right| \leq \xi \tag{2.53}$$

It follows from (2.53) and C3) that

$$\begin{aligned}
&\sum_{s=1}^{k}\left(B^s e^{-a(t-t_{k-s+1})}|\pi_{\sigma(t_{k-s})}(f_{\sigma(t_{k-s})}(t_{k-s+1})) - \pi_{\sigma(t_{k-s+1})}(f_{\sigma(t_{k-s+1})}(t_{k-s+1}))|\right)\\
&\quad \leq \xi B^{N_0}\sum_{s=1}^{k} e^{-\lambda(t-t_{k-s+1})}\\
&\quad \leq \xi B^{N_0} e^{\ln k - \lambda(t-t_k)}\\
&\quad \leq \xi B^{N_0} e^{-a^* t}
\end{aligned} \tag{2.54}$$

By substituting (2.50) and (2.54) into (2.49), we conclude that $x - \pi_{\sigma(t_k)}(f_{\sigma(t_k)})$ converges to zero $\forall t \geq t_k$, $\forall x(0) \in \mathscr{X}$ and $\forall f_i \in \mathscr{F}$. The result follows. □

2.3 Multiple Observers Method

2.3.1 Problem Formulation

Differently from sections 2.1-2.2, we address a class of HS with both continuous faults and discrete faults in this section. The system takes the form

$$\dot{x}(t) = A_\sigma x(t) + g_\sigma(x(t),t) + B_\sigma u_\sigma(t) + E_\sigma f_\sigma^c(t) \tag{2.55}$$

$$y(t) = Cx(t) \tag{2.56}$$

where $x(t) \in \Re^n$ is the non measured state, $y(t) \in \Re^p$ is the output, $u_\sigma(t) \in \Re^m$ is the control. A_σ, B_σ, E_σ and C are real constant matrices of appropriate dimensions. (A_σ, B_σ) is controllable. $g_\sigma(x(t),t)$ is a continuous Lipschitz function, i.e., $|g_\sigma(x_1,t) - g_\sigma(x_2,t)| \leq L_\sigma|x_1 - x_2|$, where $L_\sigma > 0$ is called the Lipschitz constant. Moreover, $g_\sigma(0,t) = 0$.

The *continuous actuator fault* is modelled by a "fault pattern" as in Chapter 2.1. Suppose that there exists two constants f_σ^0 and f_σ^1 such that $|f_\sigma^c| \leq f_\sigma^0$, $|\dot{f}_\sigma^c| \leq f_\sigma^1$. Such fault model covers all faults that result in a deviation of the control signal from normal.

Define $Q = \{1,2,\ldots,N\}$, where N is the number of modes. $\sigma(t) : [t_0,\infty) \to Q$ denotes the *switching function* as in sections 2.1-2.2. Denote t_j as the jth switching instant of the system (2.55)(2.56). At t_j, the system switches to mode k, where $k \in Q$, $j = 1,2,\ldots$.

The switching property is considered as in [29]: a) the switching sequence is fixed. b) there is a series of prescribed dwell periods between each switching. We also assume that the states do not jump at the switching instants.

The *discrete fault* is represented by the faulty switching function $\sigma_f(t)$, that forces the system to switch to a mode which is not the prescribed successor at the switching instant. Similarly, $\sigma_H(t)$ denotes the healthy switching function. If $\sigma(t) = \sigma_H(t)$, then there is no discrete fault in the current mode.

The FTC problem in this section can be described as: *Keep the states of system (2.55)-(2.56) always bounded and make them converge to a small closed set in spite of continuous and discrete faults.*

Different from sections 2.1-2.2, the FTC of discrete faults must be taken into account as in [132] and [145]. Since the current mode after each switching time may be unknown due to discrete faults, some identifying work must be applied for a short period. Some related work can be seen in [129], [68], [48] and [20]. Whatever method used, the necessary time period in which mode is identified (due to computation time, decision time) may cause instability. How to overcome this finite delay is a problem to be addressed.

The main idea is as follows: 1) For the continuous faults in each mode, an adaptive observer technique is proposed to provide the rapid fault estimation, based on which the FTC law is designed. 2) For the discrete faults, a novel model-free sliding mode observer is designed, which together with a series of observers related to system modes, can identify the current mode quickly while guaranteeing the stability of the system during each transition period. 3) The above two FTC strategies are combined with the average dwell time scheme such that the states of the overall hybrid system are always bounded and converge to a small closed set.

2.3.2 FTC for Continuous Faults

In this subsection, only $f^c_\sigma(t)$ is addressed. We introduce the *input-to-state practical stability* and a lemma that will be used later.

Definition 2.5. *[113] A system $\dot{x} = f(x,u)$ is said to be* input-to-state practically stable *(ISpS) over $[0,t)$ w.r.t. u if there exist functions $\beta \in \mathscr{KL}$, $\alpha, \gamma \in \mathscr{K}_\infty$, and a constant $\varsigma > 0$, such that for any bounded input u and any initial condition $x(0)$, we have*

$$\alpha(|x(t)|) \le \beta(|x(0)|,t) + \gamma(\|u\|_{[0,t)}) + \varsigma, \quad \forall t \ge 0$$

Note that when $\varsigma = 0$, ISpS becomes input-to-state stability (ISS) [114] (see also Definition 4.1 in Chapter 4).

It has been proven in Section VI of [113] that the following property holds.

Lemma 2.5. *If there exist $\alpha_1, \alpha_2, \alpha_3, \gamma_1 \in \mathscr{K}_\infty$, $\varsigma_1 > 0$ and a smooth function $V : \Re^n \to \Re_{\ge 0}$ such that*

$$\alpha_1(|x|) \le V(x) \le \alpha_2(|x|) \tag{2.57}$$

$$\dot{V}(x) \le -\alpha_3(|x|) + \gamma_1(|u|) + \varsigma_1 \tag{2.58}$$

Then the system $\dot{x} = f(x,u)$ is ISpS over $[0,t)$ w.r.t. u.

If $\varsigma_1 = 0$, then V is called *ISS Lyapunov function*[114], and the system is ISS under (2.57) and (2.58) with the state x and the input u (see Lemma 2.14 in [114]).

Now let us consider the system (2.55)(2.56) with $\sigma(t) = k$ for some $k \in Q$ starting from $t = t_j$

$$\dot{x}(t) = A_k x(t) + g_k(x(t),t) + B_k u_k(t) + E_k f_k^c(t) \tag{2.59}$$

$$y(t) = Cx(t) \tag{2.60}$$

Assumption 2.6. *There exists a matrix K_k such that $G_k(s) = C[sI - (A_k - K_k C)]^{-1} E_k$, is strictly positive real (SPR) :*

$$\forall \omega > 0 : Re(G_k(j\omega)) > 0 \tag{2.61}$$

Moreover

$$\min_{\omega \in R^+} \sigma_{\min}(A_k - K_k C - j\omega I) > L_k \tag{2.62}$$

where $\sigma_{\min}(M)$ is the smallest singular value of M.

Remark 2.5. *Assumption 2.6 is a restriction on the triple (A_k, C, E_k) in terms of the fault to residual transfer of the observer-based residual generator associated with the linear part of the system. A known necessary condition for $G_k(s)$ to be SPR is that (A_k, C) is observable and CE_k is of full column rank. It should be noted that CE_k being of full column rank is a standard assumption in fault isolation problem [10].*

Under Assumption 2.6, it has been proven in [104] that for any given matrix $Q_k \in \Re^{n\times n} > 0$ and scalar $\varepsilon > 0$, there exist two matrices $P_k \in \Re^{n\times n} > 0$ and $R_k \in \Re^{r\times q}$ such that

$$P_k E_k = C^\top R_k \tag{2.63}$$

and

$$(A_k - K_k C)^\top P_k + P_k (A_k - K_k C) + \varepsilon L_k^2 I_n + \frac{P_k^2}{\varepsilon} + Q_k \leq 0 \tag{2.64}$$

The FD scheme for mode k is designed as

$$\dot{\hat{x}} = A_k \hat{x} + g_k(\hat{x},t) + B_k u_k + E_k \hat{f}_k^c + K_k(y - \hat{y}) \tag{2.65}$$

$$\dot{\hat{f}}_k^c = \Gamma_k R_k^\top (y - \hat{y}) - \vartheta_k \Gamma_k \hat{f}_k^c \tag{2.66}$$

$$\hat{y} = C\hat{x} \tag{2.67}$$

where $\hat{x}(t), \hat{f}_k^c(t), \hat{y}(t)$ are the estimates of $x(t), f_k^c(t), y(t)$. The weighting matrix $\Gamma_k = \Gamma_k^\top > 0$, and the constant $\vartheta_k > 0$ are chosen such that $\vartheta_k - \lambda_{\max}(\Gamma_k^{-1}) > 0$.

Remark 2.6. *The diagnostic scheme (2.65)-(2.67) plays an important role to diagnose the f_k^c. Our goal is to stabilize the system, we neither care about when the fault occurs nor design a so-called* detection observer *as in [58] to detect the fault. The diagnostic scheme (2.65)-(2.67) always works no matter the mode k is faulty or not (i.e., the normal condition can be treated as a special faulty case where $f_k^c = 0$).*

Denote $e_x(t) = x(t) - \hat{x}(t)$, $e_y(t) = y(t) - \hat{y}(t)$, $e_f(t) = f_k^c(t) - \hat{f}_k^c(t)$, we have the following lemma:

Lemma 2.6. *[57] Define a set S_k as*

$$S_k \triangleq \left\{ (e_x, e_f) \middle| \lambda_{\min}(P_k)|e_x|^2 + \lambda_{\min}(\Gamma_k^{-1})|e_f|^2 \leq \frac{\beta_k}{\alpha_k} \right\}$$

where

$$\beta_k \triangleq \lambda_{\max}(\Gamma_k^{-1})(f_k^1)^2 + \sigma_k(f_k^0)^2, \quad \alpha_k \triangleq \frac{\min(c_{k1},\ c_{k2})}{\max[\lambda_{\max}(P_k),\ \lambda_{\max}(\Gamma_k^{-1})]}$$
$$c_{k1} \triangleq \lambda_{\min}(Q_k) > 0, \quad c_{k2} \triangleq \vartheta_k - \lambda_{\max}(\Gamma_k^{-1}) > 0 \tag{2.68}$$

Then under Assumption 2.6, the fault diagnostic scheme (2.65)-(2.67) guarantees that (e_x, e_f) of mode k converges to S_k exponentially at a rate greater than $e^{-\alpha_k t}$.

The following lemma gives a relation between e_x and e_f.

Lemma 2.7. *Under Assumption 2.6, the fault diagnostic scheme (2.65)-(2.67) guarantees that e_x is ISS w.r.t. e_f, i.e., there exist $\beta_{ek} \in \mathcal{KL}$, $\alpha_{ek}, \gamma_{ek} \in \mathcal{K}_\infty$ such that*

$$\alpha_{ek}(|e_x(t))| \leq \beta_{ek}(|e_x(t_j)|, t) + \gamma_{ek}(\|e_f\|_{[t_j,t)}), \quad \forall t \geq t_j \tag{2.69}$$

Proof: From (2.59), (2.60), (2.65) and (2.67), we have

$$\dot{e}_x = (A_k - K_k C)e_x + g_k(x,t) - g_k(\hat{x},t) + E_k e_f \tag{2.70}$$

Choose a Lyapunov candidate $\Theta_k = e_x^\top P_k e_x$, its derivative w.r.t. time along (2.70) is

$$\dot{\Theta}_k = e_x^\top [P_k(A_k - K_k C) + (A_k - K_k C)^\top P_k] e_x$$
$$+ 2e_x^\top P_k(g_k(x,t) - g_k(\hat{x},t)) + 2e_x^\top P_k E_k e_f$$

Note that, for two vectors $\mathbf{a_1}$, $\mathbf{a_2}$, it holds that $2\mathbf{a_1}^\top \mathbf{a_2} \leq \frac{1}{\varepsilon}\mathbf{a_1}^\top \mathbf{a_1} + \varepsilon \mathbf{a_2}^\top \mathbf{a_2}$ for $\varepsilon > 0$. Similarly, we can show that

$$2e_x^\top P_k(g_k(x,t) - g_k(\hat{x},t)) \leq e_x^\top \frac{P_k^2}{\varepsilon} e_x + \varepsilon L_k^2 e_x^\top e_x \tag{2.71}$$

From (2.64), we have

$$\dot{\Theta}_k \leq -e_x^\top Q_k e_x + 2e_x^\top P_k E_k e_f$$
$$\leq (-\lambda_{\min}(Q_k) + \varepsilon_1)|e_x|^2 + \frac{|P_k E_k|^2}{\varepsilon_1}|e_f|^2 \tag{2.72}$$

where $\varepsilon_1 > 0$ is chosen such that $-\lambda_{\min}(Q_k) + \varepsilon_1 < 0$. Inequality (2.72) implies that Θ_k is an ISS-Lyapunov function with the state e_x and the input e_f. From Lemma 2.5, the result follows.

Moreover, we have

$$\dot{\Theta}_k \leq \frac{-\lambda_{\min}(Q_k)+\varepsilon_1}{\lambda_{\max}(P_k)}\Theta_k + \frac{|P_k E_k|^2}{\varepsilon_1}|e_f|^2 \triangleq \iota_1\Theta_k + \iota_2|e_f|^2$$

Using the differential inequality theory (see Chapter 2 in [84]), we can obtain

$$\begin{aligned}
\Theta_k &\leq e^{\iota_1(t-t_j)}\Theta_k(t_j) + \int_{t_j}^{t} e^{\iota_1(t-\tau)}\iota_2|e_f(\tau)|^2 d\tau \\
&\leq e^{\iota_1(t-t_j)}\Theta_k(t_j) + \sup_{\tau\in[t_j,t)}\{\iota_2|e_f(\tau)|^2\}\int_{t_j}^{t} e^{\iota_1(t-\tau)} d\tau \\
&\leq \underbrace{e^{\iota_1(t-t_j)}\lambda_{\max}(P_k)|e_x(t_j)|^2}_{\beta_{ek}(|e_x(t_j)|,t)} + \underbrace{\frac{1}{-\iota_1}\sup_{\tau\in[t_j,t)}\{\iota_2|e_f(\tau)|^2\}}_{\gamma_{ek}(\|e_f\|_{[t_j,t)})}
\end{aligned} \tag{2.73}$$

Define $\alpha_{ek}(\cdot) = \lambda_{\min}(P_k)(\cdot)^2$, which, together with β_{ek}, γ_{ek} in (2.73), leads to (2.69). This completes the proof. □

Supposed that $e_f(t)$ is norm bounded in each $[t_j, t_{j+1})$. Inequality (2.69) means that given an initial $|e_x(t_j)|$ (or a bound of $|e_x(t_j)|$, the value of $|e_x|$ can be estimated. Define

$$e_x(t)_{est} \triangleq \alpha_{ek''}^{-1}\circ\beta_{ek''}(|e_x(t_j)|,t) + \alpha_{ek''}^{-1}\circ\gamma_{ek''}(\|e_f(t)\|_{[t_j,t)}), \quad t_j \leq t \leq t_{j+1} \tag{2.74}$$

$e_x(t)_{est}$ is the estimates of $|e_x(t)|$. It follows that $|e_x(t)| \leq e_x(t)_{est}$.

Now we are ready to design the FTC law. Since (A_k, B_k) is controllable, let $W_k = W_k^\top > 0$ be associated with a given symmetric positive definite matrix H_k by the Riccati equation

$$A_k^\top H_k + H_k A_k - 2H_k B_k B_k^\top H_k + W_k = 0 \tag{2.75}$$

The design of the proposed fault-tolerant controller makes use of the two following assumptions.

Assumption 2.7. *Given a solution H_k of (2.75), there exists a bounded function $\eta_k(x,t) > 0$ such that*

$$|x^\top H_k g_k(x,t)| \leq \eta_k(x,t)|x^\top H_k B_k| \tag{2.76}$$

Assumption 2.8. $rank(B_k, E_k) = rank(B_k)$.

Remark 2.7. *Inequality (2.76) is not restrictive. Since $g_k(0,t) = 0$, from the Lipschitz condition, one has $|g_k(x,t)| \leq L_k|x|$ and $|x^\top H_k g_k(x,t)| \leq L_k|x^\top H_k||x|$. Since (A_k, B_k) is controllable, the ratio $|x^\top H_k|/|x^\top H_k B_k|$ is homogeneous and its maximal value is found by solving* $\max(|x^\top H_k|)$ *under the constraint $|x^\top H_k B_k| = 1$ providing some bounded solution x^*. Assumption 2.8 is naturally satisfied for the actuator faulty case. Indeed, $rank(B_k) = rank(B_k, E_k) \Leftrightarrow Im(E_k) \subseteq Im(B_k)$ which is equivalent to the existence of B_k^* such that $(I - B_k B_k^*)E_k = 0$.*

The fault-tolerant controller is constructed as

$$u_k(\hat{x}) = u_{k1}(\hat{x}) + u_{k2}(\hat{x}) \tag{2.77}$$

where

$$u_{k1}(\hat{x}) \triangleq -B_k^\top H_k\hat{x} - B_k^* E_k \hat{f}_k^c, \tag{2.78}$$

$$u_{k2}(\hat{x}) \triangleq -\frac{\eta_k(\hat{x},t)}{|\phi_k(\hat{x})| + \varepsilon/2}\phi_k(\hat{x}), \quad \phi_k(\hat{x}) \triangleq \eta_k(\hat{x},t)B_k^\top H_k\hat{x} \tag{2.79}$$

with ε an arbitrarily small positive scalar.

Lemma 2.8. *Suppose that assumptions 2.6-2.8 are satisfied, under the feedback control (2.77)-(2.79), mode k in (2.59)(2.60) is ISpS over* $[t_j,t)$ *w.r.t.* e_x, e_f *and a constant* $\varsigma_k > 0$.

Proof: Applying the control (2.77) to (2.59) results in the closed-loop dynamics

$$\dot{x} = (A_k - B_k B_k^\top H_k)x + B_k B_k^\top H_k e_x + E_k e_f + g_k(x,t) + B_k u_{k2}(\hat{x}) \tag{2.80}$$

Consider a Lyapunov candidate $V_k(x) = x^\top H_k x$, where $H_k > 0$ is defined by (2.75). Its derivative along the system is

$$\begin{aligned}\dot{V}_k \leq & -\lambda_{\min}(W_k)|x|^2 + 2|H_k B_k B_k^\top H_k| \cdot |x| \cdot |e_x| \\ & +2|H_k E_k| \cdot |x| \cdot |e_f| + 2x^\top H_k[B_k u_{k2}(\hat{x}) + g_k(x,t)]\end{aligned} \tag{2.81}$$

From (2.79), one has

$$\begin{aligned}& 2x^\top H_k[B_k u_{k2}(x) + g_k(x,t)] \\ = & \frac{-2\eta_k^2(x,t)|x^\top H_k B_k|^2 + 2x^\top H_k g_k(x,t)\eta_k(x,t)|x^\top H_k B_k| + \varepsilon x^\top H_k g_k(x,t)}{\eta_k(x,t)|x^\top H_k B_k| + \varepsilon/2}\end{aligned} \tag{2.82}$$

Substituting (2.76) into (2.82) yields

$$2x^\top H_k[B_k u_{k2}(x) + g_k(x,t)] < \frac{\varepsilon|x^\top H_k g_k(x,t)|}{\eta_k(x,t)|x^\top H_k B_k| + \varepsilon/2} < \varepsilon \tag{2.83}$$

Assumption 2.7 guarantees that the control $u_{k2}(x)$ is continuous and locally bounded. There always exists a number $\delta_k > 0$ such that $|u_{k2}(\hat{x}) - u_{k2}(x)| \leq \delta_k|e_x|$ for a small $|e_x|$. Due to the convergence of the estimation in Lemma 2.6, it follows that

$$2x^\top H_k[B_k(u_{k2}(\hat{x}) - u_{k2}(x)] \leq 2|H_k B_k| \cdot \delta_k|e_x| \tag{2.84}$$

where $\delta_k > 0$. It also holds that

$$2|H_k B_k B_k^\top H_k| \cdot |x| \cdot |e_x| \leq \varepsilon_2|x|^2 + \frac{|H_k B_k B_k^\top H_k|^2}{\varepsilon_2}|e_x|^2$$

$$2|H_kE_k|\cdot|x|\cdot|e_f| \leq \varepsilon_3|x|^2 + \frac{|H_kE_k|^2}{\varepsilon_3}|e_f|^2$$

where $\varepsilon_2, \varepsilon_3 > 0$ are chosen such that $-\lambda_{\min}(W_k) + \varepsilon_2 + \varepsilon_3 < 0$. Substituting two inequalities above and (2.83), (2.84) into (2.81), one can further obtain

$$\begin{aligned}\dot{V}_k \leq & (-\lambda_{\min}(W_k) + \varepsilon_2 + \varepsilon_3)|x|^2 \\ & + \frac{|H_kB_kB_k^\top H_k|^2}{\varepsilon_2}|e_x|^2 + 2|H_kB_k|\cdot\delta_k|e_x| + \frac{|H_kE_k|^2}{\varepsilon_3}|e_f|^2 + \varepsilon\end{aligned}$$

From Lemma 2.5, the result follows. □

Based on previous analysis for single mode, now we consider the HS (2.55)(2.56). It can be obtained from Lemma 2.8 that there exist continuously differentiable functions $V_k : \mathfrak{R}^n \to \mathfrak{R}_{\geq 0}$, $k \in Q$ and $\bar{\gamma}_1(\cdot)$, $\bar{\gamma}_2(\cdot) \in \mathscr{K}_\infty$, such that $\forall p, q \in Q$

$$\bar{\alpha}_1|x|^2 \leq V_p(x) \leq \bar{\alpha}_2|x|^2 \tag{2.85}$$
$$\dot{V}_p(x) \leq -\lambda_0 V_p(x) + \bar{\gamma}_1(|e_x|) + \bar{\gamma}_2(|e_f|) + \varsigma_0 \tag{2.86}$$
$$V_p(x) \leq \mu V_q(x) \tag{2.87}$$

where constants $\bar{\alpha}_1$, $\bar{\alpha}_2$, λ_0, $\varsigma_0 > 0$, $\mu \geq 1$. The existence of μ is automatically guaranteed for the quadratic Lyapunov functions, e.g., $\mu = \bar{\alpha}_2/\bar{\alpha}_1$.

Since no discrete fault is considered, the system follows the prescribed switching sequence at each switching instant. The observer is modified for the overall system as follows:

- The fault diagnostic scheme is switched according to the current mode at each switching instant.
- The initial states $\hat{x}$ of the current observer are chosen as the final states of the previous observer. The fault estimates $\hat{f}_k^c$ are set to zero at each switching instant.

The following theorem provides a FTC strategy for the overall system with continuous faults.

Theorem 2.5. *Consider the HS (2.55)(2.56) with an initial $x(0)$, each mode satisfies assumptions 2.6-2.8. Let the switching function σ have an ADT τ_a. If $\tau_a > \frac{\ln\mu}{\lambda_0}$, where μ and λ_0 are chosen from (2.86)-(2.87), and $e_x(t^j(k+1))_{est} < e_x(t^j(k))_{est}$ where $t^j(k)$ denotes the time instant that mode j is activated for the kth time, then under the diagnostic scheme (2.65)-(2.67) and controller (2.77)-(2.79), the states of the overall switched system are always bounded and converge to a small closed set.*

Proof: Define $G_a^b(\lambda) = \int_a^b e^{\lambda s}\Phi ds$, where $\Phi \triangleq \bar{\gamma}_1(|e_x|) + \bar{\gamma}_2(|e_f|) + \varsigma_0$. Let $T > 0$ be an arbitrary time. Denote by $t_1, \ldots, t_{N_\sigma(T,0)}$ the switching instants on the interval $(0, T)$, where $N_\sigma(T, 0)$ is defined in (2.45). Similar to [125], consider the function

$$W(s) \triangleq e^{\lambda_0 s}V_{\sigma(s)}(x(s)) \tag{2.88}$$

Since $\sigma(s)$ is constant on each interval $s \in [t_j, t_{j+1})$, from (2.86), we have $\dot{W}(s) \leq e^{\lambda_0 s}\Phi, \forall s \in [t_j, t_{j+1})$. Integrating both sides of the foregoing inequality from t_j to t_{j+1}^- and from (2.87), we obtain $W(t_{j+1}) \leq \mu(W(t_j) + G_{t_j}^{t_{j+1}}(\lambda_0))$. Iterating the foregoing inequality from 0 to $N_\sigma(T,0)$, we get

$$W(T^-) \leq \mu^{N_\sigma(T,0)}\Big(W(0) + \sum_{j=0}^{N_\sigma(T,0)} \mu^{-j} G_{t_j}^{t_{j+1}}(\lambda_0)\Big) \tag{2.89}$$

where T^- denotes the time instant just before T.

Pick $\lambda \in (0, \lambda_0 - \frac{\ln \mu}{\tau_a})$, we have $\tau_a \geq \frac{\ln \mu}{(\lambda_0 - \lambda)}$. Based on (2.45), we have

$$\begin{aligned} \mu^{N_\sigma(T,0)-j} &\leq \mu^{N_0 + \frac{T}{\tau_a} - j + 1 - 1} \\ &\leq \mu^{1+N_0} e^{\tau_a(\lambda_0-\lambda)(\frac{T}{\tau_a} - 1 - j)} \leq \mu^{1+N_0} e^{(\lambda_0-\lambda)(T - t_{j+1})} \end{aligned} \tag{2.90}$$

and

$$G_{t_j}^{t_{j+1}}(\lambda_0) = \int_{t_j}^{t_{j+1}} e^{\lambda_0 s}\Phi ds \leq e^{(\lambda_0-\lambda)t_{j+1}} G_{t_j}^{t_{j+1}}(\lambda) \tag{2.91}$$

Substituting (2.90), (2.91) into (2.89) yields

$$\begin{aligned} W(T^-) &\leq \mu^{N_\sigma(T,0)} W(0) + \sum_{j=0}^{N_\sigma(T,0)} \mu^{1+N_0} e^{(\lambda_0-\lambda)T} G_{t_j}^{t_{j+1}}(\lambda) \\ &\leq \mu^{1+N_0} e^{-\lambda T}\Big(e^{\lambda_0 T - (\lambda_0-\lambda)\tau_a} W(0) + \sum_{j=0}^{N_\sigma(T,0)} e^{\lambda_0 T} G_{t_j}^{t_{j+1}}(\lambda)\Big) \\ &\leq \mu^{1+N_0} e^{-\lambda T} e^{\lambda_0 T}\Big(W(0) + G_0^T(\lambda)\Big) \end{aligned}$$

It follows that

$$\begin{aligned} \bar{\alpha}_1 |x(T)|^2 &\leq \mu^{1+N_0} e^{-\lambda T}(\bar{\alpha}_2 |x(0)|^2 + G_0^T(\lambda)) \\ &\leq \mu^{1+N_0} e^{-\lambda T} \bar{\alpha}_2 |x(0)|^2 + \mu^{1+N_0}\frac{1}{\lambda}\Big(\bar{\gamma}_1(\|e_x\|_{[0,T)}) + \bar{\gamma}_2(\|e_f\|_{[0,T)})\Big) + \bar{\varsigma} \end{aligned}$$

where $\bar{\varsigma} \triangleq (\mu^{1+N_0} \cdot \varsigma_0)/\lambda$.

This implies that the HS is ISpS w.r.t. e_x, e_f and a constant $\bar{\varsigma} > 0$. On the other hand, the inequality $e_x(t^j(k+1))_{est} < e_x(t^j(k))_{est}$ guarantees the global convergence of e_x, which together with the boundness of e_f leads to convergence of the states of the overall HS to a small closed set. This completes the proof. □

Roughly speaking, Theorem 2.5 shows that, if the average dwell time is large enough, then the overall HS is stable and the states are bounded whenever the continuous actuator faults occur in each dwell period.

2.3.3 FTC for Discrete Faults

Since the discrete faults violate the prescribed switching sequence, one would naturally try to first identify the current mode at the beginning of each time interval $[t_j, t_{j+1})$ using a short time period $\Delta t_j \ll t_{j+1} - t_j$, and then control the identified mode in the rest of the time interval.

In this section, a model-free sliding mode observer is proposed to estimate the states of current unknown mode, which together with a series of observers according to system modes, can identify the current mode quickly while guaranteeing the stability of the system in each Δt_j.

In each Δt_j, the control signal is set to zero, thus no continuous fault signal appears in Δt_j.

The system (2.59)-(2.60) without input can be written as

$$\dot{x}(t) = A_{k'}x(t) + g_{k'}(x(t),t), \quad y(t) = Cx(t) \tag{2.92}$$

where $k' \in Q$ is unknown. The system (2.92) is rewritten as

$$\dot{x}(t) = \bar{A}x(t) + F_{k'}(x(t),t), \quad y(t) = Cx(t) \tag{2.93}$$

where $F_{k'}(x,t) \triangleq A_{k'}x + g_{k'}(x,t) - \bar{A}x$, $\bar{A}$ is a matrix such that the pair $(\bar{A}, C)$ is observable. There exists a matrix $\bar{L}$ such that $\bar{A} - \bar{L}C$ is Hurwitz stable. Denote $\bar{P}$ as the symmetric positive definite solution of the Lyapunov equation $(\bar{A} - \bar{L}C)^\top \bar{P} + \bar{P}(\bar{A} - \bar{L}C) = -\bar{Q}$ with a given symmetric positive definite matrix $\bar{Q}$.

A model-free sliding mode observer is designed as

$$\dot{\bar{x}}(t) = \bar{A}\bar{x}(t) + S(\bar{e}_x(t), \rho_j) + L(y(t) - \bar{y}(t)), \quad \bar{y}(t) = C\bar{x}(t) \tag{2.94}$$

where $\bar{e}_x \triangleq x - \bar{x}$, and

$$S(\bar{e}_x(t), \rho_j) \triangleq \frac{\bar{P}^{-1}C^\top C\bar{e}_x(t)}{|C\bar{e}_x(t)|}\rho_j$$

with a constant $\rho_j > 0$ which will be designed later.

From (2.93) and (2.94), we have

$$\dot{\bar{e}}_x(t) = (\bar{A} - \bar{L}C)\bar{e}_x(t) - S(\bar{e}_x(t), \rho_j) + F_{k'}(x(t),t) \tag{2.95}$$

Assumption 2.9. *There exists a bounded function $h_{k'}(x,t)$, $|h_{k'}(x,t)| < \rho|x|$ for $\rho > 0$ such that*

$$F_{k'}(x,t) = -\bar{P}^{-1}C^\top h_{k'}(x,t) \tag{2.96}$$

Remark 2.8. *Eq.(2.96) is not hard to be satisfied if $F_{k'}(x,t)$ is bounded. It is clear that there exists a constant $\bar{F} > 0$ such that $|F_{k'}(x,t)| \leq \bar{F}|x|$. If x is bounded in Δt_j (which will be shown later), then $|F_{k'}(x,t)|$ is naturally bounded.*

Lemma 2.9. *Under Assumption 2.9, there exists a $\rho_j > 0$ such that, if the states in each Δt_j are bounded, then the origin of the system (2.95) is asymptotically stable.*

Proof: Consider a Lyapunov function candidate $\bar{V}(\bar{e}_x) = \bar{e}_x^\top \bar{P}\bar{e}_x$. Its derivative along the system (2.95) is

$$\begin{aligned}\dot{\bar{V}} &= -\bar{e}_x^\top \bar{Q}\bar{e}_x + 2\bar{e}_x^\top \bar{P}F_{k'}(x,t) - 2|C\bar{e}_x|\rho_j \\ &\leq -\bar{e}_x^\top \bar{Q}\bar{e}_x + 2|C\bar{e}_x|\cdot|x|\rho - 2|C\bar{e}_x|\rho_j\end{aligned} \tag{2.97}$$

If $|x|$ is always bounded in Δt_j, then we can choose a ρ_j large enough such that $\dot{\bar{V}} < -\bar{e}_x^\top \bar{Q}\bar{e}_x$ in Δt_j. This completes the proof. □

In order to identify the current mode, a series of following observers are also needed

$$\text{observer } i: \ \dot{\hat{x}}_i = A_i\hat{x}_i + g_i(\hat{x}_i,t) + K_i(y-\hat{y}_i),\ \hat{y}_i = C\hat{x}_i,\ i \in Q \tag{2.98}$$

which are the same as (2.65)-(2.67) without u_i and $\hat{f}_i^c$. e_{xi} denotes the state estimation error using observer i.

The sliding mode observer in (2.94) and all observers in (2.98) are invoked to estimate the current mode simultaneously in Δt_j. Set the initial states of observers to $\hat{x}(t_j^-)$ at $t = t_j$. It is supposed that all modes are *discernable* [20], i.e., for mode i without input, $|e_{xi}|$ converges faster than $|e_{xj}|, \forall j \in Q, j \neq i$. This is a quite general condition for switching control problem as for instance in [20],[129] and [68]. Roughly speaking, it means that all the modes are not overlapping.

The identifiability is analyzed in the following lemma.

Lemma 2.10. *The current mode k' can be identified at time instants $t_j + \Delta t_j$, where Δt_j can be made arbitrarily small.*

Proof: It is evident that $|e_{xk'}| - |\bar{e}_x| \leq |\bar{x} - \hat{x}_{k'}| \leq |e_{xk'}| + |\bar{e}_x|$, one has

$$|\bar{x} - \hat{x}_i| - |\bar{x} - \hat{x}_{k'}| \geq \chi, \ \ \forall i \in Q, i \neq k'$$

where $\chi \triangleq |e_{xi}| - 2|\bar{e}_x| - |e_{xk'}|$. All observers share the same initial states at $t = t_j$, so $\chi(t_j) < 0$. From Lemmas 2.7, 2.8, and (2.98), it follows that if the current mode is mode k', then $|e_{xk'}|$ converges to zero at a given rate depending on $K_{k'}$ and $Q_{k'}$. Lemma 2.9 ensures $|\bar{e}_x|$ also converges to zero at a given rate. Note that all modes are discernable, there always exist $K_{k'}$, $Q_{k'}$, $\bar{L}$, $\bar{Q}$ and ρ_j such that $\chi(t) > 0 \ \forall t \geq t_j + \Delta t_j$ with arbitrarily small Δt_j. It follows that $|\bar{x} - \hat{x}_{k'}|$ is minimal $\forall t \geq t_j + \Delta t_j$. This implies that mode k' can be identified. □

The work of identifier is to find $\hat{x}_{k'}$ that is most similar to $\bar{x}$. Although Δt_j can be made arbitrarily small as in Lemma 2.10, a small delay is necessary to overcome the possible overshoot of the state trajectories. Since $\hat{x}_i$, $\hat{x}_{k'}$ and $\bar{x}$ are all continuous and measurable, in the real implementation of the identifier, high order time derivatives of the signals can help to find the similarity (as using 1-order time derivative of signals in the simulation).

The following assumption is imposed to avoid that the system states escape into infinity or a large region before a proper controller is invoked.

Assumption 2.10. *The Δt_j determined by Lemma 2.10 is always within the following set*

$$\Omega_{\Delta t_j} \triangleq \{\Delta t_j | \Delta t_j < t_{j+1} - t_j \text{ and } |\bar{x}(t_j + \Delta t_j)| \leq \xi |\bar{x}(t_j)|\} \tag{2.99}$$

where $\xi \geq 1$, $\forall k' \in Q$, $j = 1, 2 \ldots$.

Remark 2.9. *The selection of ξ depends on system dynamics. Assumption 2.10 is not hard to be satisfied, since Δt_j can be made arbitrary small (due to Lemma 2.10). If the system without control is still stable or divergent slowly (this is the ideal case), then it is also possible that $|\bar{x}(t_j + \Delta t_j)| < \xi |\bar{x}(t_j)|$ when the current mode is detected at $t + \Delta t_j$.*

From (2.99), lemmas 2.7 and 2.9, we have

$$\begin{aligned}
|x(t_j + \Delta t_j)| &\leq |\bar{x}(t_j + \Delta t_j)| + |\bar{e}_x(t_j + \Delta t_j)| \\
&\leq |\bar{x}(t_j + \Delta t_j)| + \sqrt{\bar{\alpha}_3 e^{\bar{\alpha}_4 \Delta t_j}} |\bar{e}_x(t_j)| \\
&\leq |\bar{x}(t_j + \Delta t_j)| + \sqrt{\bar{\alpha}_3} e_x(t_j)_{est} \\
&\leq \xi |\bar{x}(t_j)| + \varepsilon_j
\end{aligned} \tag{2.100}$$

where $\bar{\alpha}_3 > 0$, $\bar{\alpha}_4 < 0$ are determined by $\bar{P}, \bar{Q}$. k'' denotes the mode activated in $[t_{j-1}, t_j)$. Note that $\varepsilon_j > 0$ can be calculated from the estimates $e_x(t_j)_{est}$ in (2.74). The main contribution of inequality (2.100) is that it provides a bound of $|x(t)|$ in Δt_j, which can be used to design ρ_j in (2.94).

The proposed identifier in this section has three good properties:

- It can provide accurate state estimates after each Δt_j .
- It is not affected by continuous actuator faults since no control signal are applied in Δt_j.
- It avoids the large transient overshoot of states in Δt_j.

2.3.4 FTC Framework

Based on the analysis in sections 2.3.2-2.3.3, the FTC problem for both continuous and discrete faults is discussed in this section. Fig.2.2 shows the block diagram of the framework, where the plant is connected with three parts: a series of observers and controllers, a model-free observer, and an identifier. The *fault tolerant control framework* works as the following procedure:

1) At switching instant t_j, stop the fault diagnostic scheme (2.65)-(2.67), set control signals and fault estimates to zero.

2) Invoke the model free observer (2.94), a series of observers (2.98), initialize all observers at t_j with the same states $\hat{x}(t_j^-)$.

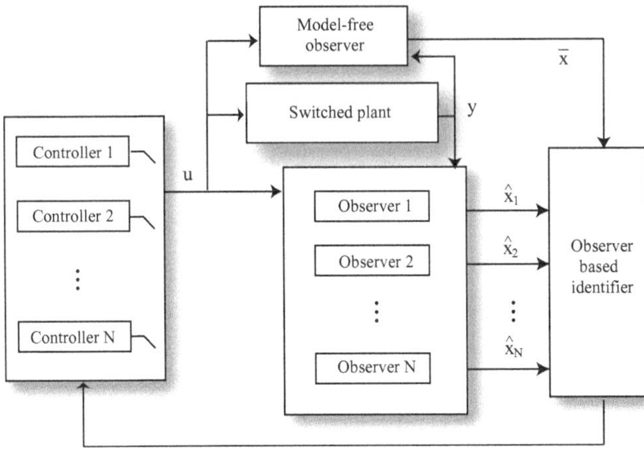

Fig. 2.2 The FTC framework

3) Choose ρ_j by (2.97) and (2.100), invoke the identifying scheme in Lemma 2.10 into the system.
4) Determine Δt_j based on Lemma 2.10.
5) At $t_j + \Delta t_j$, stop the identifier, apply the fault diagnostic scheme (2.65)-(2.67) and controller (2.77)-(2.79) into the system according to the current mode.
6) At switching instant t_{j+1}, go to 1).

The following theorem is given to guarantee the stability of overall system with both continuous and discrete faults.

Theorem 2.6. *Consider the HS (2.55)(2.56) with an initial $x(0)$ satisfying assumptions 2.9, 2.10, with each mode satisfying assumptions 2.6-2.8. Let the switching function σ have an ADT τ_a. If $\tau_a > \frac{\ln \mu}{\lambda_0}$, and $e_x(t^j(k+1))_{est} < e_x(t^j(k))_{est}$, then the proposed FTC framework guarantees that the states of the HS are always bounded and converge to a small closed set.*

Proof: Following the result of Theorem 2.5, we have

$$W(t_{j+1}) \leq \mu(W(t_j + \Delta t_j) + G_{t_j+\Delta t_j}^{t_{j+1}}(\lambda_0)) \tag{2.101}$$

If the current mode is mode k', then

$$W(t_j + \Delta t_j) = e^{\lambda_0(t_j+\Delta t_j)} V_{k'}(x(t_j + \Delta t_j)) \tag{2.102}$$

From Lemma 2.9 and (2.99), we have

$$\begin{aligned} |x(t_j + \Delta t_j)| &\leq |\bar{e}_x(t_j + \Delta t_j)| + |\bar{x}(t_j + \Delta t_j)| \\ &\leq \sqrt{\bar{\alpha}_3 e^{\bar{\alpha}_4 \Delta t_j}} |\bar{e}_x(t_j)| + \xi |\bar{x}(t_j) - x(t_j) + x(t_j)| \end{aligned}$$

$$\leq (\sqrt{\bar{\alpha}_3 e^{\bar{\alpha}_4 \Delta t_j}} + \xi)|\bar{e}_x(t_j)| + \xi|x(t_j)| \tag{2.103}$$

From (2.126), we further have

$$\begin{aligned} V_{k'}(x(t_j+\Delta t_j)) &\leq \bar{\alpha}_2 |x(t_j+\Delta t_j)|^2 \\ &\leq 2\bar{\alpha}_2(\sqrt{\bar{\alpha}_3 e^{\bar{\alpha}_4 \Delta t_j}} + \xi)|\bar{e}_x(t_j)|^2 + \frac{2\bar{\alpha}_2 \xi^2}{\bar{\alpha}_1} V_{k'}(x(t_j)) \end{aligned} \tag{2.104}$$

Define $\psi(t_j) \triangleq 2\bar{\alpha}_2(\sqrt{\bar{\alpha}_3 e^{\bar{\alpha}_4 \Delta t_j}} + \xi)|\bar{e}_x(t_j)|^2 e^{\lambda_0 \Delta t_j}$, $\Delta_j \triangleq \frac{(2\bar{\alpha}_2 \xi^2)}{\bar{\alpha}_1} e^{\lambda_0 \Delta t_j}$. In each period Δt_j, there are no input and continuous actuator fault, so $e_f(t) = 0$ $\forall t \in [t_j, t_j + \Delta t_j)$, and it is natural that $G_{t_j+\Delta t_j}^{t_{j+1}}(\lambda_0) \leq G_{t_j}^{t_{j+1}}(\lambda_0)$. Iterating the inequality (2.101) from 0 to N_σ together with (2.104), where N_σ denotes $N_\sigma(T,0)$, we get

$$\begin{aligned} W(T^-) &\leq \Big(\mu^{N_\sigma} \prod_{s=0}^{N_\sigma - 1} \Delta_s\Big) W(0) + \sum_{i=1}^{N_\sigma - 1} \Big(\mu^{N_\sigma - i+1} e^{\lambda_0 t_i} \psi(t_{i-1}) \prod_{\bar{s}=i}^{N_\sigma - 1} \Delta_{\bar{s}}\Big) \\ &+ \mu e^{\lambda_0 T} \psi(t_{N_\sigma - 1}) + \sum_{j=1}^{N_\sigma - 1} \Big(\mu^{N_\sigma - j+1} G_{t_{j-1}}^{t_j}(\lambda_0) \prod_{l=j}^{N_\sigma - 1} \Delta_l\Big) + \mu G_{t_{N_\sigma - 1}}^{\mathrm{T}}(\lambda_0) \end{aligned}$$

Since Δt_j is a bounded small time period, there exists a constant $\bar{\Delta} > 0$ such that $\prod_{s=i}^{N_\sigma - 1} \Delta_s \leq \bar{\Delta}$ $\forall i \in \{1, 2, \ldots, N_\sigma - 1\}$. Note that $e^{\lambda_0 t_i} \leq e^{\lambda_0 T}$, one has

$$\begin{aligned} W(T^-) &\leq \mu^{N_\sigma} \bar{\Delta} W(0) + e^{\lambda_0 T} \bar{\Delta} \sum_{i=1}^{N_\sigma - 1} (\mu^{N_\sigma - i+1} \psi(t_{i-1})) + \mu e^{\lambda_0 T} \psi(t_{N_\sigma - 1}) \\ &+ \bar{\Delta} \sum_{j=1}^{N_\sigma - 1} (\mu^{N_\sigma - j+1} G_{t_{j-1}}^{t_j}(\lambda_0)) + \mu G_{t_{N_\sigma - 1}}^{\mathrm{T}}(\lambda_0) \end{aligned} \tag{2.105}$$

From (2.90) and (2.91), we get $\mu^{N_\sigma - j+1} G_{t_{j-1}}^{t_j}(\lambda_0) \leq \mu^{1+N_0} e^{(\lambda_0 - \lambda)T} G_{t_{j-1}}^{t_j}(\lambda)$, for $0 < \lambda < \lambda_0$. Taking the forgoing inequality into (2.105), and following the same way as in Theorem 2.5, we can finally obtain

$$\begin{aligned} \bar{\alpha}_1 |x(T)|^2 &\leq \beta_a(|x(0)|, t) + \gamma_{\bar{e}}(\|\bar{e}_x(t_j)\|_{[0,T)}) \\ &+ \gamma_{ex}(\|e_x\|_{[0,T)}) + \gamma_{ef}(\|e_f\|_{[0,T)}) + \bar{\varsigma}_2 \end{aligned} \tag{2.106}$$

where $\beta_a \in \mathcal{KL}$, $\gamma_{\bar{e}}$, γ_{ex}, $\gamma_{ef} \in \mathcal{K}_\infty$, $\bar{\varsigma}_2 \geq 0$ are determined from (2.105).

The inequality (2.106) implies the ISpS of HS w.r.t. $e_x(t)$, $e_f(t)$, $\bar{e}_x(t_j)$ and a constant $\bar{\varsigma}_2 > 0$, where $j = 1, 2 \ldots$. which, together with $e_x(t^j(k+1))_{est} < e_x(t^j(k))_{est}$ and the boundness of e_f guarantees the global convergence of the states of the system to a small closed set. □

Remark 2.10. *Note that $\bar{e}_x(t_j)$ is a discrete vector, since its value is captured only at each switching instant. Moreover, it has been shown that $|\bar{e}_x(t_j)|$ $\forall k \in Q$ is bounded. Theorem 2.6 also implies that the value of $\Delta t_j \in \Omega_{\Delta t_j}$ does not change the system's*

ISpS property. Appropriate selection of Δt_j can reduce the bound of x in the sense of ISpS in (2.106).

Remark 2.11. *Switching the input between the nominal control strategy and zero value has been shown to be an efficient way for performance-based FTC [103]. It is natural for HS that, at each t_j, the controller is switched on according to the next mode. Setting the input to zero during a short period after each switching is reasonable.*

Example 2.2: [132] A $\bar{m}$-phase switched reluctance motor (SRM) system is employed to illustrate a potential application field of the approach. $x = [\theta_m, \ \omega_m]^\top$ is the state, where θ_m, ω_m denote the angular position and velocity of the motors.

The simplified system model is expressed as follows:

$$\begin{aligned}\dot{\theta}_m &= \omega_m \\ \dot{\omega}_m &= -\frac{\kappa_e}{J_m}\sin(\theta_m) - \frac{b_i}{J_m}\omega_m + \frac{c_i}{J_m}u_i, \quad i = 1,2,\dots,\bar{m}\end{aligned}$$

where J_m denotes the inertia of the motor. $\kappa_e > 0$ is the elasticity constant. u_i is the voltage applied to the motor of phase i, with b_i and c_i being the related viscous friction and the amplifier gain. In the simulation, $\bar{m} = 3$ is considered. The parameters are $J_m = 0.935\, kgm^2$, $\kappa_e = 0.311\, Nm/rad$, $b_1 = 1.17\, Nms/rad$, $b_2 = 2.23\, Nms/rad$, $b_3 = 0.54\, Nms/rad$, $c_1 = 20.196\, Nm/V$, $c_2 = 35.31\, Nm/V$, $c_3 = 12.44\, Nm/V$. We further describe the model by the general form (2.55)-(2.56) with

$$A_1 = \begin{bmatrix} 0 & 1 \\ 0 & -1.2513 \end{bmatrix}, \ A_2 = \begin{bmatrix} 0 & 1 \\ 0 & -2.385 \end{bmatrix}, \ A_3 = \begin{bmatrix} 0 & 1 \\ 0 & -0.5775 \end{bmatrix}$$

$$B_1 = \begin{bmatrix} 0 \\ 21.6 \end{bmatrix}, \ B_2 = \begin{bmatrix} 0 \\ 37.765 \end{bmatrix}, \ B_3 = \begin{bmatrix} 0 \\ 13.305 \end{bmatrix}, \ g(x) = \begin{bmatrix} 0 \\ -0.333\sin x_2 \end{bmatrix}$$

The position of the motor phase can be measured via the shaft position sensor, while the motor velocity is often estimated by timing the interval between phase commutations of SRM. A coupled output signal of the angular position and velocity is obtained shared by all phases, the output matrix $C = [1\ 2]$.

The *continuous actuator fault* is considered only in mode 1 with $E_1 = [0\ \ -12.5]^\top$. The matrix K_1 and Q_1 are chosen as

$$K_1 = \begin{bmatrix} 3 \\ -1.8 \end{bmatrix}, \ Q_1 = \begin{bmatrix} 0.1105 & -0.0007 \\ -0.0007 & 0.0986 \end{bmatrix}$$

Solving Eqs.(2.63)-(2.64), we obtain $R_1 = 0.3225$ and

$$P_1 = \begin{bmatrix} 0.0157 & 0.0258 \\ 0.0258 & 0.0516 \end{bmatrix}$$

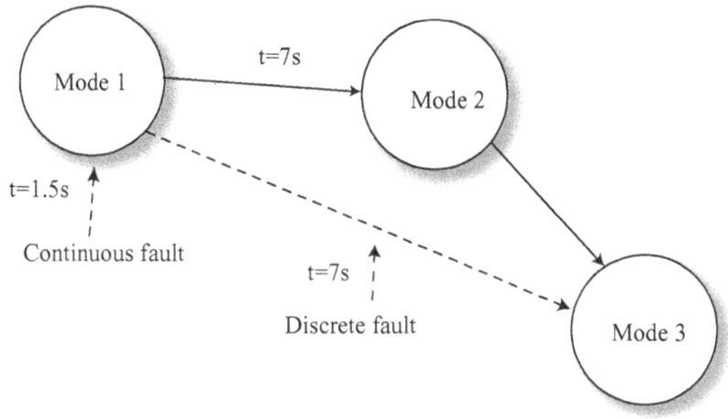

Fig. 2.3 An illustration of system's behavior

On the other hand, by choosing $W_1 = I_{2\times 2}$, we obtain the matrix H_1 from (2.75) as

$$H_1 = \begin{bmatrix} 1.0330 & 0.0327 \\ 0.0327 & 0.0325 \end{bmatrix}$$

The bounded function $\eta_1(x,t)$ is selected from (2.76) as

$$\eta_1(x,t) = \frac{0.333|0.0151x_1 - 0.0377x_2|}{|0.3266x_1 - 0.8150x_2|}$$

Take $\Gamma_1 = 20$, $\vartheta_1 = 8$, $\varepsilon = 0.01$. The related parameters of modes 2 and 3 can be obtained following the same way as for mode 1, which is omitted.

The considered switching sequence is: mode 1$\rightarrow$ mode 2$\rightarrow$ mode 3 as shown in Fig. 2.3. $N_0 = 0$. From (2.126)-(2.87), choose $\mu = 35$, $\lambda_0 = 0.8$. The switching instants are prescribed as $t_1 = 7s$, $t_2 = 14s$, which satisfy the ADT scheme in theorems 2.3 and 2.4. The system is initialized in mode 1 with $x(0) = [0.05\ \ 0.2]^\top$.

f_1^c is assumed to occur at $t = 1.5s$ as

$$f_1^c(t) = \begin{cases} 0, & 0s \leq t < 1.5s \\ 0.5 + 0.3\sin(4\pi t), & 1.5s \leq t < 7s \end{cases}$$

which corresponds to an increase in the friction of the motor, that makes the voltage deviates from normal situation. Fig. 2.4 shows the fault estimation performance, from which we can see that $\hat{f}_1^c$ follows f_1^c rapidly with a very small overshoot.

The *discrete fault* occurs at $t = t_1 = 7s$, which represents the abnormal switching behavior of the motor phase that makes mode 1 switch to mode 3 as in Fig. 2.3. At $t = 7s$, the identifier scheme is invoked. The parameter of the model free observer in (2.94) is designed as

$$\bar{A} = \begin{bmatrix} 0 & 1 \\ 0 & -1 \end{bmatrix}, \quad \bar{L} = \begin{bmatrix} 2.8 \\ -1.6 \end{bmatrix}, \quad \bar{P} = \begin{bmatrix} 2.7055 & 4.5351 \\ 4.5351 & 9.0703 \end{bmatrix}$$

where $\bar{P}$ is obtained with $\bar{Q} = \begin{bmatrix} 0.6384 & 0.6540 \\ 0.6540 & 1.8141 \end{bmatrix}$. There exists a $h(x,t)$ with ρ selected as 3. The speed of the rotor can cause an increase of the current after the corresponding voltage control has been switched off. As a consequence, such residual current can have an adverse effect on torque production at each switching instant. To avoid an unexpected oscillation of rotor, we select $\xi = 2$. From (2.97) and (2.100), we can also choose $\rho_1 = 5$. A boundary layer compensator technique [150] is used with a bound number 0.02 to eliminate the chattering.

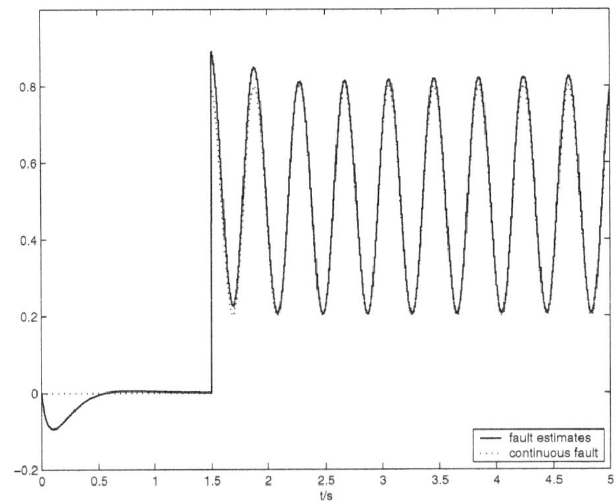

Fig. 2.4 Fault diagnosis

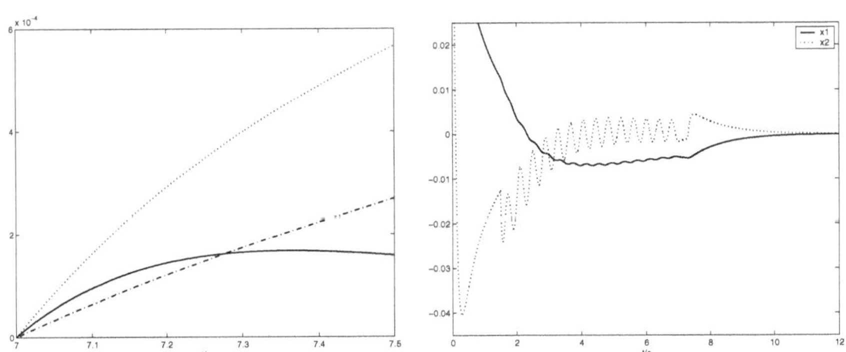

(a) Identifier performance ($|\bar{x}-\hat{x}_3|$: solid; $|\bar{x}-\hat{x}_2|$: dotted; $|\bar{x}-\hat{x}_1|$: dash-dotted)

(b) States trajectories

Fig. 2.5 FTC performance

Fig. 2.5(a) shows the performance of the identifier. Although $|\bar{x}-\hat{x}_1|$ is minimal at the beginning, $|\bar{x}-\hat{x}_3|$ is minimal and decreases at $7.35s$, while $|\bar{x}-\hat{x}_1|$ and $|\bar{x}-\hat{x}_2|$ still diverge. This implies that Mode 3 and consequently the discrete fault can be identified with $\Delta t_1 = 0.35$. The controller and fault diagnostic scheme for mode 3 are invoked into the system at $t = 7.35s$. The state trajectories throughout the system process is shown in Fig. 2.5(b), it can be seen that the states are always bounded.

2.4 Global Passivity

In sections 2.1-2.3, we designed FTC law in each faulty mode such that it is stable, then applied the standard stability results for HS. In the following two sections, we will research directly the stability of HS without reconfiguring the controller in each mode. We introduce, for the first time, the passivity theory into the FTC analysis of HS.

2.4.1 Passivity and Fault Diagnosis

Passivity theory, that provides a bridge between achievable system performances and energy-like considerations, has been widely used to analyze stability of nonlinear systems, where systems can not store more energy than that supplied by the environment outside [127]. Passivity concept has also been adopted for switched and HS [156], [151], where each mode is assumed to be passive.

We shall introduce the passivity theory into the FTC design for HS where each mode is passive in the healthy situation, and may be not passive due to the fault.

Consider the affine nonlinear system

$$\begin{aligned} \dot{x} &= f(x) + g(x)u + \Delta(x) \\ y &= h(x) \end{aligned} \tag{2.107}$$

where $x \in X \subset \Re^n$ are measurable states, $u \in U \subset \Re^m$ are inputs, $y \in Y \subset \Re^m$ are outputs. The fault is modelled by an unknown function $\Delta(x) \in \Re^n$, which effectively represents the process faults [10], and occurs at an unknown time. f, g, h and Δ are smooth functions.

Definition 2.6. *[14] A system (2.107) with $\Delta \equiv 0$ is* passive *if there exists a nonnegative function $V : X \to \Re$, which satisfies $V(0) = 0$, called the* storage function, *and a* supply rate $y^\top u$, *such that for all initial states $x(0) \in X$, $u \in U$, and $t \geq 0$*

$$\underbrace{V(x(t)) - V(x(0))}_{\textit{stored energy}} \leq \underbrace{\int_0^t y^\top(s)u(s)ds}_{\textit{supplied energy}} \tag{2.108}$$

where $x(t)$ are the states at time t.

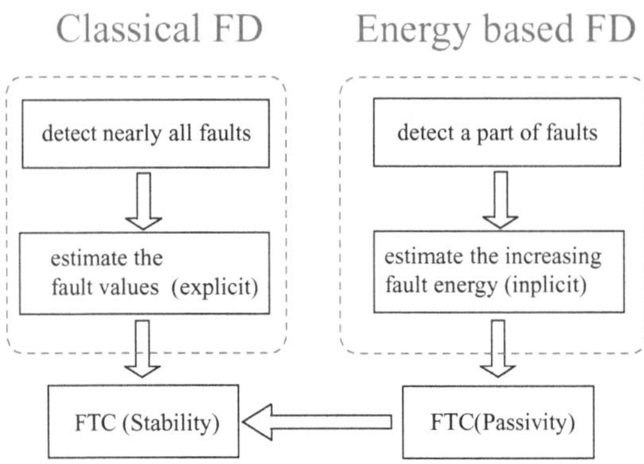

Fig. 2.6 Comparison of FD methods

The inequality (2.108) is called dissipativity inequality [127], which formalizes the property that the increase in stored energy is never greater than the amount of energy supplied by the environment. A passive system is easy to control, choosing $u = -\phi(y)$, where $\phi : U \to Y$ is a smooth function and $\phi(0) = 0$, such that $y^\top \phi(y) > 0$ for each nonzero y leads to Lyapunov stability [14].

Now we address the FD problem. As shown in Fig. 2.6, most classical methods [36, 18] are designed such that the explicit values of faults can be estimated. Here we develop a novel energy based FD technique that is concerned with the energy analysis and has its root in the passivity. Under the passivity framework, we show that only a part of faults needs to be detected and estimated implicitly.

In the following, we assume that V is a $\mathscr{C}^1$ function. The passivity property is equivalent to

$$\left[\frac{\partial V}{\partial x}(x)\right]^\top [f(x) + g(x)u] \leq y^\top u \tag{2.109}$$

Once a fault occurs, the constraint (2.108) may be violated. Adding $\Delta(x)$ into (2.109) and integrating both sides yields

$$V(x(t)) - V(x(0)) \leq \int_0^t y^\top(s)u(s)ds + \underbrace{\int_0^t \left[\frac{\partial V}{\partial x}(x)\right]^\top \Delta(x(s))ds}_{\text{fault energy } E_f} \tag{2.110}$$

As indicated in (2.110), the energy dissipativity property changes due to the fault. The fault may help to dissipate the stored energy ($E_f < 0$) or increase the stored energy ($E_f > 0$). We only care about the faults that result in $V(x(t)) - V(x(0)) > \int_0^t y^\top(s)u(s)ds$. A diagnosis threshold can be designed as

$$V(x(t)) - V(x(0)) = \int_0^t y^\top(s)u(s)ds \tag{2.111}$$

This is also called *lossless* property [14]. Note that the faults with $E_f < 0$ are not necessary to be detected since they do not change the energy dissipativity. Once the left side of (6.6) becomes larger than the right side, the fault is detected. We estimate such fault value implicitly as $V(x(t)) - V(x(0)) - \int_0^t y^\top(s)u(s)ds$. More precisely, we estimate the energy that increases due to the fault and check whether the system is still passive or not. This information will be used for fault tolerance analysis [141].

2.4.2 Fault Tolerance Analysis of Hybrid Systems

The hybrid system takes the form

$$\begin{aligned} \dot{x} &= f_\sigma(x) + g_\sigma(x)u_\sigma + \Delta_\sigma(x) \\ y &= h_\sigma(x) \end{aligned} \tag{2.112}$$

where $x \in X \subset \Re^n$ is continuous everywhere, $u_\sigma \in \Re^{m_\sigma}$, $h_\sigma \in \Re^{m_\sigma}$. All f_σ, g_σ, h_σ and Δ_σ are smooth functions. $\sigma(t) : [t_0, \infty) \to Q = \{1, 2, \ldots, N\}$ denotes the *switching function*. We denote by t_k, $k = 1, 2, \ldots$ the kth switching time. $N_{\sigma(t)}$ represents the number of switchings in $[0, t)$. t_{kj}, $k = 1, 2, \ldots$, $j \in Q$ denotes the kth switching time that mode j is activated. Suppose that there exists N non-negative storage functions $V_p(x)$, and α_1^p, $\alpha_2^p \in \mathscr{K}_\infty$, $\forall p \in Q$ that satisfy

$$\alpha_1^p(|x|) \le V_p(x) \le \alpha_2^p(|x|) \tag{2.113}$$

such that mode p is passive with $V_p(x)$ in the healthy situation.

In this work, we neither reconfigure the controller u_σ nor adjust the switching law σ. We analyze fault tolerance of the HS (2.112) under the original u_σ and σ. It will be shown that under the *global energy dissipativity*, the stability of the HS can be achieved in spite of non passive modes.

Definition 2.7. *A switched system (2.112) is* globally passive *if there exists nominal controllers* u_1, u_2, ...,u_N, *such that for all initial states* $x(0) \in X$, *and* $T \ge 0$

$$V_{\sigma(T)}(x(T)) - V_{\sigma(0)}(x(0)) - E_{tr}(x(0)) \le \int_0^T W(s)ds \tag{2.114}$$

where $W(s) \le 0$ *is defined as*

$$\int_0^T W(s) \triangleq \sum_{k=0}^{N_{\sigma(T)}} \int_{t_k}^{t_{k+1}} \Big(y^\top(s)u_{\sigma(s)}(s) + \Big[\frac{\partial V_{\sigma(s)}}{\partial x}(x)\Big]^\top \Delta_{\sigma(s)}(x(s)) \Big) ds \tag{2.115}$$

and $E_{tr} = \sum_{k=1}^{N_{\sigma(T)}} \left[V_{\sigma(t_k)} - V_{\sigma(t_k^-)} \right]$ *is bounded by a constant and tends to zero as* $x(0)$ *goes to origin.*

The left side of (2.114) represents the sum of stored energies of all modes, which could also be written as $\sum_{k=0}^{N_{\sigma(T)}}\left[V_{\sigma(t_{k+1}^{-})}-V_{\sigma(t_k)}\right]$ where $t_0=0$, $t_{N_{\sigma(T)}+1}=T$. The formulation of (2.114) is consistent with the standard passivity inequality, E_{tr} denotes the total transient energy. As shown later, E_{tr} may be eliminated under some conditions.

It is clear from (2.115) that the right hand of (2.114) denotes the total supplied energy and "fault" energy. Since $W(s)\le 0$, it follows that under the nominal controllers u_1, u_2, ...,u_N, the sum of the supplied energy during $[0,T)$ can compensate the increasing energy due to faults. This means that the total stored energies still dissipative in spite of faults.

Global passivity balances the total energy throughout the overall process, while no individual passivity of each mode is required. We shall prove that the global passivity includes the passivity property proposed in [156] as in the following proposition.

Proposition 2.1. *If each mode of a HS (2.112) is passive as in (2.108), and there exist functions $\omega_k^{k+1}(t)$, called cross supply rates such that $\omega_k^{k+1}(t)\le\phi_k^{k+1}(t)$ where $\phi_k^{k+1}(t)\in\mathscr{L}_1$ and*

$$V_q(x(t_{q(k+1)}))-V_q(x(t_{qk}))\le\int_{t_{qk}}^{t_{q(k+1)}}\omega_k^{k+1}(s)ds \tag{2.116}$$

then the system (2.112) is globally passive.

Proof: The passivity of each mode leads to the fact that each energy is non-decreasing when the related mode is activated. Suppose mode q is activated at the time T, from (2.116), we obtain

$$V_q(x(T))-V_q(x(t_{q1}))-\Theta(x(0))\le\int_0^T W(s)ds \tag{2.117}$$

where $W(s)\le 0$, $\Theta(x(0))$ is a constant and tends to zero as $x(0)$ goes to the origin. This constant is obtained from the fact that $\sum_{k=1}^{\infty}\int_{t_{qk}}^{t_{q(k+1)}}\omega_k^{k+1}(s)ds$ is bounded, since $\phi_k^{k+1}(t)\in\mathscr{L}_1$. On the other hand, for any $x(0)$, $V_{\sigma(0)}(x(0))$ is bounded, there exists a constant $\Phi\triangleq V_{\sigma(0)}(x(0))-V_q(x(t_{q1}))$, which together with (2.117), leads to the result. □

Global passivity implies the stability as shown below.

Theorem 2.7. *If a HS (2.112) is globally passive, then the origin of the system is stable in spite of faults.*

Proof: For a given arbitrary $\varepsilon>0$, since V_i is continuous and $V_i(0)=0$, based on (2.113), we can choose $\varepsilon_2^i>0$ such that $V_i<\varepsilon_2^i$ leads to $(\alpha_1^i)^{-1}(V_i)<\varepsilon$. Pick $\varepsilon_3=\min_i[\varepsilon_2^i]$, since E_{tr} tends to zero as $x(0)$ goes to the origin, we can choose ε_4 such that $|x(0)|<\varepsilon_4$ results in $\max_i[\alpha_2^i(|x(0)|)+E_{tr}(x(0))]<\varepsilon_3$. Thus, followed by (2.114),

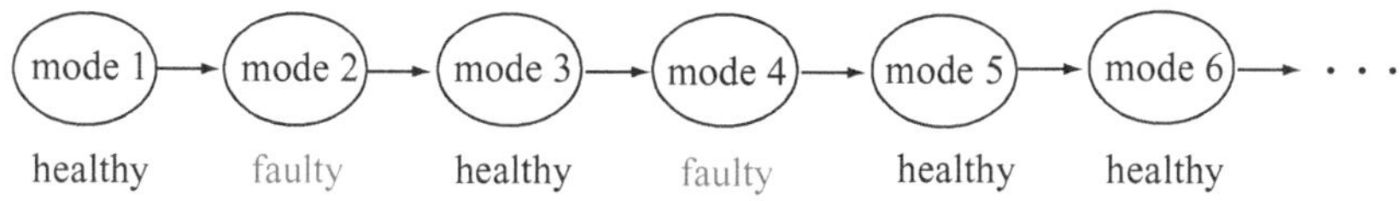

Fig. 2.7 Switching sequence

we find that if the system starts in $B(\varepsilon_4)$, we will stay within $B(\varepsilon)$. This completes the proof. □

Theorem 2.7 provides us a method to check the fault tolerance, which is equivalent to check the global passivity. However, when we use (2.114) to check the fault tolerance at any instant T, one obstacle appears since we are not sure whether there is a constant bound of the total transient energy for all $t \geq T$. This motivates the following result.

Proposition 2.2. *If a HS (2.112) is globally passive, and $V_{\sigma(t)}(x(t)) \leq V_{\sigma(t^-)}(x(t))$ at each switching instant t, then (2.114) holds with $E_{tr} = 0$.*

Proof: The result follows the fact that

$$\begin{aligned}
&\sum_{k=0}^{N_{\sigma(T)}} \left[V_{\sigma(t_{k+1}^-)}(x(t_{k+1})) - V_{\sigma(t_k)}(x(t_k))\right] \\
&\quad = V_{\sigma(T)}(x(T)) - V_{\sigma(t_{N_{\sigma(T)}})}(x(t_{N_{\sigma(T)}})) + \cdots \\
&\quad + V_{\sigma(t_{k+1}^-)}(x(t_{k+1})) - V_{\sigma(t_k)}(x(t_k)) + \cdots + V_{\sigma(t_1^-)}(x(t_1)) - V_{\sigma(0)}(x(0)) \\
&\quad \geq V_{\sigma(T)}(x(T)) - V_{\sigma(0)}(x(0))
\end{aligned} \tag{2.118}$$

Thus, from (2.115), we have $V_{\sigma(T)}(x(T)) - V_{\sigma(0)}(x(0)) \leq \int_0^T W(s)ds$. □

The condition in Proposition 2.2 guarantees that the energy in the current mode at switching time is always larger than that of the next mode. In this case, the transient energy is negative.

To further overcome the obstacle in (2.114), and allow the increase of energy at switching time, we provide a stronger version of global passivity, named "periodic fault tolerant passivity". We first define some mode sets:

- $Q_1 \subset Q$ denotes the set of healthy modes.
- $Q_2 \subset Q_1$ denotes the set of healthy modes that may be activated as the initial mode or after a healthy mode.
- $Q_3 \subset Q_1$ denotes the set of healthy modes that are activated after a faulty mode, meanwhile, are followed by a healthy mode or are the final mode.

The relation of above several sets is illustrated by Fig.2.7, from which we see that $\{\ 1,\ 3,\ 5,\ 6\ \} \in Q_1$. $\{\ 1,\ 6\ \} \in Q_2$. $5 \in Q_3$. Note that Mode 3 is activated between two faulty modes. Thus $3 \in Q_1 \setminus (Q_2 \cup Q_3)$.

Definition 2.8. *A HS (2.112) is* periodically fault tolerant passive *if there exist nominal controllers u_1, u_2, ...,u_N, such that for all initial states $x(0) \in X$, and $T \geq 0$, the following inequalities hold:*

- $\forall i \in Q_2$
$$V_i(x(t_{(k+1)i})) - V_i(x(t_{ki})) \leq 0 \tag{2.119}$$
where $0 \leq t_{(k)i} < t_{(k+1)i} \leq T$.
- $\forall i \in Q_2$, $j \in Q_3$, *such that mode* j *is the first mode of set* Q_3 *activated after mode* i. *Denote by* T_e, T_s *the end time of mode* j *and the start time of mode* i *respectively*
$$V_j(x(T_e)) - V_i(x(T_s)) \leq \int_{T_s}^{T_e} W_1(s)ds \tag{2.120}$$
where $W_1(s) \leq 0$.
- *For the case that the initial mode* i *is faulty, and there exists* $j \in Q_3$ *such that mode* j *is the first mode of set* Q_3 *activated after initial mode and is ended at* T_e
$$V_j(x(T_e)) - V_i(x(0)) \leq \int_{0}^{T_e} W_2(s)ds \tag{2.121}$$
where $W_2(s) \leq 0$.
- *For the case that the final mode* i *is faulty, and there exists* $j \in Q_2$ *such that mode* j *is the last mode of set* Q_2 *activated before the final mode and is started at* T_s
$$V_j(x(T)) - V_i(x(T_s)) \leq \int_{T_s}^{T} W_3(s)ds \tag{2.122}$$
where $W_3(s) \leq 0$.
- *For the case that no mode of the set* $Q_2 \cup Q_3$ *is activated*
$$V_{\sigma(T)}(x(T)) - V_{\sigma(0)}(x(0)) \leq \int_{0}^{T} W_4(s)ds \tag{2.123}$$
where $W_4(s) \leq 0$.

Definition 2.8 is illustrated in Fig. 2.8, from which we can see that the energy is dissipative in each small period that includes the faulty modes. Two advantages result from this property, that is 1) Inequalities (2.120)-(2.123) are not hard to justify. 2) We can check the fault tolerance in a short period after the fault occurs.

Theorem 2.8. *If a HS (2.112) is periodic fault tolerant passive, then the origin of the system is stable in spite of faults.*

Proof: We consider four cases as follows:

- Case 1: The initial and final modes are not faulty. Note that each healthy mode is passive. Inequalities (2.120)-(2.122) imply that every time when we start in the mode of the set Q_2, the energy is non-increasing until the next mode of set Q_2 is activated. The stability follows from Theorem 2.3 in [13] and Theorem 2.7.

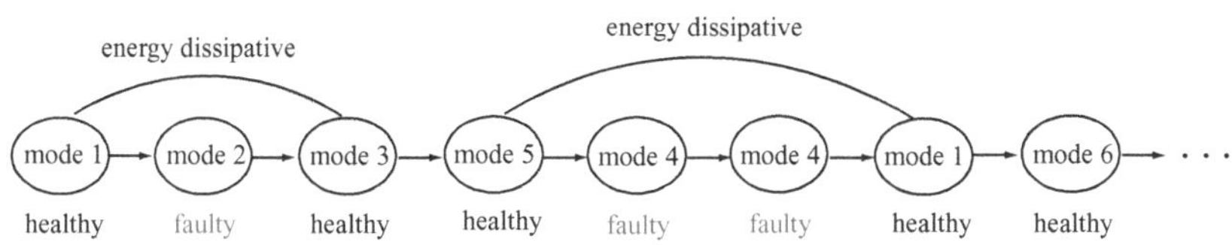

Fig. 2.8 Switching sequence

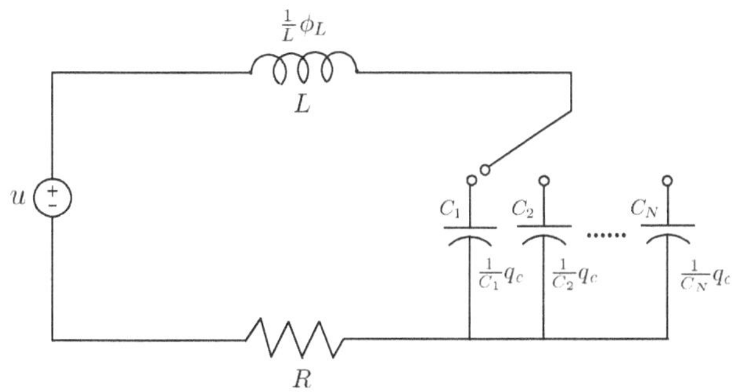

Fig. 2.9 A switched RLC circuit

- Case 2: No mode of the set $Q_2 \cup Q_3$ is activated. The stability is achieved from (2.123) and Theorem 2.7.
- Case 3: The initial mode is healthy, and the final mode is faulty. It follows from (2.122) that after the last mode of set Q_2 before final mode is activated, the energy is non-increasing. The stability is achieved from Theorem 2.3 in [13] and Theorem 2.7.
- Case 4: The initial mode is faulty, and the final mode is healthy. Similarly to Case 3, the result can be obtained from (2.121). □

Example 2.3: A switched RLC circuit that is widely employed in order to perform low-frequency signal processing in integrated circuits is taken as an example to illustrate the results. As shown in Fig. 2.9, the circuit consists of an input power source, a resistance, an inductance and N capacitors that could be switched between each other. The two measurable state variables are the charge in the capacitor and the flux in the inductance $x = [q_c, \phi_L]^\top$. The input u is the voltage.

The dynamic equations are given by

$$\begin{cases} \dot{x}_1 = \frac{1}{L}x_2 \\ \dot{x}_2 = -\frac{1}{C_i}x_1 - \frac{R}{L}x_2 + u \\ y = \frac{1}{L}x_2, \quad i = 1, 2, \ldots, N \end{cases}$$

where C_i denotes the ith capacitor. The energy function of each mode is given as

$$V_i = \frac{1}{2C_i}x_1^2 + \frac{1}{2L}x_2^2$$

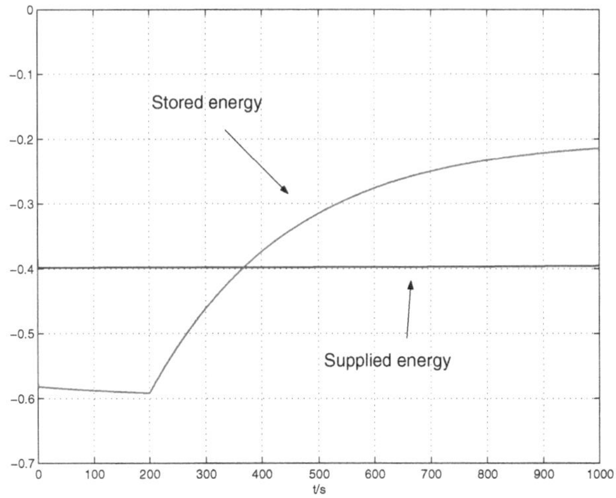

Fig. 2.10 Diagnosis performance (N=1)

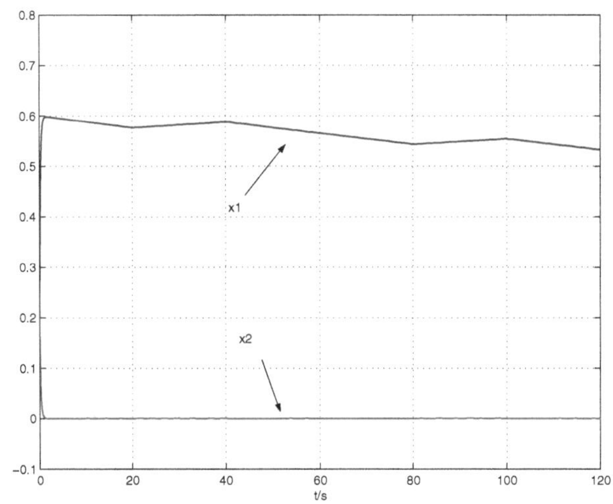

Fig. 2.11 System performance (N=3)

Let us first consider the case $N = 1$, this RLC circuit is also discussed in [91]. In the healthy situation, it can be obtained that $\dot{V} = -\frac{R}{L}x_2^2 + yu$ which satisfies the passivity. The nominal control is chosen as $u = u_n = -y$. Now we consider a leakage fault that occurs in the capacitor at $t = 200s$, the dynamic equation of $\dot{x}_2$ is changed into

$$\dot{x}_2 = -\frac{1}{C}x_1 - \frac{R}{L}x_2 + \frac{k}{C}x_1 + u \tag{2.124}$$

where $k > 0$ is an unknown faulty parameter. It follows that $\dot{V} = -\frac{R}{L}x_2^2 - \frac{k}{LC}x_1x_2 + yu$. If $-\frac{R}{L}x_2^2 \leq \frac{k}{LC}x_1x_2$, then such fault does not affect the passivity. Otherwise, the fault would be diagnosed. Set $k = -200, L = 0.1H, C = 100\mu F, R = 1\Omega$, the initial states are $[0.2, 0.2]^\top$. Fig. 2.10 shows the diagnosis performance, we can see that once the threshold is reached at nearly 370s, the fault is detected.

Suppose that $N = 3$, i.e., the system is switched among three capacitors. C_1 is activated in $[t_{3n}, t_{3n+1})$, C_2 is in $[t_{3n+1}, t_{3n+2})$, and C_3 is in $[t_{3n+2}, t_{3n+3})$, $n = 0, 1, \ldots$. The nominal input is $u_i = -\frac{1}{L}x_2$. The fault occurs in C_2 as (2.124) with $k = -200$, which violates the passivity of mode 2. It is clear that $1 \in Q_2$, $3 \in Q_3$. In the simulation, set $L = 0.1H$, $C_1 = 50\mu F$, $C_2 = 100\mu F$, $C_3 = 20\mu F$ and $R = 1\Omega$. Assume that the dwell period $t_{3n+3} - t_{3n+2} = 20s$, $t_{3n+2} - t_{3n+1} = 20s$, and $t_{3n+1} - t_{3n} = 20s$. We can check that each period $[t_{3n}, t_{3n+3})$ satisfies (2.120), and mode 1 satisfies (2.119). Thus the system is periodic fault tolerant passive. Fig. 2.11 shows the state trajectory, the system is still stable in spite of the fault.

2.5 General Stability Results in HS

Motivated by the fact that some modes may be unstable due to faults, in this section, we establish a new sufficient stability condition named " gain technique" for HS with unstable mode, and provide novel stabilizing switching laws such that the stability is guaranteed and each mode can be activated following any prescribed sequence whatever it is stable or not.

2.5.1 Preliminaries

The considered switched system takes the general form

$$\dot{x}(t) = f_{\sigma(t)}(x(t)) \tag{2.125}$$

where $x \in X \subset \Re^n$ are the states. f_σ is a nonlinear smooth function. Define $Q = \{1, 2, \ldots, N\}$, where N is the number of modes. $\sigma(t) : [0, \infty) \to Q$ denotes the *switching function*, which is assumed to be a piecewise constant function continuous from the right. f_i, $i \in Q$ are smooth functions with $f_i(0) = 0$, hence, the origin is an equilibrium point. We denote by t_j, $j = 1, 2, \ldots$ the jth switching instant, $t_0 = 0$. Let t_{ik}, $i \in Q$, $k = 1, 2, \ldots$ be the kth time when mode i is switched on. $N_{\sigma(t)}$ represents the number of switchings in $[0, t)$. In this work, we only consider nonZeno sequences (i.e., sequences that switch at most a finite number of times in any finite time interval). However, the developed theory allows infinite switchings in infinite time interval. We also assume that the states do not jump at the switching instants.

Specially, we define a class $\mathscr{GKL}$ function as in [135] $\gamma : [0, \infty) \times [0, \infty) \to [0, \infty)$ if $\gamma(\cdot, t)$ is of class $\mathscr{K}$ for each fixed $t \geq 0$ and $\gamma(s, t)$ increases as $t \to \infty$ for each fixed $s \geq 0$.

Denote $Q_s \subset Q$ as the set of stable modes and $Q_{us} \subset Q$ the set of unstable ones. $Q = Q_s \cup Q_{us}$, $Q_s \cap Q_{us} = \emptyset$ and $Q_s \neq \emptyset$. Suppose that there exist continuous non-negative functions $V_p : \Re^n \to \Re_{\geq 0}$, α_1^p, $\alpha_2^p \in \mathscr{K}_\infty$, $\forall p \in Q$, and $\phi_p \in \mathscr{K}\mathscr{L}$ $\forall p \in Q_s$, $\phi_p \in \mathscr{G}\mathscr{K}\mathscr{L}$ $\forall p \in Q_{us}$ that satisfy for $k = 1,2,\ldots$

$$\alpha_1^p(|x|) \leq V_p(x) \leq \alpha_2^p(|x|), \quad \forall p \in Q \tag{2.126}$$

$$V_p(x(t)) \leq \phi_p(V_p(x(t_{pk})), t - t_{pk}), \ \forall p \in Q_s, \ \phi_p \in \mathscr{K}\mathscr{L}, \ t \geq t_{pk} \tag{2.127}$$

$$V_p(x(t)) \leq \phi_p(V_p(x(t_{pk})), t - t_{pk}), \ \forall p \in Q_{us}, \ \phi_p \in \mathscr{G}\mathscr{K}\mathscr{L}, \ t \geq t_{pk} \tag{2.128}$$

Formulations (2.126)-(2.128) include various converging and diverging forms (e.g., the exponential decay form [47], the constant gain form [155]). For each stable mode, V_p in (2.127) is more general than a classic Lyapunov function since a bounded increase is allowed. For unstable modes, inequality (2.128) implies that V_p may increase infinitely as described by a $\mathscr{G}\mathscr{K}\mathscr{L}$ function if $t \to \infty$. $\mathscr{G}\mathscr{K}\mathscr{L}$ function is more general than the Lyapunov-like function in [148] since we do not impose an upper bound on V_p. Note that (2.127)-(2.128) are properties satisfied by functions of each mode, and do not depend on the switching sequence. V_p ($\forall p \in Q$) is not required to be differentiable.

Definition 2.9. *Given a switching function $\sigma(t)$, the origin of a switched system (2.125) is said to be* stable *under σ if for any $\varepsilon > 0$, there exists a $\delta > 0$ such that $|x(t)| \leq \varepsilon$, $t \geq 0$, whenever $|x(0)| \leq \delta$.*

Definition 2.9 describes the stability w.r.t. a given switching function $\sigma(t)$. The objectives of this section is *to propose switching laws that stabilize the system (2.125) satisfying (2.126)-(2.128) by determining the switching instants according to any given switching sequence.*

2.5.2 Stabilization of Switched Systems

In the following, we first establish a stability condition for the considered switched systems in the finite time interval with finite numbers of switchings (Lemma 2.11). Based on such stability criterion, a stabilizing switching law will be constructed (Theorem 2.9).

Lemma 2.11. *Consider a switched system (2.125) satisfying (2.126)-(2.128). Under $\sigma(t)$, if there exists a constant $\beta > 0$ such that*

$$\sum_{k=0}^{N_{\sigma(t_s,t)}} \Big(\prod_{i=k}^{N_{\sigma(t_s,t)}} \frac{\phi_{\sigma(t_i)}^{t_{i+1}-t_i}}{V_{\sigma(t_i)}^{t_i}} \Big) \leq \beta, \quad t > t_s \geq 0, \quad \text{where } t_{N_{\sigma(t_s,t)}+1} \triangleq t, \ N_{\sigma(t_s,t)} \text{ is finite} \tag{2.129}$$

Then x is bounded in $[t_s, t)$. Moreover, for any bounded $x(t_s)$, the upper bound of $|x(t)|$ can be estimated.

Remark 2.12. *Note that $\frac{\phi_{\sigma(t_i)}^{t-t_i}}{V_{\sigma(t_i)}^{t_i}}$ for $t \geq t_i$ is the bound of the gain of function $V_{\sigma(t_i)}$ when mode $\sigma(t_i)$ is activated. Condition (2.129) gives a relation among the gains of* *each activated mode and its activating period. More precisely, x is bounded in $[t_s, t)$ if the product of gains from each activated mode to the terminated mode is bounded, and the sum of these products values is also bounded. It deserves to point out that for a switched system with unstable modes, even in the finite time interval with finite switching times, x may escape to infinity under inappropriate switching law.*

Proof of Lemma 2.11: For the sake of clearness, suppose that $t_s = t_0 = 0$. Denote $N_{\sigma(t)} \triangleq N_{\sigma(0,t)}$.

Consider $t \in [0, t_1)$, we have $V^t_{\sigma(0)} \le \frac{\phi^t_{\sigma(0)}}{V^0_{\sigma(0)}} V^0_{\sigma(0)}$. Condition (2.129) ensures that $\frac{\phi^t_{\sigma(0)}}{V^0_{\sigma(0)}} \le \beta$. It follows from (2.126)-(2.128) that

$$|x(t_1)| \le \underbrace{(\alpha_1^{\sigma(0)})^{-1} \circ \beta \circ \alpha_2^{\sigma(0)}}_{\vartheta_{t_1}}(|x(0)|) \tag{2.130}$$

for $\vartheta_{t_1} \in \mathcal{K}_\infty$. According to (2.126), one has

$$V^{t_1}_{\sigma(t_1)} \le V^{t_1}_{\sigma(t_1^-)} + \alpha_2^{\sigma(t_1)}(\vartheta_{t_1}(|x(0)|)) - \alpha_1^{\sigma(t_1^-)}(\vartheta_{t_1}(|x(0)|)) \tag{2.131}$$

Define $\alpha_{t_1} = \max[\alpha_2^{\sigma(t_1)} \circ \vartheta_{t_1}, \alpha_1^{\sigma(t_1^-)} \circ \vartheta_{t_1}]$. Since $\alpha_2^{\sigma(t_1)}, \alpha_1^{\sigma(t_1^-)}, \vartheta_{t_1} \in \mathcal{K}_\infty$, it is clear that $\alpha_{t_1} \in \mathcal{K}_\infty$ and

$$\alpha_{t_1}(|x(0)|) \ge \alpha_2^{\sigma(t_1)}(\vartheta_{t_1}(|x(0)|)) - \alpha_1^{\sigma(t_1^-)}(\vartheta_{t_1}(|x(0)|)) \tag{2.132}$$

Substituting (2.132) into (2.131) results in

$$V^{t_1}_{\sigma(t_1)} \le V^{t_1}_{\sigma(t_1^-)} + \alpha_{t_1}(|x(0)|) \tag{2.133}$$

For $t \in [t_1, t_2)$, we have

$$\begin{aligned} V^t_{\sigma(t)} &\le \frac{\phi^{t-t_1}_{\sigma(t_1)}}{V^{t_1}_{\sigma(t_1)}} V^{t_1}_{\sigma(t_1)} \le \frac{\phi^{t-t_1}_{\sigma(t_1)}}{V^{t_1}_{\sigma(t_1)}} \left[V^{t_1^-}_{\sigma(t_1^-)} + \alpha_{t_1}(|x(0)|) \right] \\ &\le \frac{\phi^{t-t_1}_{\sigma(t_1)}}{V^{t_1}_{\sigma(t_1)}} \frac{\phi^{t_1}_{\sigma(0)}}{V^0_{\sigma(0)}} V^0_{\sigma(0)} + \frac{\phi^{t-t_1}_{\sigma(t_1)}}{V^{t_1}_{\sigma(t_1)}} \alpha_{t_1}(|x(0)|) \end{aligned} \tag{2.134}$$

Note that $V^0_{\sigma(0)}$ is bounded and $\alpha_{t_1} \in \mathcal{K}_\infty$. Condition (2.129) ensures that $\frac{\phi^{t-t_1}_{\sigma(t_1)}}{V^{t_1}_{\sigma(t_1)}} \frac{\phi^{t_1}_{\sigma(0)}}{V^0_{\sigma(0)}} \le \beta$ and $\frac{\phi^{t-t_1}_{\sigma(t_1)}}{V^{t_1}_{\sigma(t_1)}} \le \beta$. It follows from (2.126)-(2.128) and (2.134) that

$$|x(t_2)| \leq \underbrace{(\alpha_1^{\sigma(0)})^{-1} \circ \beta \circ \left(\alpha_2^{\sigma(0)}(|x(0)|) + \alpha_{t_1}(|x(0)|)\right)}_{\vartheta_{t_2}(|x(0)|)} \tag{2.135}$$

for $\vartheta_{t_2} \in \mathscr{K}_\infty$. One further has

$$V_{\sigma(t_2)}^{t_2} \leq V_{\sigma(t_2^-)}^{t_2} + \alpha_2^{\sigma(t_2)}(\vartheta_{t_2}(|x(0)|)) - \alpha_1^{\sigma(t_2^-)}(\vartheta_{t_2}(|x(0)|)) \tag{2.136}$$

Define $\alpha_{t_2} = \max[\alpha_2^{\sigma(t_2)} \circ \vartheta_{t_2}, \alpha_1^{\sigma(t_2^-)} \circ \vartheta_{t_2}]$. Since $\alpha_2^{\sigma(t_2)}, \alpha_1^{\sigma(t_2^-)}, \vartheta_{t_2} \in \mathscr{K}_\infty$, it follows that $\alpha_{t_2} \in \mathscr{K}_\infty$ and

$$\alpha_{t_2}(|x(0)|) \geq \alpha_2^{\sigma(t_2)}(\vartheta_{t_2}(|x(0)|)) - \alpha_1^{\sigma(t_2^-)}(\vartheta_{t_2}(|x(0)|)) \tag{2.137}$$

Substituting (2.137) into (2.136) results in

$$V_{\sigma(t_2)}^{t_2} \leq V_{\sigma(t_2^-)}^{t_2} + \alpha_{t_2}(|x(0)|) \tag{2.138}$$

for $\alpha_{t_2} \in \mathscr{K}_\infty$.

By induction, we find that under condition (2.129) there exists a function $\alpha \in \mathscr{K}_\infty$ such that at each switching instant $t_i > 0$, $i = 1, 2, \ldots, N_{\sigma(t)}$

$$V_{\sigma(t_i)}(x(t_i)) \leq V_{\sigma(t_i^-)}(x(t_i)) + \alpha(|x(0)|) \tag{2.139}$$

where $\alpha(|x(0)|) \triangleq \sup_{i=1,2,\ldots,N_{\sigma(t)}}[\alpha_{t_i}(|x(0)|)]$.

Denote $j = N_{\sigma(t)}$ for $t \geq 0$, $j \geq 0$, it follows from (2.127)-(2.128) that

$$\begin{aligned}
V_{\sigma(t)}(x(t)) &\leq \phi_{\sigma(t_j)}^{t-t_j} = \frac{\phi_{\sigma(t_j)}^{t-t_j}}{V_{\sigma(t_j)}^{t_j}} V_{\sigma(t_j)}^{t_j} \\
&\leq \frac{\phi_{\sigma(t_j)}^{t-t_j}}{V_{\sigma(t_j)}^{t_j}} \left[V_{\sigma(t_j^-)}^{t_j^-} + \alpha(|x(0)|)\right] \\
&\leq \frac{\phi_{\sigma(t_j)}^{t-t_j}}{V_{\sigma(t_j)}^{t_j}} \phi_{\sigma(t_{j-1})}^{t_j - t_{j-1}} + \frac{\phi_{\sigma(t_j)}^{t-t_j}}{V_{\sigma(t_j)}^{t_j}} \alpha(|x(0)|) \\
&\leq \frac{\phi_{\sigma(t_j)}^{t-t_j}}{V_{\sigma(t_j)}^{t_j}} \frac{\phi_{\sigma(t_{j-1})}^{t_j - t_{j-1}}}{V_{\sigma(t_{j-1})}^{t_{j-1}}} V_{\sigma(t_{j-1}^-)}^{t_{j-1}^-} + \left[\frac{\phi_{\sigma(t_j)}^{t-t_j}}{V_{\sigma(t_j)}^{t_j}} \frac{\phi_{\sigma(t_{j-1})}^{t_j - t_{j-1}}}{V_{\sigma(t_{j-1})}^{t_{j-1}}} + \frac{\phi_{\sigma(t_j)}^{t-t_j}}{V_{\sigma(t_j)}^{t_j}}\right] \alpha(|x(0)|) \\
&\qquad \vdots \\
&\leq \prod_{s=0}^{N_{\sigma(t)}} \frac{\phi_{\sigma(t_s)}^{t_{s+1}-t_s}}{V_{\sigma(t_s)}^{t_s}} V_{\sigma(0)}(x(0)) + \sum_{k=1}^{N_{\sigma(t)}} \left(\prod_{i=k}^{N_{\sigma(t)}} \frac{\phi_{\sigma(t_i)}^{t_{i+1}-t_i}}{V_{\sigma(t_i)}^{t_i}}\right) \alpha(|x(0)|)
\end{aligned} \tag{2.140}$$

Based on (2.126) and (2.139), since $\alpha \in \mathscr{K}_\infty$, there exists a $\mathscr{K}_\infty$ function $\bar{\alpha}$ such that

$$\bar{\alpha}(|x(0)|) = \max\left[\alpha_2^{\sigma(0)}(|x(0)|), \alpha(|x(0)|)\right] \tag{2.141}$$

Substituting (2.141) into (2.140), together with (2.129), yields

$$V_{\sigma(t)}(x(t)) \leq \sum_{k=0}^{N_{\sigma(t)}} \Big(\prod_{i=k}^{N_{\sigma(t)}} \frac{\phi_{\sigma(t_i)}^{t_{i+1}-t_i}}{V_{\sigma(t_i)}^{t_i}} \Big) \bar{\alpha}(|x(0)|) \leq \beta \bar{\alpha}(|x(0)|) \tag{2.142}$$

From (2.126), we finally obtain

$$|x(t)| \leq (\alpha_1^{\sigma(t)})^{-1} \beta \bar{\alpha}(|x(0)|) \tag{2.143}$$

Since $\beta > 0$ is a constant, $\alpha_1^{\sigma(t)}, \bar{\alpha} \in \mathscr{K}_\infty$, the stability result follows.

From above procedures, one can find that under condition (2.129), given any $x(t_s)$, β and switching sequence, each $\alpha_{t_i}(|x(t_s)|)$ can be calculated which is independent from the switching instants. Thus, for any bounded $x(t_s)$, we can find a function $\Omega(\cdot)$ such that $|x(t)| \leq \Omega(|x(t_s)|)$. This completes the proof. □

Remark 2.13. *The main contributions of Lemma 2.11 are twofold: 1) Both stable and unstable modes are allowed in the switched nonlinear system; 2) The "μ" condition is removed by introducing a difference $\alpha(|x(0)|)$ among functions V_p $\forall p \in \mathscr{M}$. However, the condition (2.129) is independent from $\alpha(|x(0)|)$. 3) The upper bound of $|x(t)|$ can be estimated without the information of switching instants in $[0,t)$. This property will be very useful in switching law design.*

Remark 2.14. *The condition (2.129) is valid since V_σ is a non-negative function and is impossible to become zero unless a stronger finite time stability [9] is achieved. For the case that finite time stability is achieved, (2.129) is available if we take j instead of $N_{\sigma(t)}$ where $V_{\sigma(t)}^t > 0$ for $t < t_{j+1}$.*

Remark 2.15. *It is often not easy to verify (2.129) on-line, which relies on the solutions of the system. However, this condition can help to construct a stabilizing switching law as shown below. The proposed stabilization scheme will automatically guarantee the validation of (2.129).*

Unlike the usual design methods that adjust both the switching sequence and switching instants [155], [130], we only redesign the switching instants such that the origin of switched system is always stable under any given switching sequence where each prescribed mode can be activated.

Assumption 2.11. *there exists a known constant $\chi \geq 1$ such that*

$$\chi = \max_{j \in \mathscr{M}, k=1,2\ldots} \frac{\phi_j(V_j(x(t_{jk})), 0)}{V_j(x(t_{jk}))} \tag{2.144}$$

Remark 2.16. *Assumption 1 means that the initial gain of function V_j is bounded when the corresponding mode j is just switched on at $t = t_{jk}$. In some situations, $\phi_j(V_j(x(t_{jk})),0)$ is affine w.r.t. $V_j(x(t_{jk}))$, e.g. the exponential decay form [47], the constant gain form [155]. In these cases, χ can be easily obtained* a priori.

Without loss of generality, suppose that at for a given sequence, at most m unstable modes (m is finite) are activated one by one without being interrupted by stable modes.

Choose a constant $\beta > \max[m(1+\chi)\chi^m, m(m+1)\chi^{m+1}]$, where χ is defined in (6.37). Given any required upper bound ε of $|x(t)|$ and switching sequence, the switching law is designed as:

Switching law $\mathscr{S}$ (with a given ε and a switching sequence)

1. *Let $i = 0$, choose $x(0)$ such that $(\alpha_1^{\sigma(0)})^{-1}\phi_{\sigma(0)}(V_{\sigma(0)}(x(0),0)) \leq \varepsilon$*
2. *If* (**C1**) *mode $\sigma(t_i)$ is stable and mode $\sigma(t_{i+1})$ is stable, then go to 3; Else, go to 5.*
3. *Choose t_{i+1} such that $(\alpha_1^{\sigma(t_{i+1})})^{-1}\phi_{\sigma(t_{i+1})}(V_{\sigma(t_{i+1})}(x(t_{i+1}),0)) \leq \varepsilon$.*
4. *Let $i = i+1$, go to 2.*
5. *If* (**C2**) *mode $\sigma(t_i)$ is stable and mode $\sigma(t_{i+1})$ is unstable, and there exist $h-1$ unstable modes ($h \leq m$) activated successively after mode $\sigma(t_{i+1})$, then go to 6; Else, go to 9.*
6. *Determine the bound $\Omega(|x(t_{i+1})|)$ satisfying $|x(t_{i+h+1})| \leq \Omega(|x(t_{i+1})|)$ using (2.143) in Lemma 2.11, choose t_{i+1} such that*

$$(\alpha_1^{\sigma(t_{i+h+1})})^{-1}\phi_{\sigma(t_{i+h+1})}(\alpha_2^{\sigma(t_{i+h+1})}(\Omega(|x(t_{i+1})|)),0)) \leq \varepsilon$$

let $s = 0$.

7. *Choose t_{i+2+s} such that*

$$\sum_{k=0}^{i+1+s}\Big(\prod_{j=k}^{i+1+s}\frac{\phi_{\sigma(t_j)}^{t_{j+1}-t_j}}{V_{\sigma(t_j)}^{t_j}}\Big) \leq \frac{\beta}{(h+1-s)\chi^{h+1-s}} - 1$$

8. *Let $s = s+1$; If $s \neq h$, then go to 7; Else, let $i = i+h$, go to 2.*
9. *If* (**C3**) *the initial mode $\sigma(0)$ is unstable, and there exist $h-1$ unstable modes ($h \leq m$) activated successively after mode $\sigma(0)$, then go to 10.*
10. *Determine the bound $\Omega(|x(0)|)$ satisfying $|x(t_h)| \leq \Omega(|x(0)|)$ using (2.143) in Lemma 2.11, choose $x(0)$ such that*

$$(\alpha_1^{\sigma(t_h)})^{-1}\phi_{\sigma(t_h)}(\alpha_2^{\sigma(t_h)}(\Omega(|x(0)|)),0)) \leq \varepsilon$$

let $s = 0$.

11. *Choose t_{1+s} such that $\sum_{k=0}^{s}\Big(\prod_{j=k}^{s}\frac{\phi_{\sigma(t_j)}^{t_{j+1}-t_j}}{V_{\sigma(t_j)}^{t_j}}\Big) \leq \frac{\beta}{(h+1-s)\chi^{h+1-s}} - 1$.*
12. *Let $s = s+1$; If $s \neq h$, then go to 11; Else, let $i = h$, go to 2.* ■

The main idea behind $\mathscr{S}$ is that for current stable mode $\sigma(t_i)$, if next mode $\sigma(t_{i+1})$ is stable, we let mode $\sigma(t_i)$ be activated until t_{i+1} such that $x(t_{i+1})$ results in $|x(t)| \leq \varepsilon$ during mode $\sigma(t_{i+1})$'s working period $[t_{i+1}, t_{i+2})$ (step 3). When we predict that h unstable modes will be activated after stable mode $\sigma(t_i)$, we let mode $\sigma(t_i)$ be activated long enough until t_{i+1} such that $x(t_{i+1})$ results in $|x(t)| \leq \varepsilon$ for $t \in [t_{i+1}, t_{i+h+2})$, i.e. the total activating periods of all h unstable modes and stable mode $\sigma(t_{i+h+1})$ (step 6). This can be achieved because the upper bound $\Omega(|x(t_{i+1})|)$ can be obtained without the information of switching instants $t_{i+1}, ..., t_{i+h+1}$. The switching scheme among unstable modes is based on Lemma 2.11 (steps 7, 8, 11, 12). For initial stable/unstable modes, the initial states $x(0)$ are also chosen in different ways (steps 1 and 10).

Theorem 2.9. *Consider a switched system (2.125) satisfying (2.126)-(2.128) and Assumption 2.11. For any given $\varepsilon > 0$ and any switching sequence where at most m unstable modes are activated one by one, under the switching law $\mathscr{S}$, there exist an initial states $x(0)$ and a series of switching instants satisfy $0 < t_1 < t_2 < ...$, such that the origin is stable and $|x(t)| \leq \varepsilon\ \forall t \geq 0$.*

Proof: In the step 1 of $\mathscr{S}$, choosing $x(0)$ satisfying

$$(\alpha_1^{\sigma(0)})^{-1} \phi_{\sigma(0)}(V_{\sigma(0)}(x(0), 0)) \leq \varepsilon$$

which leads to $|x(0)| \leq \varepsilon$ when mode $\sigma(0)$ is just activated. If mode $\sigma(0)$ is stable, we have from (2.126)-(2.127) that $|x(t)| \leq \varepsilon$ for $t \in [0, t_1)$. We will consider respectively three cases **C1-C3** in $\mathscr{S}$.

For **C1**, since mode $\sigma(t_i)$ is stable, it follows from (2.126)-(2.127) that there always exists a time instant $t_{i+1} > t_i$ satisfying

$$(\alpha_1^{\sigma(t_{i+1})})^{-1} \phi_{\sigma(t_{i+1})}(V_{\sigma(t_{i+1})}(x(t_{i+1}), 0)) \leq \varepsilon$$

this implies that $|x(t_{i+1})| \leq \varepsilon$ when mode $\sigma(t_{i+1})$ is just activated. Since mode $\sigma(t_{i+1})$ is also stable, we have $|x(t)| \leq \varepsilon$ for $t \in [t_{i+1}, t_{i+2})$.

For **C2**, switching on mode $\sigma(t_{i+2})$ at $t = t_{i+2}$ results in

$$\frac{\phi_{\sigma(t_{i+2})}(V^{t_{i+2}}_{\sigma(t_{i+2})}, 0)}{V^{t_{i+2}}_{\sigma(t_{i+2})}} \Big(\sum_{k=0}^{i+1} \Big(\prod_{j=k}^{i+1} \frac{\phi^{t_{j+1}-t_j}_{\sigma(t_j)}}{V^{t_j}_{\sigma(t_j)}} \Big) + 1 \Big) \leq \frac{\beta}{(h+1)\chi^h}$$

Since $\beta > m(m+1)\chi^{m+1}$, $h \leq m$, we have $\frac{\beta}{(h+1)\chi^h} < \frac{\beta}{h\chi^h} - 1$. Thus we can choose $t_{i+3} > t_{i+2}$ such that

$$\frac{\phi^{t_{i+3}-t_{i+2}}_{\sigma(t_{i+2})}}{V^{t_{i+2}}_{\sigma(t_{i+3})}} \Big(\sum_{k=0}^{i+1} \Big(\prod_{j=k}^{i+1} \frac{\phi^{t_{j+1}-t_j}_{\sigma(t_j)}}{V^{t_j}_{\sigma(t_j)}} \Big) + 1 \Big) \leq \frac{\beta}{h\chi^h} - 1$$

By induction, for $s = 1,2,...,h-1$ we have $\frac{\beta}{(h+1-s)\chi^{h-s}} < \frac{\beta}{(h-s)\chi^{h-s}} - 1$. Choose t_{i+3+s} as $\mathscr{S}$, we obtain

$$\frac{\phi_{\sigma(t_{i+2+s})}^{t_{i+3+s}-t_{i+2+s}}}{V_{\sigma(t_{i+2+s})}^{t_{i+2+s}}}\Big(\sum_{k=0}^{i+1+s}\Big(\prod_{j=k}^{i+1+s}\frac{\phi_{\sigma(t_j)}^{t_{j+1}-t_j}}{V_{\sigma(t_j)}^{t_j}}\Big)+1\Big) \leq \frac{\beta}{(h-s)\chi^{h-s}} - 1$$

Finally, we verify condition (2.129) with $t = t_{i+1+h}$ and $t_s = t_{i+1}$. There are finite numbers of switchings occurring in $(t_{i+1}, t_{i+1+h}]$, it follows from Lemma 2.11 that we can find a bound $\Omega(|x(t_{i+1})|)$ satisfying $|x(t_{i+h+1})| \leq \Omega(|x(t_{i+1})|)$ using (2.143). Since this bound is independent from the switching instants, we can determine it before h unstable modes are switched into.

Note that mode $\sigma(t_i)$ is stable, we can find a time instant $t_{i+1} > t_i$ such that

$$(\alpha_1^{\sigma(t_{i+h+1})})^{-1}\phi_{\sigma(t_{i+h+1})}(\alpha_2^{\sigma(t_{i+h+1})}(\Omega(|x(t_{i+1})|)),0)) \leq \varepsilon$$

This guarantees that $|x(t)| \leq \varepsilon$ for $t \in [t_{i+1}, t_{i+h+1}]$. Mode $\sigma(t_{i+h+1})$ is also stable, we further have $|x(t)| \leq \varepsilon$ for $t \in [t_{i+h+1}, t_{i+h+2})$.

For **C3**, note that $\beta > m(1+\chi)\chi^m$ and $\chi \geq 1$, which results in $\chi < \frac{\beta}{h\chi^h} - 1$. We can choose t_1 such that $\frac{\phi_{\sigma(0)}^{t_1}}{V_{\sigma(0)}^0} \leq \frac{\beta}{h\chi^h} - 1$, the rest of the proof follows the same procedure as in **C2**, thus is omitted here. We finally obtain (2.129) with $t = t_h$ and $t_s = 0$.

Based on above analysis, one finds that for a switched system with any given switching sequence, finite or infinite numbers of switchings and both stable and unstable modes, the switching law $\mathscr{S}$ maintains the stability of the origin, and $|x(t)| \leq \varepsilon$ for $t \geq 0$. This completes the proof. □

Remark 2.17. *Roughly speaking, $\mathscr{S}$ lets the activating periods of stable modes large enough and lets the activating periods of unstable modes small enough such that the state trajectory is bounded under a given switching sequence. Such idea is similar to that of dwell-time schemes in [136], [32] where an aggregated system is considered including stable modes and consequently activated unstable ones. This aggregated system would be stable if the total activating periods of stable modes are sufficient large. However, $\mathscr{S}$ provides an alternative way to approach stability in the absence of the "μ" condition.*

Example 2.4: Consider a numerical example with three modes. Let $\mathscr{M} = \{1,2,3\}$, $x = [x_1, x_2]^\top$, the modes take the following forms

$$f_1 = \begin{bmatrix} -x_1 + 4x_2^3 \\ -x_1 - x_2 \end{bmatrix}, \quad f_2 = \begin{bmatrix} x_1 - x_2 \\ x_2 + x_1^3 \end{bmatrix}, \quad f_3 = \begin{bmatrix} x_1 - 3x_2 \\ x_1 + x_2 \end{bmatrix}$$

The prescribed switching sequence is

$$\text{mode } 1 \rightarrow \text{mode } 2 \rightarrow \text{mode } 3 \rightarrow \text{mode } 1 \rightarrow \cdots\cdots$$

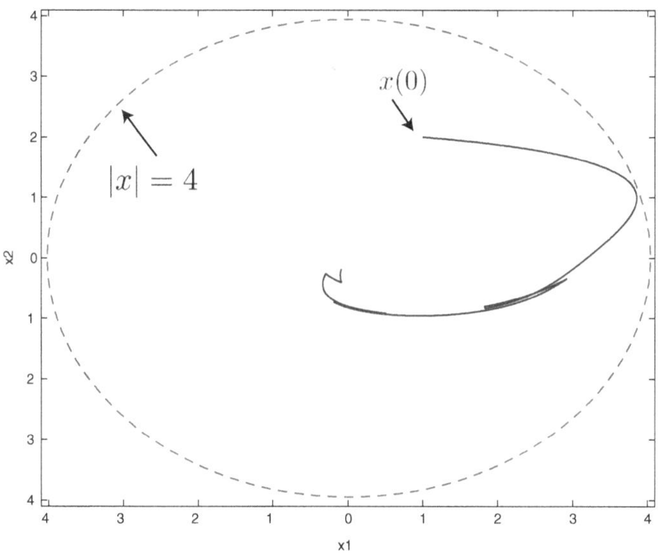

Fig. 2.12 State trajectory

For mode 1, it is not easy to find a quadratic Lyapunov function. However the origin is still stable, we choose a polynomial Lyapunov function $V_1 = x_1^2 + 2x_2^4$, this results in $V_1(x(t)) < e^{-2t}V_1(x(0))$ for $t \geq 0$. Both mode 2 and mode 3 are unstable, applying V_1 to modes 2 and 3 yields

$$\frac{dV_1(x)}{dx} f_2(x) \leq V_1^{0.5}(x) + 7V_1(x) + 4V_1^{1.5}(x) + 4V_1^3(x) \tag{2.145}$$

$$\frac{dV_1(x)}{dx} f_3(x) \leq V_1^{0.5}(x) + 11V_1(x) + 2V_1^{1.5}(x) \tag{2.146}$$

It can be seen that a common Lyapunov function is hard to impose here because inequalities (2.145)-(2.146) do not satisfy the general Lyapunov function formulation in dwell-time scheme [48]. The method in [88] is also not easy to be implemented since the right sides of (2.145) and (2.146) are polynomial forms of V_1 rather than $aV_1^m(x)$ for $a, m > 0$ in [88], and the exponents larger and smaller than 1 exist simultaneously.

We choose $V_2 = x_1^4 + 2x_2^2$, $V_3 = x_1^2 + x_2^2$. It follows that $V_2(x(t)) < e^{4t}V_2(x(0))$, $V_3(x(t)) < e^{2t}V_3(x(0))$, for $t \geq 0$. Note that MLFs techniques are difficult to be applied since the state trajectories in unstable modes are not bounded and Lyapunov-like functions are not easy to find. The "μ" condition is also hard to impose here, because V_1 and V_2 are non-quadratic.

Set $\varepsilon = 4$ which means that $|x(t)| \leq 4$ must hold for all $t \geq 0$. The prescribed switching sequence is

$$\text{mode 1} \rightarrow \text{mode 2} \rightarrow \text{mode 3} \rightarrow \text{mode 1} \rightarrow \cdots\cdots$$

Now we design the switching instants according to $\mathscr{S}$. Mode 1 is stable, choose $x(0) = [1,\ 2]^\top$ from step 1 of $\mathscr{S}$ such that $|x(t)| \leq 4$ for $t \in [0, t_1)$. Since both mode 2 and mode 3 are unstable, the switching scheme based on Lemma 1 is applied after t_1. It can be obtained from (6.37) that $\chi = 1$. $m = 2$ due to two unstable modes. Choose $\beta = 6.3 > 2(2+1)$. The activating periods of modes 2 and 3 can be calculated from step 7 of $\mathscr{S}$: $0.0059s$ for mode 2; $0.2602s$ for mode 3. Choose $t_1 = 0.9s$ from step 6 of $\mathscr{S}$ such that $|x(t)| \leq 4$ for $t \in [0, t_4)$. Consequently, choose $t_2 = 0.9059s$, $t_3 = 1.1661s$. The activating period of mode 1 is set to be $0.9s$ in the following switching process, i.e., $t_4 = 2.0661s$. Although our theory allows infinite switchings in infinite time interval, in the numerical simulation, a finite time interval $[0s,\ 4s]$ is considered. Other switching instants can be obtained straightly. Fig.2.12 shows the state trajectory, from which we can see that the stability is achieved and $|x| \leq 4$ always holds.

2.6 Conclusion

In this chapter, several FTC methods have been proposed for HS with time dependent switching. The known switching instants bring much convenience to FTC design. In sections 2.1-2.3, FTC objective has been achieved via designing the stabilizing controller in each faulty mode and a switching scheme. Sections 2.4-2.5 researched directly the stability of HS without reconfiguring the controller in each mode. It can be found that even some faulty modes are unstable, the stability of overall HS is still maintained under appropriate switching schemes.

Chapter 3
Hybrid Systems with State-Dependent Switching

In this chapter, a class of HS with state dependent switching and without full state measurement are investigated. The considered switching occurs whenever the states reach some given domains which are defined through a set of inequalities called guard set. Such kind of switchings appear widely in applications, e.g., flow control, temperature control. Two FTC methods are proposed for linear and nonlinear HS respectively.

3.1 Preliminaries

The main difference between the considered systems and that in Chapter 2 is that the switching instant is not known *a priori* or can be designed. The considered switching occurs whenever the states reach some given domains which are defined through a set of inequalities called guard set. The challenges of observer-based FTC for such systems are twofold [134]:

1) to distinguish the effects of the continuous faults and mode transitions (may include discrete faults) on the system. From the abnormal change of state estimates provided by the observer, we should first identify whether continuous faults in the current mode occur or another mode is switched into, then treat the system with different control strategies.

2) to maintain the stability of overall HS in spite of these two kinds of faults.

As for challenge 1), a natural idea is to design an observer *whose estimation error is not affected by (or robust to) continuous faults and sensitive to mode transitions.* Challenge 2) could also be solved if the accurate continuous state estimates are obtained in Challenge 1). This idea will be followed throughout this chapter.

3.2 Hybrid Linear Systems

In this section, we face the above challenges 1)-2), and propose an observer-based FTC method for a class of hybrid linear systems. The main work is outlined as follows:

H. Yang et al.: Fault Tolerant Control Design for Hybrid Systems, LNCIS 397, pp. 59–90.
springerlink.com

1. Under some mild structure conditions, each mode of HS is transformed into a new form which is friendly for the design of the observer and FTC law.
2. A novel observer is proposed for each mode of new form whose estimation error is not affected directly by continuous faults and sensitive to mode transitions. Based on such observer, a time varying threshold is proposed to detect rapidly each switching once it occurs in spite of discrete faults.
3. An observer-based FTC law is developed for each mode to guarantee the asymptotical stability of the origin. Moreover, sufficient conditions are given such that the overall HS can be stabilized in the sense of LaSalle invariance principle.

3.2.1 FTC for Linear Systems

Consider a linear system

$$\begin{aligned} \dot{x}(t) &= Ax(t) + Bu(t) + Ef^c(t), \quad |u| \le u^{\max}, \ x \in D \\ y(t) &= Cx(t) = [C^1 \ \ \mathbf{0}_{r\times n-r}]x(t) \end{aligned} \tag{3.1}$$

where $x \in D \subseteq \Re^n$ are unmeasurable states, D is a physical domain of x. $y \in \Re^r$ are outputs, C^1 is an $r \times r$ nonsingular matrix. (C,A) is observable and (A,B) is controllable, $u \in \Re^p$ are inputs with $u^{\max} > 0$ as its magnitude constraint, $f^c \in \Re^q$, with $q < r$, denote actuator faults, the $n \times q$ constant matrix E denotes fault distribution. Since the system inputs are bounded, it is reasonable to assume that actuator faults are also bounded, i.e., $|f^c| \le \bar{f}^c$, where $\bar{f}^c > 0$.

Assumption 3.1. *Rank* $(CE) = q$

Assumption 3.1 guarantees that the matrix CE is of full column rank, which implies that the effects of faults on outputs are independent.

Define a transformation $x = N^{-1}z$, where

$$N = \begin{bmatrix} C^1 & 0 \\ 0 & I \end{bmatrix} \tag{3.2}$$

Then the system (3.1) can be transformed into

$$\begin{aligned} \dot{z} &= \bar{A}z + \bar{B}u + \bar{E}f^c = \begin{bmatrix} \bar{A}_1 \\ \bar{A}_2 \\ \bar{A}_3 \end{bmatrix} z + \begin{bmatrix} \bar{B}_1 \\ \bar{B}_2 \\ \bar{B}_3 \end{bmatrix} u + \begin{bmatrix} \bar{E}_1 \\ \bar{E}_2 \\ \bar{E}_3 \end{bmatrix} f^c \\ y &= Cx = \bar{C}z = \begin{bmatrix} I_{(r-q)\times(r-q)} & 0 & 0 \\ 0 & I_q & 0 \end{bmatrix} z \end{aligned} \tag{3.3}$$

where $\bar{A} = NAN^{-1}$, $\bar{B} = NB$, $\bar{E} = NE$. z can be represented as

$$z = [z_1 \ \ z_2 \ \ z_3]^\top = [y_1 \ \ y_2 \ \ z_3]^\top$$

where $z_3 \in \Re^{n-r}$. We just need to estimate z_3. It follows from Assumption 3.1 that $\text{Rank}\begin{bmatrix}\bar{E}_1\\ \bar{E}_2\end{bmatrix} = q$, it is also assumed that $\bar{E}_2$ is nonsingular.

Define

$$S = \begin{bmatrix} I & -\bar{E}_1\bar{E}_2^{-1} & 0 \\ 0 & I & 0 \\ 0 & -\bar{E}_3\bar{E}_2^{-1} & I \end{bmatrix} \tag{3.4}$$

Left-multiplying (3.4) into (3.3), we have

$$\begin{bmatrix} \dot{y}_1 - \bar{E}_1\bar{E}_2^{-1}\dot{y}_2 \\ \dot{y}_2 \\ \dot{z}_3 - \bar{E}_3\bar{E}_2^{-1}\dot{y}_2 \end{bmatrix} = \begin{bmatrix} \bar{A}_1 - \bar{E}_1\bar{E}_2^{-1}\bar{A}_2 \\ \bar{A}_2 \\ \bar{A}_3 - \bar{E}_3\bar{E}_2^{-1}\bar{A}_2 \end{bmatrix} z + \begin{bmatrix} \bar{B}_1 - \bar{E}_1\bar{E}_2^{-1}\bar{B}_2 \\ \bar{B}_2 \\ \bar{B}_3 - \bar{E}_3\bar{E}_2^{-1}\bar{B}_2 \end{bmatrix} u + \begin{bmatrix} 0 \\ \bar{E}_2 \\ 0 \end{bmatrix} f^c \tag{3.5}$$

The advantage of the form (3.5) is that the first and third blocks of (3.5) are not affected directly by any fault, an observer can be designed for these two blocks to estimate z_3. This estimates is decoupled from the faults, thus can be used to diagnose the faults in the second block of (3.5).

Define

$$\varpi_j = \bar{A}_j - \bar{E}_j\bar{E}_2^{-1}\bar{A}_2, \;\; H_j = \bar{B}_j - \bar{E}_j\bar{E}_2^{-1}\bar{B}_2$$

where $j = 1, 3$. Partitioning ϖ_j as

$$\varpi_j = \begin{bmatrix}\varpi_{j1} & \varpi_{j2} & \varpi_{j3}\end{bmatrix} \tag{3.6}$$

then the first and third block rows of system (3.5) can be written as

$$\dot{z}_3 = \varpi_{33}z_3 + s, \;\; v = \varpi_{13}z_3 \tag{3.7}$$

where

$$\begin{aligned} s &= \varpi_{31}y_1 + \varpi_{32}y_2 + \bar{E}_3\bar{E}_2^{-1}\dot{y}_2 + H_3u \\ v &= \dot{y}_1 - \bar{E}_1\bar{E}_2^{-1}\dot{y}_2 - \varpi_{i11}y_1 - \varpi_{12}y_2 - H_1u \end{aligned}$$

To estimate z_3, an observer can be designed as

$$\dot{\hat{z}}_3 = \varpi_{33}\hat{z}_3 + s + \zeta(v - \varpi_{13}\hat{z}_3) \tag{3.8}$$

assume $(\varpi_{33}, \varpi_{13})$ is an observable pair, the observer gain ζ can be chosen to make $(\varpi_{33} - \zeta\varpi_{13})$ stable.

From the above discussion, we can let $\hat{x} = N^{-1}\hat{z} = N^{-1}\begin{bmatrix}y_1 & y_2 & \hat{z}_3\end{bmatrix}^\top$, where $\hat{z}_3$ is obtained in (3.8). Denote $\tilde{z}_3 = z_3 - \hat{z}_3$, $e(t) = x - \hat{x}$, one has

$$|e(t)| = |z_3 - \hat{z}_3| \leq \mu(\lambda^*)|\tilde{z}_3(0)|\exp(-\lambda^* t) \tag{3.9}$$

where $\lambda^* > 0$, $\mu(\lambda^*) = M\lambda^{*l}$ is polynomial in λ^* for $M, l > 0$.

The second block row in (3.5) can be written as

$$\dot{y}_2 = \bar{A}_{21}y_1 + \bar{A}_{22}y_2 + \bar{A}_{23}z_3 + \bar{B}_2u + \bar{E}_2f^c$$

Denote $\hat{f}^c$ as the fault estimate and $\tilde{f}^c \triangleq f^c - \hat{f}^c$. Then

$$\hat{f}^c = \bar{E}_2^{-1}(\dot{y}_2 - \bar{A}_{21}y_1 - \bar{A}_{22}y_2 - \bar{A}_{23}\hat{z}_3 - \bar{B}_2u) \quad (3.10)$$

$$\tilde{f}^c = -\bar{E}_2^{-1}\bar{A}_{23}\tilde{z}_3 \quad (3.11)$$

From (3.9) and (3.10), one can see that the observer-based FD scheme can provide rapid and accurate fault estimates, and meanwhile, also gives accurate continuous state estimates which are not affected by faults.

Now we design the bounded FTC law. Consider the Lyapunov candidate $V = x^\top Px$ for (3.1), where P is a positive definite symmetric matrix that satisfies the Riccati equation

$$A^\top P + PA - PBB^\top P = -Q \quad (3.12)$$

for some positive definite matrix Q.

V can be regarded as a *control Lyapunov function*[1] for system (3.1). Using the results in [73] (see also [28]), a continuous bounded FTC law can be designed as

$$u(\hat{x}) = -K(L^*_{Ax}V(\hat{x}),\hat{x})(L_BV)^\top(\hat{x}) \triangleq b(\hat{x}) \quad (3.13)$$

with

$$K(L^*_{Ax}V) = \frac{L^*_{Ax}V + \sqrt{(L^*_{Ax}V)^2 + (u^{\max}|(L_BV)^\top|)^4}}{|(L_BV)^\top|^2\left[1 + \sqrt{1 + (u^{\max}|(L_BV)^\top|)^2}\right]}$$

for $(L_BV)^\top \neq 0$, and $K(L^*_{Ax}V) = 0$, for $(L_BV)^\top = 0$, where $L^*_{Ax}V = L_{Ax}V + \rho V + |L_EV|\hat{f}^c$, with $L_{Ax}V = \hat{x}^\top(A^\top P + PA)\hat{x}$, $(L_EV)^\top = 2E^\top P\hat{x}$, $(L_BV)^\top = 2B^\top P\hat{x}$, $\rho > 0$.

For all initial states, the stability region of system (3.1) is defined by the set

$$\Phi \triangleq \{x \in D : L^*_{Ax}V(x) < u^{\max}|(L_BV)^\top(x)|\} \quad (3.14)$$

A common way of estimating the stability region (3.14) is by using the level sets of V (see Chapter 4 in [62]). An estimate is described by

$$Inv \triangleq \{x \in D : V(x) \leq c^{\max}\} \quad (3.15)$$

where Inv is expected to be the largest invariant set of Φ, $c^{\max}$ is the largest number for which $Inv \subseteq \Phi$. Fig. 3.1 describes a system with two states, where the relation of several sets are illustrated. The yellow region represents Inv.

[1] Recall that a positive definite radially unbounded smooth function $V : \Re^n \to \Re$ is called a *control Lyapunov function* for the system $\dot{x} = f(x) + G(x)u$, $x \in \Re^n$, if we have $\inf_{u\in\mathscr{U}}\{L_fV + L_GVu\} < 0 \ \forall x \neq 0$.

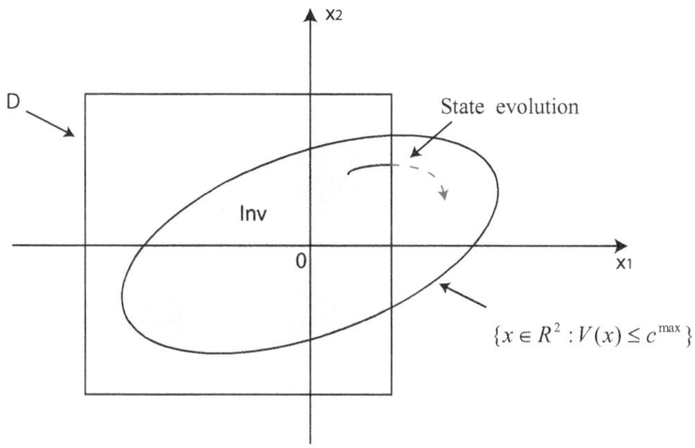

Fig. 3.1 Relations among several regions

Lemma 3.1. *Consider system (3.1), there exists a positive real number e_u, such that if $|e(t)| \le e_u, \forall t \ge 0$, and the set $\{x \in \Re^n : V(x) \le c^{\max}\} \subseteq D$, then the controller $u = b(\hat{x})$ makes the origin of the system asymptotically stable in spite of f^c.*

Proof: The time derivative of V along the closed loop trajectories is

$$\begin{aligned}\dot{V} &= L_{Ax}V + L_B V u(x) + L_E V f^c + L_B V(u(\hat{x}) - u(x)) \\ &= \frac{(L_{Ax}V + L_E V f^c)\sqrt{1 + (u^{\max}|(L_B V)^\top|)^2} + L_E V f^c}{\left[1 + \sqrt{1 + (u^{\max}|(L_B V)^\top|)^2}\right]} \\ &\quad - \frac{|L_E V|\bar{f}^c + \rho V + \sqrt{\left(L^*_{Ax}V\right)^2 + (u^{\max}|(L_B V)^\top|)^4}}{\left[1 + \sqrt{1 + (u^{\max}|(L_B V)^\top|)^2}\right]} \\ &\quad + L_B V(u(\hat{x}) - u)\end{aligned} \tag{3.16}$$

From (3.11), one has $L_E V f^c \le |L_E V|\bar{f}^c$. It is clear from (3.16) that, if $L^*_{Ax}V(x) < 0$, we have $\dot{V} < -\rho V + L_B V(u(\hat{x}) - u)$. When $0 \le L^*_{Ax}V(x) < u^{\max}|(L_B V)^\top(x)|$, we have

$$\begin{aligned}&(L_{Ax}V + L_E V f^c)\sqrt{1 + (u^{\max}|(L_B V)^\top|)^2} \\ &\quad < (L^*_{Ax}V - \rho V)\sqrt{1 + (u^{\max}|(L_B V)^\top|)^2} \\ &\quad < \sqrt{\left(L^*_{Ax}V\right)^2 + (u^{\max}|(L_B V)^\top|)^4} \\ &\qquad - \rho V\sqrt{1 + (u^{\max}|(L_B V)^\top|)^2}\end{aligned} \tag{3.17}$$

Substituting (3.17) in (3.16), we have that whenever $L^*_{Ax}V(x) < u^{\max}|(L_BV)^\top(x)|$,

$$\begin{aligned}\dot{V} &< -\rho V + L_BV(u(\hat{x}) - u) \\ &\leq -\rho^*|x|^2 + M^G|u(\hat{x}) - u| \end{aligned} \tag{3.18}$$

where $\rho^* > 0, M^G = \max_{V=c^{\max}}(|L_BV|)$, M^G exists since $|L_BV(\cdot)|$ is continuous over the region *Inv*.

Since $\{x \in \Re^n : V(x) \leq c^{\max}\} \subseteq D$, we have $Inv = \{x \in \Re^n : V(x) \leq c^{\max}\}$. Firstly, we analyze $\dot{V}$ on the boundary *Inv*. Inequality (3.18) can be written as

$$\dot{V} < -\rho c^{\max} + M^G|u(\hat{x}) - u| \tag{3.19}$$

Note that $|u(\hat{x}) - u(x)|$ is continuous $\forall t \geq 0$ and vanishes when $e = 0$, since e is always bounded which is not affected by faults, there exist two positive real numbers e_u and $\kappa(e_u)$, such that if $|e| \leq e_u$, then $|u(\hat{x}) - u(x)| \leq \kappa|e| \leq \rho c^{\max}/M^G$, which implies V is always negative on the boundary *Inv*, so $x(t) \in Inv\ \forall t \geq 0$.

Secondly, substituting the estimate $\kappa|e|$ into (3.18) yields

$$\begin{aligned}\dot{V} < -\rho^*|x|^2 + M^G\kappa|e| \leq -r\rho^*|x|^2, \\ \forall|x| \geq \sqrt{\frac{M^G\kappa|e|}{r\rho^*}} \triangleq \gamma(|e|)\end{aligned} \tag{3.20}$$

where $\gamma(\cdot)$ is a class $\mathscr{K}$ function. Based on [62], we have that, for any $x(t) \in Inv$, there exists a class $\mathscr{KL}$ function $\beta(\cdot,\cdot)$ and a class $\mathscr{K}$ function $\gamma_1(\cdot)$, such that

$$|x(t)| \leq \beta(|x(0)|,t) + \gamma_1(\sup_{\tau\geq 0}|e(\tau)|), \quad \forall t \geq 0 \tag{3.21}$$

which means that x is input-to-state stable with respect to e. Note that Eq. (3.8) and (3.9) ensure that $\lim_{t\to\infty} e(t) = 0$, which together with (3.21), leads to $\lim_{t\to\infty} x(t) = 0$. This completes the proof. □

3.2.2 FTC for Hybrid Systems

Based on the above FTC solution for linear system, we focus on the HS modeled by a hybrid automaton as defined in Definition 1.1.

The trajectories of a hybrid automaton $\mathscr{H}$ that start from some initial state $(q_0, x_0) \in Init$ consist of a sequence of continuous flows and discrete transitions. When the discrete state $q \in Q$ is maintained, the continuous state x evolves according to the differential equation $x = F_q(x, u, f^c)$ where $F_q \in \mathscr{F}$ as long as $x \in Inv(q)$. After x reaches the guard set, the system would switch into next mode. It is assumed that the states x are continuous at each switching instant.

Definition 3.1. *The system $\mathscr{H}$ is* live *if for $i, i' \in Q$*

$$\forall x \in Inv(i)\ or\ x \notin Inv(i), \exists e = (i,i'), x \in G(i,i') \tag{3.22}$$

$$\forall e = (i,i') \; and \; x \in G(i,i'), \; x \in Inv(i') \tag{3.23}$$

The liveness of HS ensures the succession of the trajectory under appropriate control input. Review Fig. 3.1, where x may escape from Inv if $\{x \in \Re^2 : V(x) \leq c^{\max}\} \nsubseteq D$. Condition (3.22) means that before or when x escapes from the invariant set of the current mode, the switching must happen. Condition (3.23) guarantees that $x \in Inv(i')$ after mode i' is switched into.

This section models the plant as a class of $\mathscr{H}$ with the following properties:

- (P1) Both normal and faulty switchings are continuous state-dependent and are not controlled by any discrete input V.
- (P2) The vector field F for mode i is of the form

$$\begin{aligned} \dot{x}(t) &= A_i x(t) + B_i u_i(t) + E_i f_i^c(t), \;\; |u_i| \leq u_i^{\max}, \; x \in D_i \\ y_i(t) &= C_i x(t) = [C_i^1 \;\; \mathbf{0}_{r\times n-r}]x(t), \;\; t \in [t_k, t_{k+1}) \end{aligned} \tag{3.24}$$

which satisfies the conditions imposed on the system (3.1). $f_i^c \in \Re^q$ denotes actuator faults for mode i, where $|f_i^c| \leq \bar{f}_i^c$, where $\bar{f}_i^c > 0$.
- (P3) The *discrete fault* $f_d \in F_d$ is such that $x \in Inv(i) \wedge x \in G(i,i',f_d)$, where the system is switched from i to i' under $G(i,i',f_d)$.
- (P4) The system is live in the heathy situation, and the switching sequence is deterministic in both healthy and faulty situations, i.e., each trajectory contains only one switching sequence for all initial $(q_0,x_0) \in Init$. No *Zeno* phenomenon occurs.

Remark 3.1. *The considered model is more practical than that in [110], [123] and [147], since it involves the strict physical bound of control signals and unmeasurable states. The mode transition takes place just when states reach the guard set G, the switching instants are not controllable by the so-called discrete inputs as in [123].*

The FTC Problem (P) for HS can be described as: *Keep the HS live as in (3.22)-(3.23), and make the origin asymptotically stable in spite of any fault in P2, P3.*

(P) is similar to the target control problem for HS in [123], where the target objective is the origin. It is supposed that the prescribed determined switching sequence can bring x to the origin under appropriate controllers in the healthy situation. Details about how to choose such sequence can be seen in [123].

The characteristics of continuous and discrete faults motivate us to consider four faulty situations:

Under continuous faults f_i^c:

Case 1: $x \notin Inv(i) \wedge x \notin G(i,i')$.

Under discrete faults $G(i,i',f_d)$:

Case 2: $x \in Inv(i) \wedge x \in G(i,i',f_d) \wedge x \in Inv(i')$.
Case 3: $x \in Inv(i) \wedge x \in G(i,i',f_d) \wedge x \notin Inv(i')$.
Case 4: $x \notin Inv(i) \wedge x \notin G(i,i',f_d)$.

In Case 1, f_i^c changes the dynamics of mode i, and makes the states escape from the invariant set before switching happens, the liveness would be violated. In Cases 2 and 3, Faulty switching happens under $G(i,i',f_d)$. Note that the system is still live in Case 2, whereas in Case 3, the faulty switching makes the state escape from the invariant set of the next mode, which also violates the liveness. In Case 4, the switching does not happen when the continuous evolution is impossible, the system is locked. All the above four cases may destroy the stability of the HS. We define two switchings: *stable switching* (in the normal case and Cases 1,2) and *unstable switching* (in Cases 3,4).

In the following discussions, we first solve Cases 1 and 2 by applying the method, then provide a *relaxed* FTC method to solve Case 3, and discuss an *active switching detection* technique for Case 4. Finally, we present a FTC framework.

Since each mode satisfies the conditions imposed on the system (3.1), the observer-based FD and FTC methods developed in Section 3.2.1 is applied to each mode. The idea of switching detection appears from the analysis of estimation error $v_i - \varpi_{i13}\hat{z}_3$ for mode i. If all the modes are not overlapping, i.e., each observer works well only when applied to its related mode, then, similar to fault detection problem [57], $v_i - \varpi_{i13}\hat{z}_3$ can be regarded as a *residual* for mode i to detect the switching, since $\lim_{t\to\infty}(v_i - \varpi_{i13}\hat{z}_3) = 0$ before switching occurs. We give a quite general assumption for switching control problem [20] as follows:

Assumption 3.2. *All modes of $\mathscr{H}$ are* discernable, *i.e., for mode i, the estimation error $|e(t)|$ is convergent as in (3.9) only under the observer (3.8) which is associated with mode i.*

Under Assumption 3.2, given an initial $(i,x(t_k)) \in Init(i)$ for mode i, any mode transition can be detected using following time varying threshold:

$$|v_i - \varpi_{i13}\hat{z}_3| \leq \varpi_{i13}\mu(\lambda_i^*)|\bar{z}_3(t_k)|\exp(-\lambda_i^*(t-t_k)) \tag{3.25}$$

Now consider the mode transition in Case 2. Two subcases of Case 2 can be given:

$$x \in Inv(i) \wedge x \in G(i,i',f_d) \cap G^c(i,i'), \quad \textbf{switch earlier}$$

$$x \in Inv(i) \wedge x \in G(i,i') \cap G_f^c(i,i'), \quad \textbf{switch earlier}$$

Since the sequence is deterministic, once the mode transition is detected, the controller (3.13) and the observer (3.8) are switched according to the next mode. The initial states $\hat{x}$ of the current observer are chosen as the final states of the previous observer.

Recall that in the normal case, and Cases 1, 2, the switching (normal or faulty) does not affect the liveness of HS, i.e., it always holds that $x \in Inv(i')$. Fig. 3.2 shows the relation of several sets and the system trajectory. The green and yellow regions denote respectively the invariant sets of two modes, $Inv(1)$, $Inv(2)$. It can be seen

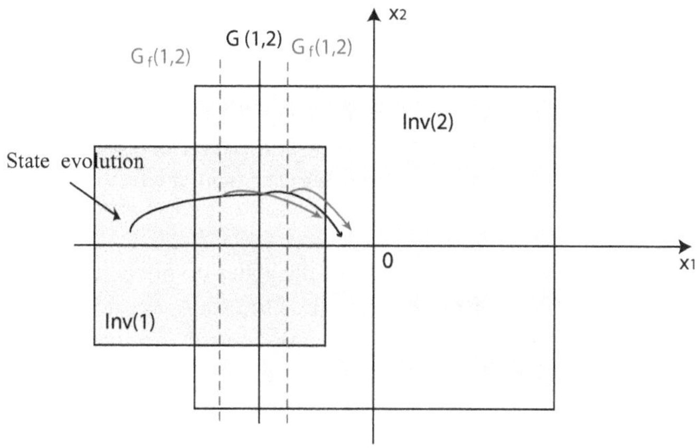

Fig. 3.2 FTC for Case 1 and Case 2

that the state trajectory starting from mode 1 is always within $Inv(2)$ under both $G(1,2)$ and $G(1,2,f_d)$.

The following theorem extends the LaSalle invariance principle to the HS under Cases 1 and 2. For the sake of simplicity, denote $\bar{G}$ as the guard set for both the normal situation and Case 2.

Theorem 3.1. *Consider a HS satisfying P1-P4 and assumptions 3.1, 3.2, the initial states* $(q_0, x(t_0)) \in Inv(q_0) \wedge x(t_0) \in G^c(q_0, q_1)$, $\hat{x}(t_0)$ *is such that* $\mu(\lambda_1^*)|e(t_0^+)| \leq \min\{e_{u,i}, \forall i \in Q\}$. *Under the controller* $u_i = b_i(\hat{x})$ *which is switched according to the mode transition, if* $\forall x \in Inv(i) \wedge x \in \bar{G}(i,i')$, *the following condition holds:*

$$V_{i'}(\hat{x}) + M_i + M_{i'} < V_i(\hat{x}) \tag{3.26}$$

where M_i *is such that* $|e| \leq e_{u,i} \rightarrow |V_i(x) - V_i(\hat{x})| \leq M_i$, *then the origin of the HS is asymptotically stable in the normal case, and cases 1, 2.*

Proof: Due to the continuity of $V_i(\cdot)$, there exists a positive real number M_i such that if $|e| \leq e_{u,i}$, then $|V_i(x) - V_i(\hat{x})| \leq M_i$. Therefore, we have

$$V_{i'}(x) - M_{i'} \leq V_{i'}(\hat{x}) \tag{3.27}$$

$$V_i(x) + M_i \geq V_i(\hat{x}) \tag{3.28}$$

Inequalities (3.27) and (3.28), together with (3.64), lead to

$$V_{i'}(x) < V_i(x) \tag{3.29}$$

Define $\Lambda_1^i = \{x \in Inv(i) \wedge x \in \bar{G}^c(i,i') : \dot{V}_i = 0\}$, $\Lambda_2^i = \{x \in Inv(i) \wedge x \in \bar{G}(i,i') : V_i = V_i'\}$. Let Λ be the largest invariant subset of $\Lambda_1^i \cup \Lambda_2^i$ $\forall i \in Q$. Λ is an invariant set to be attracted. It is clear from Lemma 3.1 that under u_i, $\dot{V}_i$ is always negative, thus $\lim_{t_{k+1} \to \infty} V_i = 0$ $\forall i \in Q$. On the other hand, based on (3.29) and Theorem IV.1

in [80], one concludes that the trajectory of $\mathscr{H}$ approaches Λ, thus the origin is asymptotical stable. □

Remark 3.2. *The switching can be detected using the threshold (3.25) with a short time delay. Due to the discernability of the modes, such delay is often very short and much less than the activating period of mode i, which would be acceptable for practical applications.*

In Case 3, a new mode is switched into while the states do not belong to its invariant set. This is very dangerous for the system since the states would escape to a large region or infinity without being limited by any control command and guard set. Indeed, continuous faults f_i^c do not always exist, a possible solution to solve Case 3 is to design a *variable* invariant set according to the time when f_i^c occurs.

Define two stability regions for mode i

$$\Phi_h(i) \triangleq \{x \in D_i : L^{\star}_{A_i x} V_i(x) < u_i^{\max} |(L_{B_i} V_i)^{\top}(x)|\}$$
$$\Phi_f(i) \triangleq \{x \in D_i : L^{\diamond}_{A_i x} V_i(x) < u_i^{\max} |(L_{B_i} V_i)^{\top}(x)|\}$$

where $L^{\star}_{A_i x} V_i = L_{A_i x} V_i + \rho_i V_i$, and $L^{\diamond}_{A_i x} V_i = L_{A_i x} V_i + \rho_i V_i + |L_{E_i} V_i| \bar{\bar{f}}_i^c + |L_{E_i} V_i \bar{E}_{i2}^{-1} \bar{A}_{i23}|$ $e^{\star 1}_{u,i}$, $\bar{\bar{f}}_i^c > 0$ is defined such that $|\hat{f}_i^c| \leq \bar{\bar{f}}_i^c$, $e^{\star 1}_{u,i} > 0$ will be given later.

Design two controllers

$$u_i(\hat{x}) = -K_i(L^{\star}_{A_i x} V_i(\hat{x}), \hat{x})(L_{B_i} V_i)^{\top}(\hat{x}) \triangleq b_h^i(\hat{x})$$
$$u_i(\hat{x}) = -K_i(L^{\diamond}_{A_i x} V_i(\hat{x}), \hat{x})(L_{B_i} V_i)^{\top}(\hat{x}) \triangleq b_f^i(\hat{x})$$

Similar to (3.15), we define $Inv_h(i)$ and $Inv_f(i)$ as invariant subsets of $\Phi_h(i)$ and $\Phi_f(i)$ respectively. We have the following Lemma:

Lemma 3.2. *Consider system (3.24) with $x(t_k) \in Inv(i) \wedge x(t_k) \in G^c(i,i')$, and $\{x \in \mathfrak{R}^n : L^{\star}_{A_i x} V_i(x) < u_i^{\max} |(L_{B_i} V_i)^{\top}(x)|\} \subseteq D_i$, $\{x \in \mathfrak{R}^n : L^{\diamond}_{A_i x} V_i(x) < u_i^{\max} |(L_{B_i} V_i)^{\top}(x)|\}$ $\subseteq D_i$. There exist two positive numbers $e^{\star 1}_{u,i}$ and $e^{\star 2}_{u,i}$ such that if $\mu(\lambda_i^*)|e(t_k^+)| \leq e^{\star 1}_{u,i}$ and $\hat{x}(t_k^f) \in Inv_f(i) \wedge |e(t_k^f)| \leq e^{\star 2}_{u,i}$, where $t = t_k^f$ is the time when f_i^c occurs, and $Inv_f(i)$ is such that $\forall |e| \leq e^{\star 2}_{u,i}$, $\hat{x} \in Inv_f(i) \to x \in Inv_f(i) \wedge x \in G^c(i,i')$, then the bounded controller*

$$u_i^{\diamond}(\hat{x}) = \begin{cases} b_h^i(\hat{x}) & t \in [t_k, t_k^f) \\ b_f^i(\hat{x}) & t \in [t_k^f, t_{k+1}) \end{cases} \tag{3.30}$$

makes the origin of mode i asymptotically stable.

Proof: Since the system is fault-free (i.e., $f_i^c = 0$) for $t \in [t_k, t_k^f)$. The time-derivative of V_i along the closed-loop trajectories is

$$\dot{V}_i = L_{A_i x} V_i + L_{B_i} V_i u_i^{\diamond}(x) + L_{B_i} V_i (u_i^{\diamond}(\hat{x}) - u_i^{\diamond}(x))$$

It can be obtained similarly to Lemma 3.1 that there exists $e^{\star 1}_{u,i}$ such that $\forall |e(t)| \leq e^{\star 1}_{u,i}$ and $x(t_k) \in Inv_h(i) \wedge x(t_k) \in G^c(i,i')$, the controller $b^i_h(\hat{x})$ makes the origin of the ith mode asymptotically stable.

At $t = t^f_k$, the faults occur and are detected, the time-derivative of V_i along the closed-loop trajectories under controller $b^i_f(\hat{x})$ is

$$\begin{aligned}\dot{V}_i &= L_{A_ix}V_i + L_{B_i}V_iu^\diamond_i(x) + L_{E_i}V_if^c_i + L_{B_i}V_i(u^\diamond_i(\hat{x}) - u^\diamond_i(x)) \\ &= \frac{(L_{A_ix}V_i + L_{E_i}V_if^c_i)\sqrt{1+(u^{\max}_i|(L_{B_i}V_i)^\top|)^2} + L_{E_i}V_if^c_i}{\left[1+\sqrt{1+(u^{\max}_i|(L_{B_i}V_i)^\top|)^2}\right]} \\ &\quad - \frac{|L_{E_i}V_i|\bar{\hat{f}}^c_i - |L_{E_i}V_i\bar{E}^{-1}_{i2}\bar{A}_{i23}|e^\star_{u,i} + \rho_iV_i}{\left[1+\sqrt{1+(u^{\max}_i|(L_{B_i}V_i)^\top|)^2}\right]} \\ &\quad + \frac{\sqrt{\left(L^\diamond_{A_ix}V_i\right)^2 + (u^{\max}_i|(L_{B_i}V_i)^\top|)^4}}{\left[1+\sqrt{1+(u^{\max}_i|(L_{B_i}V_i)^\top|)^2}\right]} + L_{B_i}V_i(u^\diamond_i(\hat{x}) - u^\diamond_i)\end{aligned}$$

From (3.11), one has

$$L_{E_i}V_if^c_i \leq |L_{E_i}V_i|\bar{\hat{f}}^c_i + |L_{E_i}V_i\bar{E}^{-1}_{i2}\bar{A}_{i23}|e^{\star 1}_{u,i}$$

The subsequent proof follows the same way as in Lemma 3.1, one can conclude that if $\hat{x}(t^f_k) \in Inv_f(i)$, then under $b^i_f(\hat{x})$, there exists a constant $e^{\star 2}_{u,i} > 0$ such that $\forall |e(t)| \leq e^{\star 2}_{u,i}$, the states will stay in the region $Inv_f(i)$, and the origin of the ith mode is asymptotically stable. □

The main contribution of Lemma 3.2 is that it relaxes the invariant set as shown in Fig.3.3. where $Inv(2) \subseteq Inv_h(2)$. The Case 3 is said to be *fault tolerable* if

$$\left(x \in Inv(i)\right) \wedge x \in G(i,i',f_d) \wedge x \in Inv_h(i') \tag{3.31}$$

Theorem 3.1 can be directly extended to Case 3 as in the following corollary without the proof.

Corollary 3.1. *Consider a HS satisfying P1-P4 and assumptions 3.1, 3.2, the initial states* $(q_0, x(t_0)) \in Inv(q_0) \wedge x(t_0) \in G^c(q_0,q_1)$, $\hat{x}(t_0)$ *is such that* $\mu(\lambda^*_1)|e(t^+_0)| \leq \min\{e_{u,i}, \forall i \in Q\}$. *Under the controller* $u^\diamond_k(\hat{x})$ *which is switched according to the mode transition, if 1) conditions in Lemma 3.2 and (3.31) hold, 2)* $\forall x \in Inv(i) \wedge x \in G(i,i',f_d)$, *the condition (3.64) holds, then the origin of the HS is asymptotically stable.*

In Case 4, the system is locked and does not respond to any control command, nothing can be said about the subsequent system's behavior, the system may be

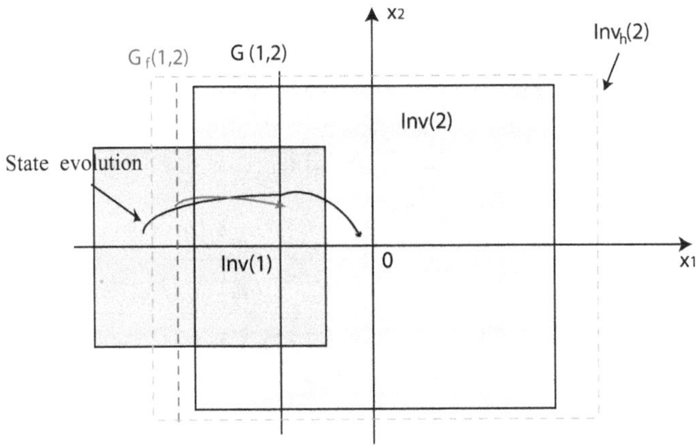

Fig. 3.3 FTC for Case 3

entirely destroyed or some new modes occur which are not included in the original HS. To avoid this phenomenon, an *active switching detection* technique is developed. The main idea is to generate a *Switching Alarm* before the normal switching (when $x \in G(i,i')$) occurs, and also generates a *Lock Alarm* when we identify that $x \in G(i,i')$ but no switching occurs. This is possible due to the structure of the observer in Section 3.2.1. After *Lock Alarm*, some emergency measures must be taken to the system by human (stop the system, or change equipments, or force the system to switch, and so on).

For mode i, define a set $G_{near}(i,i')$ and three sets as

$$\chi_s(i) \triangleq \{x \in X : x \in Inv(i) \wedge x \in G^c_{near}(i,i')\}$$
$$\chi_n(i) \triangleq \{x \in X : x \in Inv(i) \wedge x \in G^c(i,i')\}$$
$$\chi_l(i) \triangleq \{x \in X : x \in Inv(i) \wedge x \in G(i,i')\}$$

where $G_{near}(i,i')$ is close to $G(i,i')$ such that $\chi_s(i) \subseteq \chi_n(i)$.

Since the observer (3.8) always follows the system (3.7), estimation error converges to zero, similar to Lemma 3.2, we can define $\Psi_s(i)$ and $\Psi_l(i)$ such that $\forall |e| \le e^{\star 1}_{u,i}$, $\hat{x} \in \Psi_s(i) \to x \in \chi_s(i)$, $\hat{x} \in \Psi_l(i) \to x \in \chi_l(i)$. The *Active Switching Detection* strategy for mode i is in two steps as shown in Fig. 3.4:

1) When $\hat{x} \in \Psi_s(i)$, *Switching Alarm* is generated.
2) If no mode transition occurs after $\hat{x} \in \Psi_l(i)$, *Lock Alarm* is generated.

Remark 3.3. *Due to estimation errors of the observer, the Lock Alarm will be generated later for a short time delay after states reach the guard set. This delay is acceptable in most situations, only except a very special case that Case 4 occurs in this time delay (e.g., once the states reach the guard set, the continuous evolution of the current mode is impossible).*

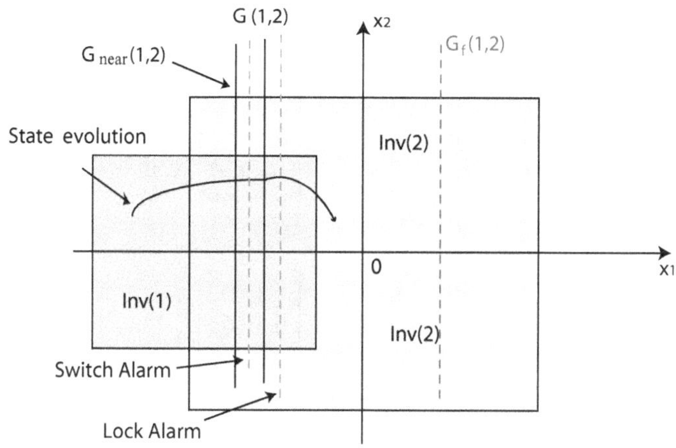

Fig. 3.4 FTC for Case 4

We are in the position to provide a FTC framework

1) *Apply the FD scheme in sections 3.2.2 and controller (3.13) for the current mode.*
2) *When the mode transition is detected before* Switching Alarm,
   If $x \in Inv(i')$, *go to 1);*
   If $x \in Inv_h(i') \cap Inv^c(i')$, *apply the controller (3.30) for current mode.*
3) *When the mode transition is detected after* Switching Alarm *and before* Lock Alarm, *go to 1).*
4) *When the mode transition does not occur after* Lock Alarm, *take some measures to the system by human.*

Example 3.1 (Example 1.1 revisited): Recall the CPU processing control system in Example 1.1. As described before, the system is modeled as a hybrid automaton with two modes: busy mode and usual mode. Fig. 3.5 shows the determined sequence. $x \in \Re^3 = [\pi,\ \rho,\ \omega]^\top$ is the state with π being the amount of CPU tasks in the buffer, ρ the CPU temperature, and ω angular velocity of a cooling fan. $c \in \Re$ and $v \in \Re$ are the clock frequency and the voltage input of a cooling fan. The FTC objectives are to make the above CPU process switch appropriately between two modes according to the guard set (liveness), and to make the origin asymptotically stable in spite of any fault. This means that the cost on the continuous states and inputs is minimized, which leads to energy saving.

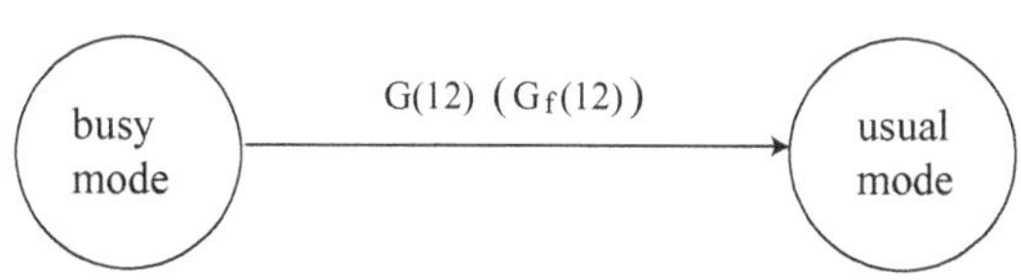

Fig. 3.5 Switching sequence

The continuous models around the equilibrium state[2] are given as:

mode 1 :

$$\begin{bmatrix}\dot{\pi}\\ \dot{\rho}\\ \dot{\omega}\end{bmatrix} = \begin{bmatrix}0 & 0 & 0\\ 0 & -0.05 & -0.5\\ 0 & 0 & -3\end{bmatrix}\begin{bmatrix}\pi\\ \rho\\ \omega\end{bmatrix} + \begin{bmatrix}-2 & 0\\ 0.2 & 0\\ 0 & 0.5\end{bmatrix}\begin{bmatrix}c\\ v\end{bmatrix} + \begin{bmatrix}2\\ -0.2\\ 0\end{bmatrix} f_1^c$$

mode 2 :

$$\begin{bmatrix}\dot{\pi}\\ \dot{\rho}\\ \dot{\omega}\end{bmatrix} = \begin{bmatrix}0 & 0 & 0\\ 0 & -0.05 & -0.5\\ 0 & 0 & -3\end{bmatrix}\begin{bmatrix}\pi\\ \rho\\ \omega\end{bmatrix} + \begin{bmatrix}-1 & 0\\ 0.1 & 0\\ 0 & 1.5\end{bmatrix}\begin{bmatrix}c\\ v\end{bmatrix} + \begin{bmatrix}1\\ -0.1\\ 0\end{bmatrix} f_2^c$$

$$C_1 = C_2 = \begin{bmatrix}1 & -1 & 0\\ 0 & 2 & 0\end{bmatrix}$$

The models satisfy Assumption 3.1, we can see that both f_1^c and f_2^c affect the clock frequency input channel. We obtain from (3.2) that $N^{-1} = \begin{bmatrix}1 & 0.5 & 0\\ 0 & 0.5 & 0\\ 0 & 0 & 1\end{bmatrix}$, and from (3.3), we have

$$\bar{A}_1 = \bar{A}_2 = \begin{bmatrix}0 & 0.025 & 0.5\\ 0 & -0.05 & -1\\ 0 & 0 & -3\end{bmatrix}, \bar{B}_1 = \begin{bmatrix}-2.2 & 0\\ 0.4 & 0\\ 0 & 0.5\end{bmatrix}, \bar{B}_2 = \begin{bmatrix}-1.1 & 0\\ 0.2 & 0\\ 0 & 1.5\end{bmatrix}$$

and $\bar{E}_1 = [2.2\ \ -0.4\ 0]^\top, \bar{E}_2 = [1.1\ \ -0.2\ 0]^\top$, then from (3.5) and (3.6), we get $\varpi_{11} = \varpi_{21} = [0\ -0.25\ -5]$, $\varpi_{13} = \varpi_{23} = [0\ 0\ -3]$, it is clear that $[\varpi_{133}, \varpi_{113}]$ and $[\varpi_{233}, \varpi_{213}]$ are observable, $H_{13} = [0\ 0.5], H_{23} = [0\ 1.5]$. The observers for modes 1 and 2 are designed from (3.8) with $\zeta_1 = 0.4$, $\zeta_2 = -0.2$ respectively. Since the observer is of 1-order, the precise threshold can be given to detect the mode transition without any delay, Assumption 3.2 is not required in this situation.

In mode 1, $D_1 = \{x \in \Re^3 : \pi + \rho \geq 8\}$, $|c| \leq 5$, $|v| \leq 10$. In mode 2, $D_2 = \{x \in \Re^3 : \pi + \rho \leq 25\}$, $|c| \leq 2$, $|v| \leq 5$. $G(1,2) = \{x \in \Re^3 : \pi + \rho \leq 10\}$. Assume $|f_1^c| \leq 2.5$, $|f_2^c| \leq 1$. Choose $\rho_1 = 0.05, \rho_2 = 0.08$, and

$$P_1 = \begin{bmatrix}0.0540 & -0.0062 & 0\\ -0.0062 & 2.1310 & 0.1646\\ 0 & 0.1646 & 2.4753\end{bmatrix}, \quad P_2 = \begin{bmatrix}1.0240 & -0.0048 & 0\\ -0.0048 & 1.6200 & 0.1246\\ 0 & 0.1246 & 1.9752\end{bmatrix}$$

$x(t_1) = x(0)$ is assumed to be $[8\ 9.5\ 9]^\top$. From Lemma 3.1, $\hat{x}(0)$ is chosen as $[8\ 9.5\ 8.85]^\top$. $f_1^c = 2 + 0.2\sin(5t)$ which occurs at $t = 0.15s$. Fig. 3.6 shows the switching detection performance using threshold (3.25), the uncontrollable

[2] The state of system when a sufficiently long time has passed after booting the system is defined as the equilibrium state of this model.

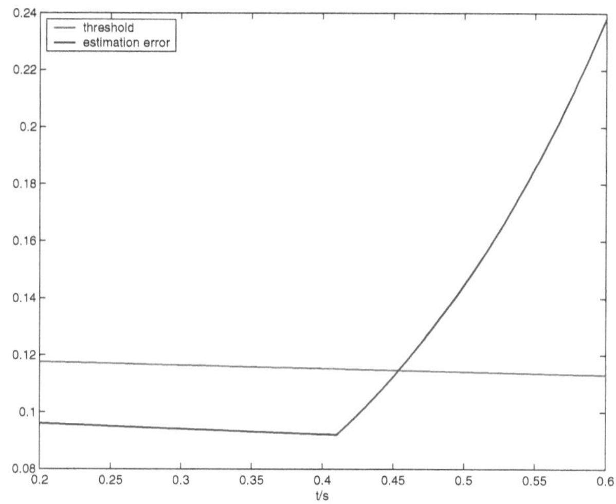

Fig. 3.6 Switching detection performance with $G(1,2)$

switching occurs at $t = 0.41s$, a short detection delay of $0.04s$ exists. Two states π and ρ are illustrated. System evolution is shown in Fig. 3.7(a) where $Inv(1)$ and $Inv(2)$ are computed via level set technique in [62]. We can see that, in the presence of f_1^c, stabilization of the HS is achieved as in Lemma 3.1 and Theorem 3.1, the switching detection delay nearly has no effect on the stability. Now consider the faulty guard set $G(1,2,f_d) = \{x \in \Re^3 : \pi + \rho \leq 10.5\}$, which implies that the mode transition occurs with larger amount of CPU tasks and higher temperature. Fig. 3.7(b) shows the stability of system.

Now we consider Case 3 with $G(1,2,f_d) = \{x \in \Re^3 : 0.8\pi + \rho \leq 12\}$, from Fig. 3.8(a), it is clear that $x \in G(1,2,f_d) \wedge (2,x) \notin Inv(2)$, however, using the relaxed

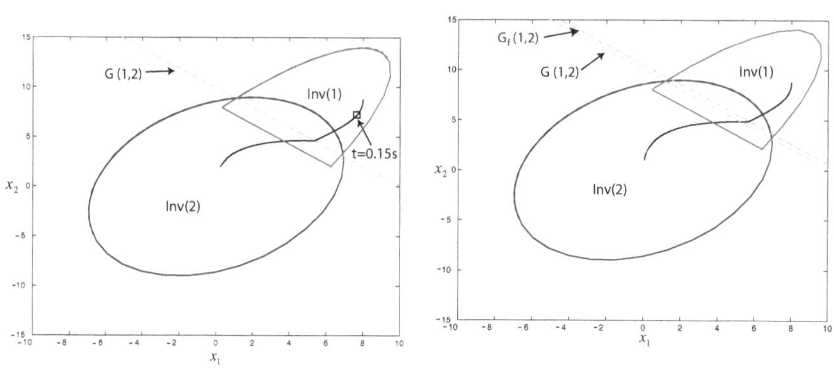

(a) FTC for f_1^c (b) FTC for f_1^c and stable $G(1,2,f_d)$

Fig. 3.7 FTC for the stable switchings

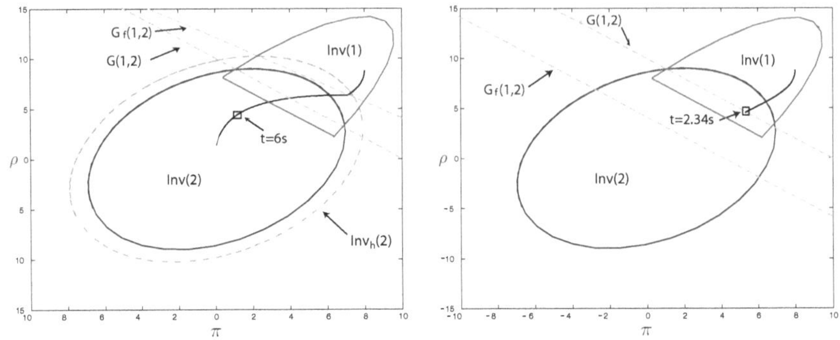

(a) Relaxed FTC for unstable $G(1,2,f_d)$ (b) Active switching detection for unstable $G(1,2,f_d)$

Fig. 3.8 FTC for the unstable switchings

method, it can be seen that $(2,x) \in Inv_h(2)$. Consider $f_2^c = 0.8$ which occurs at $t = 6s$, Fig. 3.8(a) shows that the stabilization of the system is achieved as in Corollary 3.1.

Next, we consider Case 4 with $G(1,2,f_d) = \{x \in \Re^3 : \pi + \rho \leq 4\}$. From Fig. 3.8(b), we can find that the system will be locked in mode 1 and is impossible to switch into mode 2, this is very dangerous since the temperature in CPU can not decrease. The switching alarm and lock alarm are generated at $t = 2.21s$ and $t = 2.34$ respectively, which prevents the system from being dangerous.

3.3 Hybrid Nonlinear Systems

Following the similar idea as in Section 3.2, we now focus on the output tracking problem for a class of hybrid nonlinear systems with uncontrollable state-dependent switching, parametric uncertainties, both continuous and discrete faults, and without full state measurements. Firstly, under geometric conditions, each mode of HS is transformed into a new form which is suitable for both the observer and the FTC law design. Then, a novel observer is designed for each mode whose estimation error is not affected by continuous faults and sensitive to mode transitions. Such observer leads to a time varying threshold for the switching detection of the HS. Finally, sufficient conditions are given to solve the fault tolerant tracking problem for overall HS.

3.3.1 Preliminaries

The HS that we consider takes the form

$$\begin{aligned} \dot{x} &= g_0^\sigma(x) + g^\sigma(x)u^\sigma + \phi^\sigma(x,u)\theta^\sigma + e^\sigma(x)f^\sigma \\ y &= h(x) \end{aligned} \tag{3.32}$$

where $x \in \Re^n$ are unmeasurable states, $u^\sigma \in \Re^{p^\sigma}$ are inputs, $y^\sigma \in \Re^m$ are outputs, $\theta^\sigma \in \Re^{l^\sigma}$ is an unknown constant vector representing parametric uncertainties, $|\theta^\sigma| \leq \theta_0^\sigma$, for $\theta_0^\sigma > 0$.

The *continuous fault* is modelled by a "fault pattern", which consists of the distribution matrix $e^j(x)$ and a "fault signal" $f^j \in \Re^{q^j}$.

g_0^σ, g^σ, ϕ^σ, e^σ and h^σ are smooth and known functions, and $q^\sigma < m \leq p^\sigma$ is considered. $\sigma(t) : [t_0, \infty) \to Q = \{1, 2, \ldots, N\}$ denotes the *switching function* as in Chapter 2.

Define $G : Q \times Q \to \Re^n$ as a guard condition related to two modes. The system is switched from mode i to mode j , $i, j \in Q$ if the continuous states x in mode i reach the guard set $G(i, j)$. The *discrete fault* is represented by the faulty guard set $G_f : Q \times Q \to \Re^n$ that makes the system switch under an abnormal switching condition.

It is assumed that the states x are continuous at each switching instants, and the switched sequence is prescribed and fixed in spite of faults.

The FTC problem is precisely described as: *Keep the outputs of each mode y asymptotically track the given reference signals $y_d^j = [y_{d1}^j, y_{d2}^j, \ldots, y_{d(m)}^j]^\top \in \Re^m$ during the activating period of mode j in spite of continuous and discrete faults, parametric uncertainties, meanwhile, make the continuous states bounded.*

We discuss the system transformation, observer and FTC design problems for non-hybrid systems in sections 3.3.2-3.3.3, then apply the results to hybrid system in Section 3.3.4.

3.3.2 Fault Diagnosis for Nonlinear Systems

Consider the following affine nonlinear system

$$\begin{aligned} \dot{x} &= g_0(x) + g(x)u + \phi(x,u)\theta + e(x)f \\ y &= h(x) \end{aligned} \tag{3.33}$$

where $x \in \Re^n$, $u \in \Re^p$, $y \in \Re^m$, $f \in \Re^q$, $\theta \in \Re^l$ play the same roles as in (3.32). $|\theta| \leq \theta_0$ and $q < m \leq p$, g_0, g, ϕ, e and h are smooth and known.

Definition 3.2. The FD block strict feedback form *of system (3.33) is*

$$\dot{z}_1 = Az_1 + \gamma_1(z_1, y)u + \gamma_2(z_1, y) + \psi_1(z_1, u, y)\theta \tag{3.34}$$

$$\bar{y}_1 = Cz_1 \tag{3.35}$$

$$\dot{z}_2 = \psi_0(z) + \gamma_3(z_2, y)u + \bar{e}(z)f + \psi_2(z, u)\theta \tag{3.36}$$

$$\bar{y}_2 = z_2 \tag{3.37}$$

where $z = [z_1^\top, z_2^\top]^\top$, $z_1 = [\xi_1^\top, \xi_2^\top, \ldots, \xi_{m-q}^\top]^\top \in \Re^{n-q}$, $z_2 = [\xi_{m-q+1}^\top, \ldots, \xi_m^\top]^\top \in \Re^q$ are the states of system (3.34)-(3.37), with $\xi = [\xi_1^\top, \xi_2^\top, \ldots, \xi_m^\top]^\top \in \Re^n$, $\xi_i \in \Re^{\bar{\rho}_i} = [\xi_{i1}, \ldots, \xi_{i\bar{\rho}_i}]^\top$. $y = [\bar{y}_1^\top, \bar{y}_2^\top]^\top$ with $\bar{y}_1 \in \Re^{m-q}$, $\bar{y}_2 \in \Re^q$. Moreover, $A = diag[A_1, \ldots, A_{m-q}] \in \Re^{(n-q)\times(n-q)}$, $C = diag[C_1, \ldots, C_{m-q}] \in \Re^{(m-q)\times(n-q)}$, with

$$A_i \in \Re^{\bar{\rho}_i \times \bar{\rho}_i} = \begin{bmatrix} 0 & 1 & \cdots & 0 \\ \vdots & \vdots & \ddots & \vdots \\ 0 & 0 & \cdots & 1 \\ 0 & 0 & \cdots & 0 \end{bmatrix}, C_i \in \Re^{1\times\bar{\rho}_i} = \begin{bmatrix} 1, 0, \ldots, 0 \end{bmatrix}$$

for $1 \le i \le m-q$. $\gamma_1(z_1, y) = [\bar{g}_1^\top, \bar{g}_2^\top, \ldots, \bar{g}_{m-q}^\top]^\top$ *with*

$$\bar{g}_i \in \Re^{\rho_i \times p} = \begin{bmatrix} \bar{g}_{i1}(\xi_1, \ldots, \xi_{i-1}, \xi_{i1}, y_{i+1}, \ldots, y_m) \\ \bar{g}_{i2}(\xi_1, \ldots, \xi_{i-1}, \xi_{i1}, \xi_{i2}, y_{i+1}, \ldots, y_m) \\ \vdots \\ \bar{g}_{i\bar{\rho}_i}(\xi_1, \ldots, \xi_{i-1}, \xi_i, y_{i+1}, \ldots, y_m) \end{bmatrix} \tag{3.38}$$

and $\gamma_2(z_1, y) = [\bar{\bar{g}}_1^\top, \bar{\bar{g}}_2^\top, \ldots, \bar{\bar{g}}_{m-q}^\top]^\top$ *with*

$$\bar{\bar{g}}_i \in \Re^{\bar{\rho}_i} = [0\, 0 \ldots 0, L_{g_0}^{\bar{\rho}_i} h_i(\xi_1 \ldots \xi_{i-1} \xi_i y_{i+1} \ldots y_m)]^\top \tag{3.39}$$

Remark 3.4. *The form given in Definition 3.2 is an extension of the* the block parametric strict feedback form *in [63] to the faulty case. In our model, both* ψ_1 *and* ψ_2 *terms do not required to take the certain triangular forms as in [63], since the parameter* θ *can be estimated by the observer rather than the control strategy as shown later.*

Assumption 3.3. *There exists a set of integer numbers* $\{\bar{\rho}_1, \bar{\rho}_2, \ldots, \bar{\rho}_m\}$ *such that* $\sum_{i=1}^m \bar{\rho}_i = n$ *and* $\xi = T(x) \in \Re^n$ *is a diffeomorphism where*

$$T(x) = [h_1(x), L_{g_0(x)} h_1(x), \ldots, L_{g_0(x)}^{\bar{\rho}_1 - 1} h_1(x), \\ h_2(x), \ldots, L_{g_0(x)}^{\bar{\rho}_2 - 1} h_2(x), \ldots, L_{g_0(x)}^{\bar{\rho}_m - 1} h_m(x)]^\top$$

The relative degree of the rth output y_r *of system (3.33), denoted as* ρ_r*, is such that* $\rho_r = \bar{\rho}_r = 1$, $m-q+1 \le r \le m$.

Under Assumption 3.3, $dT(x)$ is invertible $\forall x \in \Re^n$, let $r_i(x)$ be the ith column of $[dT(x)]^{-1}$ and $\mathscr{R}(i)_j := \text{span}\{r_{v_i - j}, \ldots, r_{v_i}\}$, where $v_i = \sum_{j=1}^i \bar{\rho}_j$.

Lemma 3.3. *Under Assumption 3.3, the diffeomorphism* $\xi = T(x)$ *can transform the system (3.33) into (3.34)-(3.37) if and only if*

I $[g_0(x), \mathscr{R}(i)_{\bar{\rho}_i - 2}] \subset \mathscr{R}(i)_{\bar{\rho}_i - 1} + \mathscr{R}(i+1)_{\bar{\rho}_{i+1} - 1} + \cdots + \mathscr{R}(m-q)_{\bar{\rho}_{m-q} - 1}$, *for* $2 \le i \le m-q$.
II $[g_0(x), \mathscr{R}(i)_j] \subset \mathscr{R}(i)_j + \mathscr{R}(i+1)_{\bar{\rho}_{i+1} - 1} + \cdots + \mathscr{R}(m-q)_{\bar{\rho}_{m-q} - 1}$, *for* $1 \le i \le m-q$, $0 \le j \le \bar{\rho}_i - 2$.
III $[g_0(x), \mathscr{R}(\bar{i})_{\bar{\rho}_{\bar{i}} - 2}] \subset \mathscr{R}(1)_{\bar{\rho}_1 - 2} + \mathscr{R}(2)_{\bar{\rho}_2 - 2} + \cdots + \mathscr{R}(m-q)_{\bar{\rho}_{m-q} - 2}$, *for* $m-q+1 \le i \le m$, $1 \le \bar{i} \le m-q$.
IV $L_e L_{g_0}^s h_i = 0$, *for* $1 \le s \le \bar{\rho}_i - 1$, $1 \le i \le m-q$.

Proof: We first show that the conditions I and II lead to the block triangular forms of γ_1 and γ_2 as in (3.38) and (3.39). By the construction of $\mathscr{R}(i)_j$, we have in ξ−coordinate [82]

$$\mathscr{R}(i)_j = \text{span}\{\frac{\partial}{\partial \xi_{\langle v_i \rangle}}, \ldots, \frac{\partial}{\partial \xi_{\langle v_i - j \rangle}}\} \tag{3.40}$$

where $\xi_{\langle \iota \rangle}$, $1 \leq \iota \leq n-q$ denotes the ιth element of ξ.

Define $b_{g,l}$ such that $\sum_{i=1}^{m-q} \sum_{j=1}^{\bar{\rho}_i} L_g L_{g0}^{j-1} h_i \frac{\partial}{\partial \xi_{\langle v_{i-1}+j \rangle}} = \sum_{l=1}^{n-q} b_{g,l} \frac{\partial}{\partial \xi_{\langle l \rangle}}$. Condition II can be represented in $\xi-$coordinate as

$$[\sum_{l=1}^{n-q} b_{g,l} \frac{\partial}{\partial \xi_{\langle l \rangle}}, \frac{\partial}{\partial \xi_{\langle k \rangle}}] \in \text{span}\{\frac{\partial}{\partial \xi_{\langle v_{m-q} \rangle}}, \frac{\partial}{\partial \xi_{\langle v_{m-q}-1 \rangle}}, \ldots, \frac{\partial}{\partial \xi_{\langle v_i \rangle}}, \ldots, \frac{\partial}{\partial \xi_{\langle v_i - j \rangle}}\} \tag{3.41}$$

for $\bar{\rho}_i - j \leq k \leq \bar{\rho}_i$. Note that $[\sum_{l=1}^{n-q} b_{g,l} \frac{\partial}{\partial \xi_{\langle l \rangle}}, \frac{\partial}{\partial \xi_{\langle k \rangle}}] = \sum_{l=1}^{n-q} \frac{\partial b_{g,l}}{\partial \xi_{\langle k \rangle}} \frac{\partial}{\partial \xi_{\langle l \rangle}}$, which, together with (3.41) implies that $\frac{\partial b_{g,l}}{\partial \xi_{\langle k \rangle}} = 0$, for $1 \leq i \leq m-q$, $1 \leq l \leq v_i - 1$, and $\max(l+1, v_{i-1}+2) \leq k \leq v_i$.

Similarly, define $b_{g0} \triangleq \sum_{i=1}^{m-q} L_{g0}^{\bar{\rho}_i} h_i \frac{\partial}{\partial \xi_{\langle v_i \rangle}}$, one can obtain from Condition I that $\frac{\partial b_{g0,\bar{\rho}_l}}{\partial \xi_{\langle k \rangle}} = 0$, for $2 \leq i \leq m-q$, $v_{i-1}+2 \leq k \leq v_i$, and $1 \leq l \leq i-1$. The block triangular forms of γ_1 and γ_2 follows.

Now we show that under Condition III, the term $\gamma_3(z_2, y)$ is independent on $z_1 \backslash \bar{y}_1$. Similar to (3.40), we have

$$\mathscr{R}(\bar{i})_{\bar{\rho}_{\bar{i}}-2} = \text{span}\{\frac{\partial}{\partial \xi_{\langle v_{\bar{i}} \rangle}}, \frac{\partial}{\partial \xi_{\langle v_{\bar{i}}-1 \rangle}}, \ldots, \frac{\partial}{\partial \xi_{\langle v_{\bar{i}} - \bar{\rho}_{\bar{i}}+2 \rangle}}\}$$

Condition III can be represented in $\xi-$coordinate as

$$[\sum_{i=m-q+1}^{m} L_{g0} h_i \frac{\partial}{\partial \xi_{\langle v_i \rangle}}, \frac{\partial}{\partial \xi_{\langle k \rangle}}] \in \text{span}\{\frac{\partial}{\partial \xi_{\langle v_{m-q} \rangle}}, \frac{\partial}{\partial \xi_{\langle v_{m-q}-1 \rangle}}, \ldots, \frac{\partial}{\partial \xi_{\langle v_{\bar{i}} - \bar{\rho}_{\bar{i}}+2 \rangle}} \ldots, \frac{\partial}{\partial \xi_{\langle v_1 - \bar{\rho}_1 + 2 \rangle}}\} \tag{3.42}$$

for $\bar{\rho}_{\bar{i}-1} + 2 \leq k \leq \bar{\rho}_{\bar{i}}$. Note that

$$[\sum_{i=m-q+1}^{m} L_{g0} h_i \frac{\partial}{\partial \xi_{\langle v_i \rangle}}, \frac{\partial}{\partial \xi_{\langle k \rangle}}] = \sum_{i=m-q+1}^{m} L_{g0} h_i \frac{\partial}{\partial \xi_{\langle k \rangle}} \frac{\partial}{\partial \xi_{\langle v_i \rangle}}$$

which, together with (3.42) implies that $L_{g0} h_i \frac{\partial}{\partial \xi_{\langle k \rangle}} = 0 (m-q+1 \leq i \leq m)$ followed by the property that $\gamma_3(z_2, y)$ is independent on $z_1 \backslash \bar{y}_1$.

Finally, condition VI decouples the subsystems (3.34)-(3.35) from continuous faults f. □

The form (3.34)-(3.37) that results from the transformation $T(x)$ takes several advantages:

1. The subsystem (3.34)-(3.35) is not affected by continuous faults. An observer can be designed for this subsystem to provide the estimates of z_1 and θ. These estimates are decoupled from the continuous fault, thus they can be used to diagnose the faults in the subsystem (3.36)-(3.37). Another benefit is to detect the switching of the HS as it will be discussed in Section 3.3.4.
2. Both (3.34)-(3.35) and (3.36)-(3.37) are in the block strict feedback form. Which are more friendly for FTC design than that in [34] and [56]. The back-stepping control method in [63] can be developed to achieve the fault tolerant tracking goal.

We present a novel observer for the subsystem (3.34)-(3.35), which relaxes the Lipschitz conditions as in the usual high gain observer. The observer will be constructed firstly through the following several steps as in [66].

Step 1 : Define

$$\bar{M}_i(z_1,u,y) \triangleq [C_i^\top,(C_iF_i(z_1,u,y))^\top,\ldots,(C_iF_i^{\bar{\rho}_i-1}(z_1,u,y))^\top]^\top$$

for $1\leq i\leq m-q$, where $F_i(z_1,u,y)=A_i+G_{ij}(z_1,y)u$, with $G_{ij}(z_1,y)=\partial\bar{g}_i/\partial\xi_j$ for $1\leq j\leq m-q$.

Step 2 : Let $N_i=R_i(\bar{M}_iF_i\bar{M}_i^{-1}-A_i)^\top R_i$, where $R_i=[\beta_i\ A_i\beta_i\ A_i^{\bar{\rho}_i-1}\beta_i]$ with $\beta_i=[0\ldots0\ 1]^\top$. From the construction, N_i can be decomposed into $N_i=L_iC_i$, where $L_i\in\Re^{\bar{\rho}_i}\times 1$.

Step 3 : Define

$$W_i(z_1,u,y) \triangleq [C_i^\top,(C_i\bar{A}_i(z_1,u,y))^\top,\ldots,(C_i\bar{A}^{\bar{\rho}_i-1}(z_1,u,y))^\top]^\top$$

where $\bar{A}_i=A_i+N_i$, and also define $M_i=W_i^{-1}\bar{M}_i$.

We can obtain [66]

$$\begin{aligned} M_i(z_1,u,y)F_i(z_1,u,y)M_i^{-1}(z_1,u,y) &= A_i+L_i(x,u,y)C_i \\ C_iM_i^{-1}(z_1,u,y) &= C_i \end{aligned} \tag{3.43}$$

Assumption 3.4

3.4.1 The partial derivatives of $\bar{g}_i$ w.r.t. z_1 and their respective time derivatives are bounded.

3.4.2 There exists a function $B(z_1,u,y)\in\Re^{(n-q)\times(m-q)}$ such that $\psi_1(z_1,u,y)=B\bar{\psi}_1(z_1,u,y)$, where B is Lipschitz w.r.t. z_1, $|B|\leq b_0$, and $|\bar{\psi}_1|\leq\bar{q}(z_1,u,y)\leq q_0$ for a function q and numbers $b_0,q_0>0$.

3.4.3 There exist matrices $P=P^\top\in\Re^{n\times n}=diag[P_1,\ldots,\ P_{m-q}]$ with $P_i=P_i^\top\in\Re^{\bar{\rho}_i\times\bar{\rho}_i}$ and a function $R(z_1,u,y)\in\Re$ such that

$$\Delta_\varepsilon P\Delta_\varepsilon M(z_1,u,y)B(z_1,u,y)=C^\top R(z_1,u,y)$$

$$(A_i-K_iC_i)^\top P_i+P_i(A_i-K_iC_i)=-Q_i$$

where $\Delta_\varepsilon = diag[\Delta_{\varepsilon_1}, \ldots, \Delta_{\varepsilon_{m-q}}]$, $\Delta_{\varepsilon_i} = diag[1/\varepsilon_i, \ldots, 1/\varepsilon_i^{\bar{\rho}_i}]$, *with* ε *a design parameter.* $M = diag[M_1, \ldots, M_{m-q}]$, $Q_i = Q_i^\top > 0$, $K_i \in \Re^{\bar{\rho}_i \times 1}$ *are such that* $(A_i - K_i C_i)$ *is stable.*

Remark 3.5. *Note that Conditions 3.4.1 and 3.4.2 are taken instead of Lipschitz conditions on* γ_1, γ_2 *and* ψ_1. *Condition 3.4.3 is weaker than the strict positive real (SPR) condition in [57], [56], since the term* Δ_{ε_i} *is involved. Note that if* $\bar{y}_1$ *is only single output (as in our application), the dimension of B could be relaxed as* $B \in \Re^{(n-q)\times\kappa}$ *for* $\kappa > 0$, *and R could be chosen as a vector to further relax Condition 3.4.3.*

The observer is constructed as

$$\begin{aligned}
\dot{\hat{z}}_1 &= A\hat{z}_1 + \psi_1(\hat{z}_1,u,y)\hat{\theta} + \gamma_1(\hat{z}_1,y)u + \gamma_2(\hat{z}_1,y) \\
&\quad + M^{-1}(\hat{z}_1,u,y)[L(\hat{z}_1,y) + \Delta_\varepsilon^{-1}K](\bar{y}_1 - \hat{\bar{y}}_1) \\
&\quad + \hat{B}sgn(\hat{R}^\top)[\theta_0(q_0 + q(\hat{z}_1,y))sgn(\bar{y}_1 - \hat{\bar{y}}_1)] \qquad (3.44)\\
\hat{\bar{y}}_1 &= C\hat{z}_1 \qquad (3.45)\\
\dot{\hat{\theta}} &= \Gamma \bar{\psi}_1^\top(\hat{z}_1,u,y)R^\top(\hat{z}_1,u,y)(\bar{y}_1 - \hat{\bar{y}}_1) \qquad (3.46)
\end{aligned}$$

where $L = \text{diag}[L_1, \ldots, L_{m-q}]$, and $K = \text{diag}[K_1, \ldots, K_{m-q}]$. The weighting matrix $\Gamma = \Gamma^\top > 0$. $\hat{\Xi}$ denotes $\Xi(\hat{z}_1,u,y)$. Denote $e_z = [e_1^\top, \ldots, e_{m-q}^\top]^\top$ with $e_i = \xi_i - \hat{\xi}_i$, $1 \le i \le m-q$, $e_\theta = \theta - \hat{\theta}$.

Theorem 3.2. *Under Assumption 3.4, the observer described by (3.44)-(3.45) together with the adaptive algorithm (3.46) can realize* $\lim_{t\to\infty} e_z = 0$ *and* $\lim_{t\to\infty} e_\theta = 0$ *if there exist two positive constants* σ *and* t_0 *such that for all t, the following persistent excitation condition holds:*

$$\int_t^{t+t_0} \psi_1^\top(z_1(s),y(s))\psi_1(z_1(s),y(s))ds \ge \sigma I \qquad (3.47)$$

Proof: The proof of the theorem follows the recursive way. Consider the *i*th subsystem of (3.34) and (3.44), we have

$$\begin{aligned}
\dot{e}_i &= A_i e_i + (\bar{g}_i - \hat{\bar{g}}_i)u + \bar{\bar{g}}_i - \hat{\bar{\bar{g}}}_i - (\widehat{M_i^{-1}})(\hat{L}_i + \Delta_{\varepsilon_i}^{-1}K_i)C_i e_i \\
&\quad + \psi_{1i}\theta - \hat{\psi}_{1i}\hat{\theta} - \Upsilon_i \qquad (3.48)
\end{aligned}$$

where $\psi_1 = [\psi_{11}^\top, \ldots, \psi_{1(m-q)}^\top]^\top$, $\Upsilon = [\Upsilon_1^\top, \ldots, \Upsilon_{m-q}^\top]^\top \triangleq Bsgn(R^\top)[\theta_0(q_0 + q(\hat{z}_1,y))$ $sgn(\bar{y}_1 - \hat{\bar{y}}_1)]$. Consider the transformation $\tilde{e}_i \triangleq \Delta_{\varepsilon_i}\hat{M}_i e_i$ and choose a Lyapunov candidate function $V_i = \tilde{e}_i^\top P_i \tilde{e}_i$. Based on [66], it can be shown that the time derivative of V_i along (3.48) satisfies

$$\begin{aligned}
\dot{V}_i &\le -\varepsilon_i \lambda_{\min}(Q_i)|\tilde{e}_i|^2 + \sum_{j=1}^{i} \mu_{ij}|\tilde{e}_i|^2 + \sum_{j=1}^{i-1} \mu_{ji}|\tilde{e}_j|^2 \\
&\quad + 2\tilde{e}_i^\top P_i \Delta_{\varepsilon_i}\hat{M}_i(\psi_{1i}\theta - \hat{\psi}_{1i}\hat{\theta} - \Upsilon_i)
\end{aligned}$$

where $\mu_{ij}, \mu_{ji} > 0$.

Now, consider the Lyapunov candidate function as $W(e_z, e_\theta) = V_z + V_\theta$ for the overall system, where $V_z = \sum_{i=1}^{m-q} V_i$, $V_\theta = e_\theta^\top \Gamma^{-1} e_\theta$. Denote $\tilde{e} = [\tilde{e}_1^\top, \ldots, \tilde{e}_{m-q}^\top]^\top$. The time derivative of W along (3.34), (3.44) and (3.46) is

$$\dot{W} \leq \sum_{i=1}^{m-q} \left((-\varepsilon_i \lambda_{\min}(Q_i) + \sum_{j=1}^{m-q} \mu_{ij}) |\tilde{e}_i|^2 \right)$$
$$+ \underbrace{2\tilde{e}^\top P \Delta_\varepsilon \hat{M}(\psi_1 \theta - \hat{\psi}_1 \hat{\theta}) - 2\tilde{e}^\top P \Delta_\varepsilon \hat{M} \Upsilon - 2e_\theta^\top \bar{\psi}_1^\top \hat{R}^\top C e_z}_{\Psi}$$

Based on Assumption 3.4.3 and (3.43), we have

$$\begin{aligned}
\Psi &= 2\tilde{e}^\top P \Delta_\varepsilon \hat{M} \hat{B} \hat{\bar{\psi}}_1 e_\theta - 2e_\theta^\top \hat{\bar{\psi}}_1^\top \hat{R}^\top C \Delta_\varepsilon^{-1} \tilde{e} \\
&\quad + 2\tilde{e}^\top P \Delta_\varepsilon \hat{M} \hat{B}(\bar{\psi}_1 - \hat{\bar{\psi}}_1)\hat{\theta} + 2\tilde{e}^\top P \Delta_\varepsilon \hat{M}(B - \hat{B})\bar{\psi}_1 \hat{\theta} \\
&\quad - 2\tilde{e}^\top P \Delta_\varepsilon \hat{M} \hat{B} sgn(\hat{R}^\top)[\theta_0(q_0 + q(\hat{z}_1, y)) sgn(\bar{y}_1 - \hat{\bar{y}}_1)] \\
&\leq 2\tilde{e}^\top P \Delta_\varepsilon \hat{M}(B - \hat{B})\bar{\psi}_1 \hat{\theta}
\end{aligned} \tag{3.49}$$

From Assumption 3.4.2, we further obtain

$$\dot{W} \leq -\eta |\tilde{e}|^2 \tag{3.50}$$

where ε_i, $1 \leq i \leq m - p$, is chosen such that $\eta > 0$.

Since M and Δ_ε are all bounded and nonsingular, inequality (3.50) implies the stability of the origin $e_z = 0$, $e_\theta = 0$. One can get $\lim_{t \to \infty} e_z = 0$, which, together with (3.46), the persistent condition (3.47) and the uniform boundedness of e_θ, leads to $\lim_{t \to \infty} e_\theta = 0$. □

The fault estimates can be obtained straightly from (3.36) as

$$\hat{\bar{e}} \hat{f} = \dot{\hat{\bar{y}}}_2 - \hat{\psi}_0 - \hat{\psi}_2 \hat{\theta} - \bar{\gamma}_2 u \tag{3.51}$$

which yields $\hat{\bar{e}} \hat{f} - \bar{e} f = (\psi_0 - \hat{\psi}_0) + (\psi_2 \theta - \hat{\psi}_2 \hat{\theta})$. Since $\lim_{t \to \infty} e_z = 0$, $\lim_{t \to \infty} e_\theta = 0$, due to the continuity of $\bar{e}$, ψ_0 and ψ_2, there always exist two numbers $k_z, k_\theta > 0$ such that for all bounded z, $\hat{z}$, if $|e_z|$ and $|e_\theta|$ are sufficiently small, the following inequality holds

$$|\hat{\bar{e}} \hat{f} - \bar{e} f| \leq k_z |e_z| + k_\theta |e_\theta| \tag{3.52}$$

Moreover, we have $\lim_{t \to \infty} |\hat{\bar{e}} \hat{f} - \bar{e} f| = 0$. Note that f can be estimated if $\bar{e}$ is invertible, however, it will be shown in the next section that inequality (3.52) is enough to achieve the FTC objective.

3.3.3 FTC for the Nonlinear System

The observer based fault tolerant tracking control strategy is discussed in three parts. We first analyze the *local controller* to achieve the tracking objective for

z_1 subsystem (3.34)-(3.35) and z_2 subsystem (3.36)-(3.37) respectively, then design the global fault tolerant tracking controller for the overall system.

The tracking controller for the observer (3.44)-(3.45) of z_1 subsystem is designed first, the convergence of observer implies the availability of the controller for z_1 system. To facilitate the analysis, we give the following assumption.

Assumption 3.5. *There exists a set of relative degrees $\{\rho_1, \rho_2, \ldots, \rho_m\}$ such that $\sum_{i=1}^{m} \rho_i = n$ and $\xi = T(x) \in \Re^n$ is a diffeomorphism where*

$$T(x) = [h_1(x), L_{g_0(x)} h_1(x), \ldots, L_{g_0(x)}^{\rho_1 - 1} h_1(x),\\ h_2(x), \ldots, L_{g_0(x)}^{\rho_2 - 1} h_2(x), \ldots, L_{g_0(x)}^{\rho_m - 1} h_m(x)]^\top \tag{3.53}$$

Moreover, $\rho_r = 1$, $m - q + 1 \leq r \leq m$.

Under Assumption 3.5, the structure of $\bar{g}_i$ in (3.38) is changed into

$$\bar{g}_i \in \Re^{\rho_i \times p} = [\underbrace{0,\ 0,\ \ldots,\ 0,}_{\rho_i - 1 \text{ order}}\ \bar{g}_{i\rho_i}(\xi_1, \ldots, \xi_{i-1}, \xi_i, y)]^\top \tag{3.54}$$

$\hat{\gamma}_1$ in observer (3.44) is also modified to be consistent with the system.

Remark 3.6. *If Assumption 3.5 does not hold, then the system (3.34)-(3.35) would contain the tracking dynamics and zero dynamics, the proposed method can be extended to that case if the system (3.34)-(3.35) is minimum phase, i.e., the zero dynamics are input-to-state stable w.r.t. the linearizable states as in most of related literatures [63].*

Define $\Theta = [\Theta_1^\top, \ldots, \Theta_{m-q}^\top]^\top$ with $\Theta_i \triangleq \hat{M}_i^{-1}[\hat{L}_i + \Delta_{\varepsilon_i}^{-1} K_i] C e_z$. Eq.(3.44) is rewritten as

$$\begin{aligned} \dot{\hat{\xi}}_{ij} &= \hat{\xi}_{i(j+1)} + \hat{\bar{\Delta}}_{ij}, \quad 1 \leq j \leq \rho_i - 1 \\ \dot{\hat{\xi}}_{i\rho_i} &= \upsilon_i \quad 1 \leq i \leq m - q \end{aligned} \tag{3.55}$$

where

$$\begin{aligned} \hat{\bar{\Delta}}_i &= [\hat{\bar{\Delta}}_{i1}, \ldots, \hat{\bar{\Delta}}_{i(\rho_i - 1)})]^\top \triangleq \hat{\psi}_{1i}\hat{\theta} + \Theta_i + \Upsilon_i, \hat{\bar{\Delta}}_{i\rho_i} = 0 \\ \Upsilon &= [\Upsilon_1^\top, \ldots, \Upsilon_{m-q}^\top]^\top \triangleq Bsgn(R^\top)[\theta_0(q_0 + q(\hat{z}_1, y))sgn(\bar{y}_1 - \hat{\bar{y}}_1)] \end{aligned}$$

υ_i is the *local controller* for system (3.55).

Define y_{di} as a tracking signal for y_i, where y_{di} has the bounded ρ_i-orders time derivative, i.e., $\dot{y}_{di}, \ldots, y_{di}^{(\rho_i)}$ are all bounded.

Assumption 3.6. *There exists a known smooth function $\bar{\varpi}_{ij}(\hat{\xi}_{i1}, \ldots, \hat{\xi}_{ij})$ such that $|\varpi_{ij}| \leq \bar{\varpi}_{ij}$, $1 \leq j \leq \rho_i, 1 \leq i \leq m - q$, where $\varpi_{i1} \triangleq \hat{\bar{\Delta}}_{i1}$, $\varpi_{i2} \triangleq \hat{\bar{\Delta}}_{i2} - \frac{\partial \alpha_{i1}(\hat{\xi}_{i1}, t)}{\partial \hat{\xi}_{i1}} \dot{\hat{\xi}}_{i1} - \frac{\partial \alpha_{i1}}{\partial t}$, and*

$$\varpi_{ij} \triangleq \hat{\Delta}_{ij} - \sum_{s=1}^{j-1} \frac{\partial \alpha_{i(j-1)}(\hat{\xi}_{i1}, \ldots, \hat{\xi}_{i(j-1)}, t)}{\partial \hat{\xi}_{is}} \dot{\hat{\xi}}_{is} - \frac{\partial \alpha_{i(j-1)}}{\partial t}, \quad 3 \leq j \leq \rho_i$$

α_{ij} is a fictitious controller to be chosen.

Define the tracking error as $\chi_i \in \Re^{\rho_i} = [\chi_{i1}, \ldots, \chi_{i\rho_i}]^\top$, where $\chi_{is} = \hat{\xi}_{is} - y_{di}^{(s-1)}$, $1 \leq s \leq \rho_i$, and define the transformation $\tilde{\chi}_i = [\tilde{\chi}_{i1}, \ldots, \tilde{\chi}_{i\rho_i}]^\top$ with $\tilde{\chi}_{i1} = \chi_{i1}$, and $\tilde{\chi}_{ij} = \chi_{ij} - \alpha_{i(j-1)}$.

Lemma 3.4. *Under assumptions 3.4, 3.5 and 3.6, there exists a series of* local controllers *to make the output of the observer (3.44)-(3.45) exponentially* track *the given signals $\bar{y}_d \in \Re^{m-q} \triangleq [y_{d1}, \ldots, y_{d(m-q)}]^\top$ while the continuous states are bounded.*

Proof: From (3.55) and Assumption 3.6, we have $\dot{\tilde{\chi}}_{i1} = \tilde{\chi}_{i2} + \varpi_{i1} + \alpha_{i1}$. α_{i1} is designed as

$$\alpha_{i1} = -\frac{1}{2}\bar{k}_i \tilde{\chi}_{i1} - \frac{\bar{\varpi}_{i1}^2 \tilde{\chi}_{i1}}{\bar{\varpi}_{i1}|\tilde{\chi}_{i1}| + \varepsilon_i e^{-a_i t}} \tag{3.56}$$

where $\bar{k}_i, \varepsilon_i, a_i > 0$ are designed by the user. Consider the Lyapunov candidate function $\bar{V}_{i1} = \frac{1}{2}\tilde{\chi}_{i1}^2$, it follows that $\dot{\bar{V}}_{i1} = \tilde{\chi}_{i1}(\tilde{\chi}_{i2} + \varpi_{i1} + \bar{\alpha}_{i1}) \leq -\bar{k}_i \bar{V}_{i1} + \tilde{\chi}_{i1}\tilde{\chi}_{i2} + \varepsilon_i e^{-a_i t}$. Similar procedures are introduced for $\tilde{\chi}_{ij}$, $2 \leq j \leq \rho_i$. Consider the Lyapunov candidate function $\bar{V}_{ij} = \bar{V}_{i(j-1)} + \frac{1}{2}\tilde{\chi}_{ij}^2$, we have $\dot{\bar{V}}_{ij} \leq -\bar{k}_i \bar{V}_{ij} + \tilde{\chi}_{ij}\tilde{\chi}_{i(j+1)} + j\varepsilon_i e^{-a_i t}$ with $\alpha_{ij} = -\tilde{\chi}_{i(j-1)} - \frac{1}{2}\bar{k}_i \tilde{\chi}_{ij} - \frac{\bar{\varpi}_{ij}^2 \tilde{\chi}_{ij}}{\bar{\varpi}_{ij}|\tilde{\chi}_{ij}| + \varepsilon_i e^{-a_i t}}$.

Finally, choose a Lyapunov function $V_{ti} = \bar{V}_{i(\rho_i - 1)} + \frac{1}{2}\tilde{\chi}_{i\rho_i}^2$, we obtain

$$\dot{V}_{ti} \leq -\bar{k}_i V_{ti} + (\rho_i)\varepsilon_i e^{-a_i t} \tag{3.57}$$

which results from the local controller

$$\upsilon_i = -\tilde{\chi}_{i(\rho_i - 1)} - \frac{1}{2}\bar{k}_i \tilde{\chi}_{i\rho_i} - \frac{\bar{\varpi}_{i\rho_i} +^2 \tilde{\chi}_{i\rho_i}}{\bar{\varpi}_{i\rho_i}|\tilde{\chi}_{i\rho_i}| + \varepsilon_i e^{-a_i t}} + y_{di}^{(\rho_i)}$$

This completes the proof. □

Similarly to the procedure for z_1 subsystem, define the tracking error as $\chi_i = y_i - y_{di}$, $m - q + 1 \leq i \leq m$. From (3.36)-(3.37), we have $\dot{\chi}_i = \hat{\Delta}_{fi} - \dot{y}_{di} + \upsilon_i$, $m - q + 1 \leq i \leq m$, where υ_i also denotes *the local controller*, and $\hat{\Delta}_{fi} \triangleq \psi_{0i} + \psi_{2i}\theta + \bar{e}_i f$. Define $\hat{\Delta}_i \triangleq (\psi_{0i} - \hat{\psi}_{0i}) + (\psi_{2i}\theta - \hat{\psi}_{2i}\hat{\theta}) + (\bar{e}_i f - \hat{\bar{e}}_i \hat{f})$, then based on Theorem 3.2 and (3.52), it is clear that $|\hat{\Delta}_i| \leq \bar{\varpi}_i$ for $\bar{\varpi}_i > 0$. The *local controller* can be designed as

$$\upsilon_i = -\frac{1}{2}\bar{k}_i \tilde{\chi}_i + \dot{y}_{di} - \frac{\bar{\varpi}_i^2 \tilde{\chi}_i}{\bar{\varpi}_i|\tilde{\chi}_i| + \varepsilon_i e^{-a_i t}} - \hat{\psi}_{0i} - \hat{\psi}_{2i}\hat{\theta} - \hat{\bar{e}}_i \hat{f} \tag{3.58}$$

for $m - q + 1 \leq i \leq m$, which makes the time derivative of Lyapunov function $V_{ti} = \frac{1}{2}\tilde{\chi}_i^2$ satisfy the inequality as in (3.57).

Theorem 3.3. *Suppose that all the conditions in Lemma 3.3, Theorem 3.2 and assumptions 3.5, 3.6 hold. There exists a control law for the system (3.33) to make the outputs asymptotically* track *the given signals* $y_d \in \Re^m = [\bar{y}_d^\top, \bar{\bar{y}}_d^\top]^\top = [y_{d1}, \dots, y_{dm}]^\top$ *while guaranteeing that all the states x are bounded in spite of faults and parametric uncertainties, if*

$$Rank \begin{bmatrix} \gamma_1(\hat{z}_1, y) \\ \gamma_3(z_2, y) \end{bmatrix} = m \tag{3.59}$$

Proof: If Eq.(3.59) holds, the fault tolerant tracking controller is designed as

$$u = \begin{bmatrix} \gamma_1(\hat{z}_1, y) \\ \gamma_3(z_2, y) \end{bmatrix}^\dagger [\tilde{\upsilon}_1, \dots, \tilde{\upsilon}_{m-q}, \upsilon_{m-q+1}, \dots, \upsilon_m]^\top \tag{3.60}$$

where † denotes right inverse, $\tilde{\upsilon}_i \triangleq \upsilon_i - (\hat{\psi}_{1(i\rho_i)}\hat{\theta} + \Theta_{i\rho_i} + \Upsilon_{i\rho_i} + \hat{\bar{g}}_{i\rho_i})$, $1 \le i \le m-q$. It is clear that the controller (3.60) guarantees the tracking performance and the boundedness of states for systems (3.44)-(3.45) and (3.36)-(3.37). From the convergence result of observer in Theorem 3.2, the outputs of system (3.34) also asymptotically *track* $\bar{y}_d$ and the states z_1 are bounded. This completes the proof. □

To this end, we summarize the FTC design procedure as follows:

1) *Transform the original system (3.33) into the fault diagnosis block strict feedback form (3.34)-(3.37).*
2) *Design the observer (3.44)-(3.46), and the fault diagnostic scheme (3.51).*
3) *Design the observer-based fault tolerant tracking controller (3.60).*

3.3.4 FTC for Hybrid Nonlinear System

We are now in the position to extend the result in sections 3.3.2-3.3.3 to the HS. The FTC framework, shown in Fig.3.9, consists of high level (discrete event supervisor) and low level (continuous modes). The observer estimates the current continuous states, and meanwhile, detects the switchings. Based on the information from the observer, the controller, the fault diagnostic scheme, and the observer itself are switched according to the current mode.

The idea of switching detection appears from the analysis of estimation error $\bar{y}_1^j - \hat{\bar{y}}_1^j$. If all the modes are not overlapping, i.e., each observer works well only when applied to its related mode, then, similar to fault detection problem [57], $\bar{y}_1^j - \hat{\bar{y}}_1^j$ can be regarded as a *residual* for mode j to detect the switching, since $\lim_{t\to\infty} \bar{y}_1^j - \hat{\bar{y}}_1^j = 0$ before mode transition occurs. Here, we propose a time varying threshold to detect the switching instants as in the following Lemma.

Lemma 3.5. *Suppose that all the modes of (3.32) satisfy the conditions in Theorem 3.3. If all modes are* discernable, *i.e., for mode j, the estimation error* $|e_z^j|$ *is convergent as in (3.50) only under the observer (3.44)-(3.46) associated with mode j, then all the normal switchings and faulty switchings can be detected.*

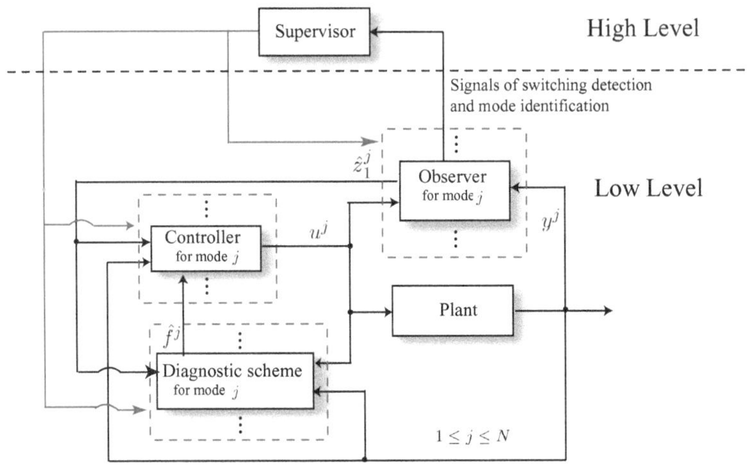

Fig. 3.9 The FTC framework for hybrid nonlinear systems

Proof: The proof follows the results of Theorem 3.2. We first consider the initial period before any switching happens. Supposed the system is initialized in mode j from $t = 0$, the inequality (3.50) can be rewritten as $\dot{V}_z^j \leq -\bar{\eta}_1^j V_z^j - \dot{V}_\theta^j$, for $\bar{\eta}_1^j > 0$. Using the differential inequality theory, we have

$$
\begin{aligned}
V_z^j(t) &\leq e^{-\bar{\eta}_1^j t} V_z^j(0) - \int_0^t e^{-\bar{\eta}_1^j (t-\tau)} \dot{V}_\theta^j(\tau) d\tau \\
&= e^{-\bar{\eta}_1^j (t)} V_z^j(0) - e^{-\bar{\eta}_1^j t} \left[V_\theta^j(\tau) e^{\bar{\eta}_1 \tau} |_0^t - \bar{\eta}_1^j \int_0^t e^{\bar{\eta}_1^j \tau} V_\theta^j(\tau) d\tau \right] \\
&\leq e^{-\bar{\eta}_1^j t} V_z^j(0) + e^{-\bar{\eta}_1^j t} \left[V_\theta^j(0) + \bar{\eta}_1^j \int_0^t e^{\bar{\eta}_1^j \tau} V_\theta^j(\tau) d\tau \right]
\end{aligned} \tag{3.61}
$$

where $\bar{\eta}_1^j \triangleq \frac{\eta_1^j}{\lambda_{max}(P^j)}$. Given an initial $z_1(0)$ or a bound of $z_1(0)$(note that the continuous state is always bounded from lemmas 3.4,3.5 and Theorem 3.3). It follows from (3.61) that $|e_z^j(t)| \leq e_{bound}^j(0)$ with

$$
\begin{aligned}
(e_{bound}^j(0))^2 \triangleq e^{-\bar{\eta}_1^j t} \Bigg[& \bar{\eta}_2^j |e_z^j(0)|^2 + \bar{\eta}_3^j \left(\theta_0^j + |\hat{\theta}^j(0)| \right)^2 \\
& + \bar{\eta}_4^j \int_0^t e^{\bar{\eta}_1^j \tau} \left(\theta_0^j + |\hat{\theta}^j(\tau)| \right)^2 d\tau \Bigg]
\end{aligned}
$$

where $\bar{\eta}_2^j \triangleq \frac{\lambda_{max}(\bar{P}^j)}{\lambda_{min}(\bar{P}^j)}$, $\bar{\eta}_3^j \triangleq \frac{\lambda_{max}((\Gamma^j)^{-1})}{\lambda_{min}(\bar{P}^j)}$, $\bar{\eta}_4^j \triangleq \bar{\eta}_1^j \bar{\eta}_3^j$, and $\bar{P} \triangleq M^\top \Delta_\varepsilon P \Delta_\varepsilon M$. Since all the modes are discernable, no matter whether the discrete fault occurs, once $|\bar{y}_1^j(t) - \hat{\bar{y}}_1^j(t)| = |Ce_z^j(t)| > \sqrt{m^j - q^j}e_{bound}^j(0)$, the switching that ends the activating period of mode j is detected.

By induction, denote $t^j(k)$ as the switching instant that activates mode j for kth time, we have

$$V_z^j(t) \leq e^{-\bar{\eta}_1^j t} V_z^j(t^j(k)) + e^{-\bar{\eta}_1^j t} \left[V_\theta^j(t^j(k)) + \bar{\eta}_1^j \int_{t^j(k)}^t e^{\bar{\eta}_1^j \tau} V_\theta^j(\tau) d\tau \right] \quad (3.62)$$

The time varying threshold is designed as

$$|e_z^j(t)| \leq e_{bound}^j(t^j(k)) \quad (3.63)$$

with

$$e_{bound}^j(t^j(k)) \triangleq e^{-\bar{\eta}_1^j t} \left[\bar{\eta}_2^j |e_z^j(t^j(k))|^2 + \bar{\eta}_3^j \left(\theta_0^j + |\hat{\theta}^j(t^j(k))| \right)^2 + \bar{\eta}_4^j \int_{t^j(k)}^t e^{\bar{\eta}_1^j \tau} \left(\theta_0^j + |\hat{\theta}^j(\tau)| \right)^2 d\tau \right]$$

Once $|\bar{y}_1^j(t) - \hat{\bar{y}}_1^j(t)| = |Ce_z^j(t)| > \sqrt{m^j - q^j}e_{bound}^j(t^j(k))$, the switching that ends the activating period of mode j is detected. □

After the switching detection, the controller (3.60) and the observer (3.44)-(3.46) related to the current mode is activated. The initial states $\hat{z}_1$ of the current observer are chosen as the final states of the previous observer. The following theorem gives the conditions to guarantee the global tracking property.

Theorem 3.4. *Suppose that the conditions in lemmas 3.6, 3.7 hold, consider the HS (3.32) under a family of controllers (3.60), diagnostic scheme (3.51) and observers (3.44)-(3.46), where all the modes satisfy the conditions in Theorem 3.3. If, at* $t = t^j(k)$*, the following inequalities hold :*

$$V_t^j(t^j(k+1)) < V_t^j(t^j(k)) \quad (3.64)$$

$$e_{bound}^j(t^j(k+1)) < e_{bound}^j(t^j(k)) \quad 1 \leq j \leq N \quad (3.65)$$

where $V_t \triangleq \sum_{i=1}^m V_{ti}$*, then* $y^j, \forall j \in Q$ *asymptotically tracks* y_d^j *during the activating period of mode j, while x is always bounded in spite of faults and uncertainties.*

Proof: Based on (3.57) in Lemma 3.4, we can further obtain that $\dot{V}_t^j \leq -\bar{k}^j V_t^j + n\varepsilon^j e^{-a^j t}$, where $\bar{k}^j = \bar{k}_i^j, a^j = a_i^j$, $1 \leq i \leq m$. Appropriate selections of ε^j and a^j can make $\dot{V}_t^j < 0, \ \forall \sigma(t) = j$. If (3.64) holds, then the Multiple Lyapunov functions (MLFs) method in [22] can be applied to conclude that the tracking error of the HS is Lyapunov stable. On the other hand, for each time $t^j(k)$ when mode j is identified, the sequence $V_t^j(t^j(k))$ is decreasing and positive, and therefore has a limit $\zeta \geq 0$. One has

$$\lim_{k\to\infty}\left[V_t^j(t^j(k+1))-V_t^j(t^j(k))\right]=\zeta-\zeta=0$$

Note that there exists a class $\mathscr{K}$ function ω such that

$$0=\lim_{k\to\infty}\left[V_t^j(t^j(k+1))-V_t^j(t^j(k))\right]\leq\lim_{k\to\infty}[-\omega(|\tilde{\chi}^j|)]\leq 0 \tag{3.66}$$

Inequality (3.66) implies that the tracking error $\tilde{\chi}^j$ converges to the origin, which combined with Lyapunov stability, leads to the asymptotic stability of $\tilde{\chi}^j$ for the HS in spite of faults and parametric uncertainty. On the other hand, Inequality (3.65) guarantees that $\lim_{t\to\infty} e_{bound}(t)=0$, which leads to the global convergence of the estimation error e_z to zero. □

Remark 3.7. *The switching detection using Lemma 3.5 may have a short time delay $t^j(k)-t^{j*}(k)$, where $t^{j*}(k)$ is the real switching instant. The effect of this delay is acceptable in the practical situation. Moreover, since e_z^j is always bounded in e_{bound}^j, continuous state estimation performance is guaranteed in the delay. For the case that x may diverge during the delay, the non-decreasing MLFs control method in [155] could be applied.*

Example 3.2: [134] A well known three-tank system is employed to illustrate the application of our approach. The schematic diagram of the system is depicted in Fig. 3.10. The system consists of three cylindrical tanks linked to each other through connecting cylindrical pipes. Two pumps control two incoming flows. The control objective is *to keep h_2 and h_3 rise or drop with given velocities and maintain the water levels in three tanks in certain regions.*

The system is modeled as a hybrid automaton with the following three modes:

Mode 1 (save water): Valves V_1, V_2 are open, V_3, V_4 are closed. Levels h_2 and h_3 rise according to given velocities.

Mode 2 (lose water): Valves V_1, V_2, V_3, V_4 are open. Levels h_2 and h_3 drop according to given velocities.

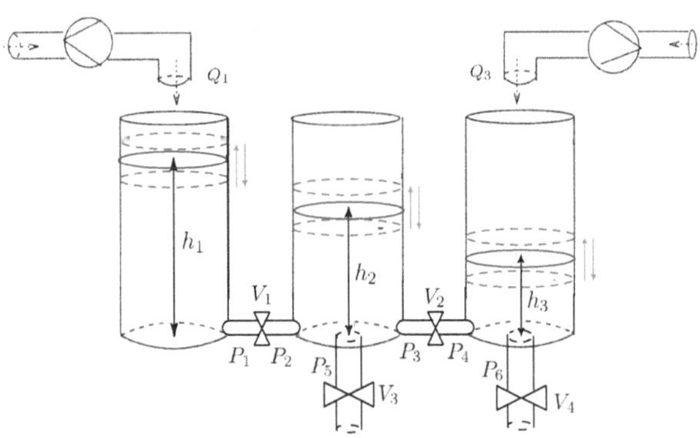

Fig. 3.10 The three tank system

h_1, h_2, h_3 are continuous states of the system with h_2 and h_3 as outputs for both modes. In the following, $(\cdot)^{\langle 1\rangle}$ and $(\cdot)^{\langle 2\rangle}$ denote the parameters for mode 1 and 2 respectively.

In the normal situation, the switching sequence is prescribed as mode 1 $\rightarrow$ mode 2 $\rightarrow$ mode 3 $\rightarrow$ mode 1, which would be affected by the faulty switching. Suppose that the coefficients of pipes P_i, $1 \le i \le 4$ are the same. For the sake of simplicity, only f and θ in mode 1 and $G_f(1,2)$ are considered.

We first verify the decomposition result of Lemma 3.3. According to the Bernoulli's principle and Toricelli's law, the analytic model of mode 1 with f and θ can be written as

$$\begin{aligned}
\begin{bmatrix} \dot{h}_1 \\ \dot{h}_2 \\ \dot{h}_3 \end{bmatrix} &= \begin{bmatrix} -a\sqrt{h_1-h_2} \\ a\sqrt{h_1-h_2} - a\sqrt{h_2-h_3} \\ a\sqrt{h_2-h_3} \end{bmatrix} + \begin{bmatrix} \frac{Q_1}{S} \\ 0 \\ \frac{Q_3}{S} \end{bmatrix} \\
&\quad + \begin{bmatrix} -a\sqrt{h_1-h_2} \\ 0 \\ a\sqrt{h_2-h_3} \end{bmatrix} f + \begin{bmatrix} -\frac{0.006\sqrt{h_2-h_3}}{\sqrt{h_1-h_2}} - \frac{0.006a}{s\sqrt{h_1-h_2}}\frac{Q_1}{S} \\ 0.006 \\ \frac{0.006\sqrt{h_2-h_3}}{\sqrt{h_1-h_2}} - \frac{0.006\sqrt{h_2-h_3}}{a} \end{bmatrix} \theta \\
y_1 &= h_2,\ y_2 = h_3
\end{aligned}$$

where the fault term corresponds to sediment deposit in P_1 and P_4, i.e., sections of P_1 and P_4 progressively change. The uncertainty term denotes the modelling error and input disturbance related to Q_1. It can be checked that Assumption 3.5 is satisfied. Indeed, according to (3.53), a diffeomorphism is chosen as $\xi_1^{\langle 1\rangle} = h_2$, $\xi_2^{\langle 1\rangle} = a(\sqrt{h_1-h_2} - \sqrt{h_2-h_3})$, $\xi_3^{\langle 1\rangle} = h_3$. Mode 1 in $\xi^{\langle 1\rangle}$-coordinate can be represented as

$$\begin{aligned}
\dot{\xi}_1^{\langle 1\rangle} &= \xi_2^{\langle 1\rangle} + 0.006\theta \\
\dot{\xi}_2^{\langle 1\rangle} &= -\frac{a^2}{2} + \frac{a^3\sqrt{y_1-y_2}}{2(\xi_2^{\langle 1\rangle} + a\sqrt{y_1-y_2})} - \frac{a\xi_2^{\langle 1\rangle}}{2\sqrt{y_1-y_2}} \\
&\quad + \frac{a^2 Q_1}{2S(\xi_2^{\langle 1\rangle} + a\sqrt{y_1-y_2})} + \frac{aQ_3}{2S\sqrt{y_1-y_2}} \\
&\quad - 0.006\Big(\frac{S^2a^3\sqrt{y_1-y_2} + a^2Q_1}{2S^2(\xi_2^{\langle 1\rangle} + a\sqrt{y_1-y_2})^2} + \frac{a}{2\sqrt{y_1-y_2}} - \frac{1}{2}\Big)\theta \\
y_1 &= \xi_1^{\langle 1\rangle}
\end{aligned}$$

$$\begin{aligned}
\dot{\xi}_3^{\langle 1\rangle} &= a\sqrt{y_1-y_2} + \frac{Q_3}{S} + a\sqrt{y_1-y_2}f \\
&\quad + \Big(\frac{0.006a\sqrt{y_1-y_2}}{\xi_2^{\langle 1\rangle} + a\sqrt{y_1-y_2}} - \frac{0.006\sqrt{y_1-y_2}}{a}\Big)\theta \\
y_2 &= \xi_3^{\langle 1\rangle}
\end{aligned}$$

which verifies Lemma 3.3. Moreover the persistent excitation condition (3.47) holds since the coefficient matrix of θ does not tend to zero.

One further have that

$$M^{\langle 1\rangle} = \begin{bmatrix} 1 & 0 \\ \frac{S^2a^3\sqrt{y_1-y_2}+a^2Q_1}{2S^2(\xi_2^{\langle 1\rangle}+a\sqrt{y_1-y_2})^2}+\frac{a}{2\sqrt{y_1-y_2}} & 1 \end{bmatrix}$$

$$L^{\langle 1\rangle} = \begin{bmatrix} -\frac{S^2a^3\sqrt{y_1-y_2}+a^2Q_1}{2S^2(\xi_2^{\langle 1\rangle}+a\sqrt{y_1-y_2})^2}-\frac{a}{2\sqrt{y_1-y_2}} & 0 \\ \frac{S^2a^3\xi_2^{\langle 1\rangle}-Sa^3Q_1}{\sqrt{y_1-y_2}(2S\xi_2^{\langle 1\rangle}+2Sa\sqrt{y_1-y_2})^2}-\frac{aQ_3-Sa\xi_2^{\langle 1\rangle}}{4S\sqrt{(y_1-y_2)^3}} & 0 \end{bmatrix}$$

The transformation of mode 2 and 3 similar to that of mode 1 is thus omitted. It can be shown that Lemma 3.3 is applied for all three modes. Also note that the dynamics of three modes are different from each other, which guarantees the discernability as in Lemma 3.5.

Table 3.1 summarizes typical values of the three-tank system.

Table 3.1 Physical parameters of the three tank system

$S_1 = S_2 = S_3 = S = 0.0171\text{m}^2$	Tank cross-section areas
$S_{c1} = S_{c2} = S_{c3} = S_{c4} = 0.0002\text{m}^2$ $S_{c5} = S_{c6} = 0.0004\text{m}^2$	Pipe cross-section areas
$a_{zi} = 0.5279 (1 \le i \le 6)$	Pipe coefficients
$h_{max} = 0.63\text{m}$	Maximum level
$Q_{max} = 5\times 10^{-4}\text{m}^3/s$	Maximum In-flow rate

where the coefficient $a_i = a_{zi}(S_{ci}/S)\sqrt{2g}(1 \le i \le 6)$ with g being the gravitational constant. The system required behavior is shown in Table 3.2.

Table 3.2 System required behavior

Mode	Reference signals	guard set
Mode 1	$y_{d1}^{\langle 1\rangle} = 0.45 - 0.08e^{-0.01(t-t^1(k))}$ $y_{d2}^{\langle 1\rangle} = 0.33 - 0.08e^{-0.01(t-t^1(k))}$	$G(1,2) = \{h_1 \ge 0.545\}$
Mode 2	$y_{d1}^{\langle 2\rangle} = 0.35 + 0.08e^{-0.01(t-t^2(k))}$ $y_{d2}^{\langle 2\rangle} = 0.23 + 0.08e^{-0.01(t-t^2(k))}$	$G(2,3) = \{h_1 \le 0.495\}$
Mode 3	$y_{d1}^{\langle 3\rangle} = 0.35 + 0.06e^{-0.008(t-t^2(k))}$ $y_{d2}^{\langle 3\rangle} = 0.23 + 0.06e^{-0.008(t-t^2(k))}$	$G(3,1) = \{h_1 \le 0.485\}$

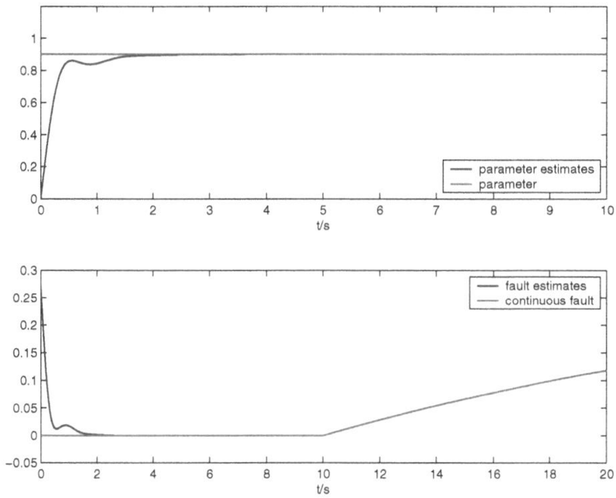

Fig. 3.11 Estimation performance

The system is initialized in mode 1, one switching and the operation of two modes are considered in the simulation. The initial levels are $[0.48\ 0.37\ 0.25]^\top$. As for mode 1, $f = 0.3 - 0.3e^{-0.05(t-10)}$ which is assumed to occur at $t = 10s$, $\theta = 0.9\text{m}/s$ is unknown, $\theta_0 = 1$. The parameters of the observer are chosen as $\Gamma^{\langle 1 \rangle} = 5$, $\varepsilon_1^{\langle 1 \rangle} = 3$, $K^{\langle 1 \rangle} = [1\ 0.25]^\top$, $P^{\langle 1 \rangle} = \begin{bmatrix} 1.25 & -0.5 \\ -0.5 & 3 \end{bmatrix}$, $R^{\langle 1 \rangle} = 0.1481$, it can be seen that Condition 2.3 is satisfied with $\bar{\psi}_1^{\langle 1 \rangle} = 0.006$, $Q^{\langle 1 \rangle} = \begin{bmatrix} 2.25 & -1 \\ -1 & 1 \end{bmatrix}$, $\hat{\theta}(0) = 0$. The parameters of the controller are designed as $\bar{\varpi}_{11}^{\langle 1 \rangle} = 0.0026$, $\bar{\varpi}_{12}^{\langle 1 \rangle} = 2.5|\hat{\xi}_2^{\langle 1 \rangle}|$, $\bar{k}^{\langle 1 \rangle} = 5$, $\varepsilon^{\langle 1 \rangle} = 0.0002$, and $a^{\langle 1 \rangle} = 5$. Related parameters of mode 2 and 3 can be obtained following the same way, which are omitted.

Fig. 3.11 shows the estimation performance, where rapid and accurate estimates of θ can be provided after $t = 2s$ which verifies Theorem 3.2. Note that $\bar{e}^{\langle 1 \rangle}$ is invertible, f is also estimated effectively. Fig. 3.12(a) shows that the switching occurs at $t = 98.28s$ when h_1 reaches the guard set, and is detected at $t = 98.298s$ with a delay of $0.018s$, the region of $h_1(98.298)$ is obtained from Lemma 3.6 as $[0.546 - 6.5 \times 10^{-6}, 0.546 + 6.5 \times 10^{-6}]$ which belongs to $Inv(2) \backslash Inv(3)$, so mode 2 can be identified as the current mode. Figures 3.12 shows the behaviors of h_1, h_2, h_3 and two inputs, from which we can see that the tracking performance is always maintained in spite of continuous faults and uncertainties, and h_1, h_2, h_3 are always in the invariant set, the switching detection delay is acceptable.

Now consider the discrete fault as $G_f(1,2) = \{h_1 \geq 0.543\}$, the switching is detected at $t = 86.572s$, and mode 2 can also be identified according to Lemma 3.6. Fig. 3.13 shows the behaviors of three levels, which implies that the faulty switching can be detected rapidly and the tracking performance is guaranteed with the continuous state staying in the invariant set.

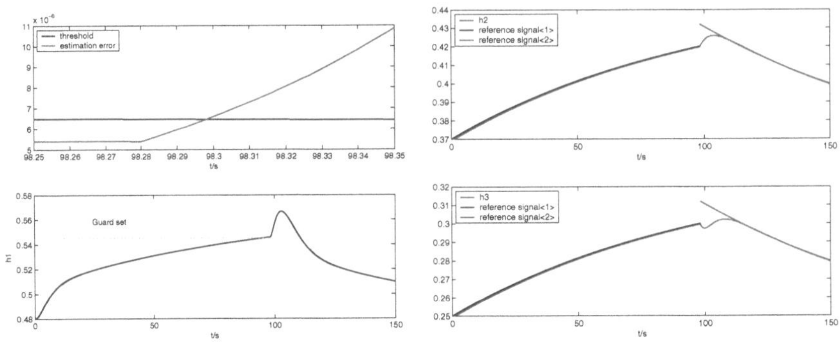

(a) Switching detection and h_1's behavior (b) The behaviors of h_2 and h_3

Fig. 3.12 FTC performance

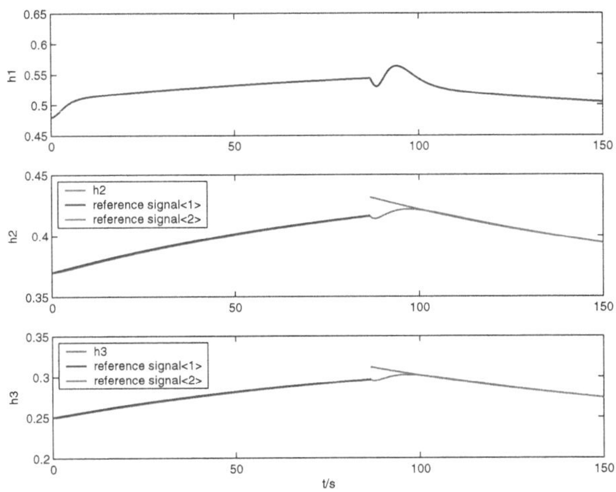

Fig. 3.13 System behavior

3.4 Conclusion

This Chapter has investigated the observer-based FTC problem of HS with uncontrollable state-dependent switching. The key idea is to design, under some structure conditions imposed on each mode, the observer for each mode whose estimation error is not affected by continuous faults and sensitive to mode transitions.

Chapter 4
Hybrid Systems with Impulsive and Stochastic Switching

In this chapter, two important classes of HS are considered that are with impulsive and stochastic switchings. For the former system, the FTC strategy is based on the trade-off between the frequency of switching, the impulsive magnitude, and the decreasing rate of Lyapunov functions along the solution of the system; Similarly, for the latter one, the FTC objective is achieved via the trade-off among the fault occurrence transition rate, the frequency of switching, and the decreasing rate of Lyapunov functions. The work in this chapter can be regarded as an extension of that in Section 2.3.

4.1 Impulsive Switching Case

Hybrid impulsive systems (HIS) represent an important type of hybrid systems [46, 136] that have gained much attention in engineering, where the continuous states abruptly change due to the impulse effect at each switching instant. Examples of HIS include some biological neural networks, frequency-modulated signal processes, flying object motions [69, 45].

In this section, we focus on the FTC problem for hybrid nonlinear impulsive systems with both continuous and discrete faults, and without full state measurements. An observer-based FTC law is designed for each mode, and two consequent cases are considered. For the case that each mode is input-to-state stable (ISS) w.r.t. the estimation error as the input, an ADT scheme is proposed such that the ISS property of the HIS is maintained in spite of faults and impulse effects. For the case that only partial modes are ISS under the FTC law, a novel *double ADT scheme* is developed to keep the overall system still ISS.

4.1.1 Preliminaries

The HIS that we consider takes the form

$$\begin{cases} \dot{x}(t) = A_{\sigma(t)}x(t) + G_{\sigma(t)}(x(t))\theta_{\sigma(t)}(t) + B_{\sigma(t)}u_{\sigma(t)}(t) \\ y(t) = C_{\sigma(t)}x(t) \qquad\qquad t \neq t_k, k \in \{1,2,...\} \end{cases} \tag{4.1}$$

H. Yang et al.: Fault Tolerant Control Design for Hybrid Systems, LNCIS 397, pp. 91–112.
springerlink.com

$$\begin{cases} x(t) = f_{\sigma(t^-),\sigma(t)}\left(x(t^-), u_{\sigma(t^-)}(t^-), \theta^d_{\sigma(t^-),\sigma(t)}(x(t^-))\right) \\ y(t) = C_{\sigma(t)}x(t) \qquad\qquad\qquad\qquad\qquad\qquad t = t_k, k \in \{1,2,...\} \end{cases} \tag{4.2}$$

where $x(t) \in \Re^n$ is the non measured state which is continuous between impulses. $y(t) \in \Re^r$ is the output, $u_\sigma(t) \in \Re^m$ is the control. A_σ, B_σ and C_σ are real constant matrices of appropriate dimensions. (A_σ, B_σ) is controllable, (A_σ, C_σ) is observable. $\theta_\sigma \in \Re^j$ is a bounded parameter, $|\theta_\sigma| \leq \bar{\theta}_\sigma$ for $\bar{\theta}_\sigma > 0$. In the fault-free case, we have $\theta_\sigma = \theta_{H\sigma}$ with $\theta_{H\sigma}$ a known constant vector. The nonlinear term $G_\sigma(x)$ is a continuous Lipschitz function, i.e., $|G_\sigma(x_1) - G_\sigma(x_2)| \leq L_\sigma |x_1 - x_2|$ for $L_\sigma > 0$. It is assumed that $G_\sigma(0) = 0$, and $|G_\sigma(x)| \leq \bar{g}_\sigma$ for $\bar{g}_\sigma > 0$.

The *continuous fault* changes the parameter θ_σ unexpectedly as in [59]. In the faulty case, $\theta_\sigma = \theta_{H\sigma} + \theta_{f\sigma}$, where $\theta_{f\sigma}$ denotes the unknown constant fault vector, $|\theta_{f\sigma}| \leq \bar{\theta}_{f\sigma}$, for $\bar{\theta}_{f\sigma} > 0$.

Define $Q = \{1, 2, \ldots, N\}$, where N is the number of modes. $\sigma(t) : [0, \infty) \rightarrow Q$ denotes the piecewise constant switching function [69]. At the kth switching instant t_k, the system (4.1) switches from mode i to mode j, where $i = \sigma(t), \forall t \in [t_{k-1}, t_k)$ and $j = \sigma(t), \forall t \in [t_k, t_{k+1})$. It is supposed that the switching can be detected at each switching instant.

The impulsive dynamics (4.2) is activated at each t_k. *The discrete fault* is considered as an abnormal impulse effect, which is represented by the unknown function $\theta^d_{\sigma(t^-),\sigma(t)}(x(t^-))$, and does not exist in the fault-free case.

There are quite a few practical systems that can be described by the HIS model (4.1)-(4.2), e.g., the biped walking robot [44], the switched reluctance motor [116], etc. The objective is to *design the FTC law u_σ and provide a sufficient condition on the switching frequency of σ such that the state x is always bounded in spite of faults and impulse effects.*

4.1.2 FTC for Single Mode

Let us first consider the system (4.1) with $\sigma(t) = j$ for some $j \in Q$ starting from $t = t_k$, and design the controller u_j such that mode j is stabilized in spite of fault θ_{fj}.

Assumption 4.1. *There exist two constant matrices $E_j, K_j \in \Re^{n\times r}$ such that $G_j(x) = E_j\bar{G}_j(x)$, and for a given matrix $Q_j \in \Re^{n\times n} > 0$, it holds that*

$$(A_j - K_jC_j)^\top P_j + P_j(A_j - K_jC_j) = -Q_j, \quad and \quad P_jE_j = C_j^\top R_j$$

for a matrix $P_j \in \Re^{n\times n} > 0$ and scalar R_j. Moreover, rank$(B_j, E_j) =$ rank(B_j).

The *FD observer* for mode j is designed as

$$\dot{\hat{x}}(t) = A_j\hat{x}(t) + G_j(\hat{x}(t))\hat{\theta}_j(t) + B_ju_j(t) + K_j(y(t) - \hat{y}(t)) \tag{4.3}$$

$$\hat{y}(t) = C_j\hat{x}(t) \tag{4.4}$$

$$\dot{\hat{\theta}}_j(t) = \Gamma_jG_j^\top(\hat{x}(t))R_j(y(t) - \hat{y}(t)) \tag{4.5}$$

where $\hat{x}(t), \hat{\theta}_j(t), \hat{y}(t)$ are the estimates of $x(t), \theta_j(t), y(t)$. The matrix $\Gamma_j = \Gamma_j^\top > 0$.

Similarly as in section 2.2, the observer (4.3)-(4.5) always diagnoses θ_j no matter the mode j is faulty or not. Denote $e_x(t) = x(t) - \hat{x}(t)$, $e_y(t) = y(t) - \hat{y}(t)$, $e_\theta(t) = \theta_j(t) - \hat{\theta}_j(t)$.

Lemma 4.1. *Under Assumption 4.1, the observer described by (4.3)-(4.5) can realize $\lim_{t\to\infty} e_x = 0$ and $\lim_{t\to\infty} e_\theta = 0$ if there exist two positive constants ρ and t_0 such that for all t, the following persistent excitation condition holds:*

$$\int_t^{t+t_0} \bar{G}_j^\top(x(s))\bar{G}_j(x(s))ds \geq \rho I \tag{4.6}$$

Proof: The proof can be obtained following the procedure in [59], which is omitted. □

Lemma 4.1 means that the observer (4.3)-(4.5) provides both the continuous state estimates $\hat{x}$ and the fault estimates $\hat{\theta}_j$, which will be used for controller design.

Definition 4.1. *[115]: A system $\dot{x} = f(x,u)$ is said to be* input-to-state stable *(ISS) w.r.t the input u if there exist functions $\beta \in \mathscr{K}\mathscr{L}$, $\alpha, \gamma \in \mathscr{K}_\infty$ such that for any initial $x(0)$, we have*

$$\alpha(|x(t)|) \leq \beta(|x(0)|, t) + \gamma(\|u\|_{[0,t)}), \quad \forall t \geq 0$$

Lemma 4.2. *[115]: If there exist α_1, α_2, α_3, $\gamma_1 \in \mathscr{K}_\infty$, and a smooth function $V: \Re^n \to \Re_{\geq 0}$ such that $\alpha_1(|x|) \leq V(x) \leq \alpha_2(|x|)$, $\dot{V}(x) \leq -\alpha_3(|x|) + \gamma_1(|u|)$ then the system $\dot{x} = f(x,u)$ is ISS w.r.t. u.*

Recall that (A_j, B_j) is controllable. Let $W_j = W_j^T > 0$ be associated with a given symmetric positive definite matrix H_j by the Riccati equation

$$A_j^T H_j + H_j A_j - 2H_j B_j B_j^T H_j + W_j = 0 \tag{4.7}$$

Since $G_j(x)$ satisfies $|G_j(x)| \leq L_j|x|$. It has been shown in [45] that there exists $\eta_j > 0$ such that

$$\theta_{Hj}^\top G_j^\top(x) H_j x \leq \eta_j x^\top H_j x \tag{4.8}$$

To this end, our fault-tolerant controller is constructed as

$$u_j(\hat{x}) = -B_j^T H_j \hat{x} - B_j^* E_j \bar{G}_j(\hat{x})(\hat{\theta}_j - \theta_{Hj}) \tag{4.9}$$

Theorem 4.1. *Suppose that Assumption 4.1 is satisfied. Under the feedback controller (4.9), mode j in (4.1) is ISS w.r.t. e_x and e_θ, i.e. there exist functions $\beta \in \mathscr{K}\mathscr{L}$, $\alpha, \gamma_1, \gamma_2 \in \mathscr{K}_\infty$ such that for any initial state $x(t_k)$ and $t \geq t_k$, we have $\alpha(|x(t)|) \leq \beta(|x(t_k)|, t) + \gamma_1(\|e_x\|_{[t_k,t)}) + \gamma_2(\|e_\theta\|_{[t_k,t)})$ if*

$$-\lambda_{\min}(W_j) + \eta_j|H_j| < 0 \tag{4.10}$$

Proof: Applying the control (4.9) to mode j in (4.1) results in the closed-loop dynamics

$$\dot{x} = (A_j - B_j B_j^\top H_j)x + B_j B_j^\top H_j e_x + G_j(x)\theta_{Hj} \\ + E_j\Big(\bar{G}_j(x)\theta_{fj} - \bar{G}_j(\hat{x})\hat{\theta}_{fj}\Big) \quad (4.11)$$

where $\hat{\theta}_{fj} \triangleq \hat{\theta}_j - \theta_{Hj}$. Consider a Lyapunov candidate $V_j(x) = x^\top H_j x$ with $H_j > 0$ defined by (4.7). Its derivative along the system (4.11) is

$$\dot{V}_j = -x^\top W_j x + 2x^\top H_j B_j B_j{}^\top H_j e_x + 2x^\top H_j G(x)_j \theta_{Hj} \\ + 2x^\top H_j E_j(\bar{G}_j(x)\theta_{fj} - \bar{G}_j(\hat{x})\hat{\theta}_{fj}) \quad (4.12)$$

It is clear that

$$\bar{G}_j(x)\theta_{fj} - \bar{G}_j(\hat{x})\hat{\theta}_{fj} = \bar{G}_j(\hat{x})(\theta_{fj} - \hat{\theta}_{fj}) + (\bar{G}_j(x) - \bar{G}_j(\hat{x}))\theta_{fj} \quad (4.13)$$

Substituting (4.8) and (4.13) into (4.12), together with the fact that there exists an arbitrary $\varepsilon > 0$ such that $2ab \le \varepsilon a^2 + b^2/\varepsilon$ for two numbers a, b, yields

$$\dot{V}_j \le (-\lambda_{\min}(W_j) + \eta_j|H_j| + \varepsilon_1 + \varepsilon_2 + \varepsilon_3)|x|^2 \\ + \Big(\frac{|H_j B_j B_j^T H_j|^2}{\varepsilon_1} + \frac{|H_j E_j L_j|^2 \bar{\theta}_{fj}^2}{\varepsilon_2}\Big)|e_x|^2 + \frac{|H_j E_j|^2 \bar{g}_j^2}{\varepsilon_3}|e_\theta|^2 \quad (4.14)$$

where $\varepsilon_1, \varepsilon_2, \varepsilon_3 > 0$, $\bar{\theta}_{fj}$ and $\bar{g}_j$ denote the norm bounds of θ_{fj} and G_j. If the condition (4.10) holds, $\varepsilon_1, \varepsilon_2, \varepsilon_3 > 0$ can be chosen small enough such that V_j satisfies Lemma 4.2, the result follows. □

If we can choose H_j and W_j such that (4.10) is satisfied, then each single mode is ISS w.r.t e_x and e_θ in spite of continuous faults, which, together with Lemma 4.1, implies that x converges to zero.

4.1.3 FTC for Hybrid Impulsive Systems

In this section, we first consider that all modes are ISS w.r.t. e_x and e_θ, then extend the result to the case that some modes may be not stabilized in the sense of ISS, because (4.10) does not hold. We will show that under some switching conditions, it is not necessary to design the stabilizing controller for each faulty mode. The stability of the overall HIS is still guaranteed.

Consider the HIS (4.1)-(4.2). Since all modes are ISS, it can be obtained from Theorem 4.1 that there exist continuously differentiable functions $V_k : \Re^n \to \Re_{\ge 0}$, $k \in Q$ and $\bar{\gamma}_1(\cdot)$, $\bar{\gamma}_2(\cdot) \in \mathscr{K}_\infty$, such that $\forall p \in Q$

$$\bar{\alpha}_1|x|^2 \le V_p(x) \le \bar{\alpha}_2|x|^2 \quad (4.15)$$

$$\dot{V}_p(x) \le -\lambda_0 V_p(x) + \bar{\gamma}_1(|e_x|) + \bar{\gamma}_2(|e_\theta|) \quad (4.16)$$

where constants $\bar{\alpha}_1$, $\bar{\alpha}_2$, $\lambda_0 > 0$.

Assumption 4.2. *There exist two numbers $\xi_1, \xi_2 \geq 0$ such that the impulsive dynamic (4.2) with discrete faults satisfies*

$$|x(t_k)| \leq \xi_1 |x(t_k^-)| + \xi_2 |e_x(t_k^-)|, \quad k \in \{1,2,...\} \tag{4.17}$$

Remark 4.1. *Assumption 4.2 is a mild condition due to the following aspects: 1) Since the impulsive dynamics includes x and $\hat{x}$, the discrete fault is also a function of x, the form of (4.17) appears naturally for the norm bound of $x(t_k)$. 2) The magnitudes of ξ_1 and ξ_2 are not restricted, and can be taken arbitrarily large. 3) Inequality (4.17) does not restrict the decay rate of the impulsive dynamics as in [69], [74], and has no relation with the continuous dynamics.*

The FD observer (4.3)-(4.4) and the controller (4.9) are switched according to the current mode at each switching instant t_k. The initial observer state of the current mode is chosen as the previous value $\hat{x}(t_k^-)$. The parameter estimates $\hat{\theta}_\sigma(t_k)$ are set to $\theta_{H\sigma(t_k)}$ at t_k.

Theorem 4.2. *Consider the HIS (4.1)-(4.2) that satisfies Assumption 4.2, and all modes are ISS w.r.t. $e_x(t)$, $e_\theta(t)$. The HIS is ISS w.r.t. $e_x(t)$, $e_\theta(t)$ in spite of any fault and any large impulse effect if the switching function σ has an ADT τ_a such that*

$$\tau_a > \frac{\ln \varpi}{\lambda_0} \tag{4.18}$$

where $\varpi \triangleq \frac{2\bar{\alpha}_2 \xi_1^2}{\bar{\alpha}_1}$ *and* $\varpi \geq 1$.

Proof: The proof can be straightly obtained following the same line as that of Theorem 2.6, thus it is omitted. □

Roughly speaking, Theorem 4.2 shows that, under a low switching frequency, the overall HIS is ISS w.r.t. e_x, e_θ.

Remark 4.2. *The condition (4.18) is similar to the condition in Theorem 2.6 for the non-impulsive systems. However, if $\varpi \leq 1$, i.e., the impulsive dynamics decreases the norm bound of x, then the HIS can switch arbitrarily without affecting the ISS. This property is unavailable for non-impulsive HS.*

Remark 4.3. *The discrete fault is hard to be detected since it appears and vanishes instantaneously, unless the impulsive dynamics satisfies some special structures such that the fault can be detected rapidly from outputs. Theorem 4.2 shows that the discrete fault detection and diagnosis is not necessary to keep the HIS ISS.*

Now consider the case that some modes are ISS while others may be not. Define two subsets of Q as $Q = Q_s \cup Q_{us}$, where Q_s (Q_{us}) denotes the set of modes that are (not) ISS.

The following two inequalities are considered instead of inequality (4.16)

$$\begin{cases} \dot{V}_p(x) \leq -\lambda_0 V_p(x) + \bar{\gamma}_1(|e_x|) + \bar{\gamma}_2(|e_\theta|) \ \forall p \in Q_s \\ \dot{V}_q(x) \leq \lambda_1 V_q(x) + \bar{\gamma}_1(|e_x|) + \bar{\gamma}_2(|e_\theta|) \quad \forall p \in Q_{us} \end{cases} \tag{4.19}$$

where $0 < \lambda_1 \triangleq \max_{j\in Q_{us}}\{-\lambda_{\min}(W_j)+\eta_j|H_j|\}$. In this case, the continuous flow in mode $p \in Q_{us}$ can potentially destroy ISS.

Define T_s (T_{us}) the dwell period of ISS (non-ISS) modes in $[t,T)$. Then we define the *double ADT* as follows, which generalizes Definition 2.4 (ADT) and provides two ADT scales for the HIS with both ISS and non-ISS modes.

Definition 4.2. *Let $N_\sigma^s(T,t)$ ($N_\sigma^{us}(T,t)$) denote the number of switchings of σ during the period T_s (T_{us}), if there exist two positive numbers τ_s and τ_{us} such that*

$$N_\sigma^s(T,t) \le N_0 + \frac{T_s}{\tau_s}, \quad N_\sigma^{us}(T,t) \le N_0 + \frac{T_{us}}{\tau_{us}}, \quad \forall T \ge t \ge 0 \tag{4.20}$$

where $N_0 > 0$, then τ_s and τ_{us} are called double ADT *of σ over (t,T).*

Consider the time interval $[0,T)$ for $T > 0$, for the sake of simplicity, in the following, we divide $[0,T) = [0,T_c^-) \cup [T_c,T)$ and focus on two cases: Case 1, $T_{us} = T_c, T_s = (T - T_c)$, i.e., non-ISS modes work in $[0,T_c^-)$ and ISS ones work in $[T_c,T)$. Case 2, $T_s = T_c, T_{us} = T - T_c$, i.e., ISS modes work in $[0,T_c^-)$ and non-ISS ones work in $[T_c,T)$. The results can be extended to the more general case. It is still assumed that $\varpi \ge 1$.

Theorem 4.3. *Consider the HIS (4.1)-(4.2) that satisfies Assumption 4.2. The HIS is ISS w.r.t. $e_x(t)$, $e_\theta(t)$ in spite of any fault and impulse effect, if the switching function σ has the double ADT τ_s, τ_{us} such that*

$$\lambda_0\tau_s > \ln\varpi, \quad T_{us} = T_c,\ T_s = (T - T_c) > 0 \tag{4.21}$$

$$\lambda_0\tau_s > \max\left\{\ln\varpi, \ln\varpi\frac{T_{us}}{\tau_{us}} + \lambda_1 T_{us}\right\}\ T_s = T_c > 0,\ T_{us} = T - T_c \tag{4.22}$$

where $T > 0$ is an arbitrary time.

Before proving Theorem 4.3, we provide some insight into the conditions (4.21)-(4.22): if the system remains in an ISS mode after the last switching instant, then the HIS with partial ISS modes is ISS under the same conditions as that with all ISS modes. In contrast, if the system stays in a non-ISS mode after the last switching instant, then (4.22) implies that

- The larger (smaller) λ_1 is, the longer (shorter) ADT of ISS modes is needed.
- The larger (smaller) λ_0 is, the shorter (longer) ADT of ISS modes is needed.
- With a frequent switching of non-ISS modes, a long ADT of ISS modes is needed.
- With a long dwell period of non-ISS modes, a long ADT of ISS modes is needed.

The above analysis reflects the trade-off among the frequency of switching, and the decreasing rate of Lyapunov functions along the solution of ISS modes and non-ISS ones. It can be seen that ISS can still be achieved if the negative effect on the ISS resulting from non-ISS modes can be compensated by the positive effect of ISS modes.

Proof of Theorem 4.3: Let $T>0$ be an arbitrary instant. The function $W(s)$ is modified as

$$W(s)=\begin{cases} e^{\lambda_0 s}V_{\sigma(s)}(x(s)) & \forall\sigma(s)\in Q_s \\ e^{-\lambda_1 s}V_{\sigma(s)}(x(s)) & \forall\sigma(s)\in Q_{us}\end{cases}\tag{4.23}$$

Then we have $\dot W(s)\le e^{\lambda_0 s}\Phi,\forall s\in T_s$, and $\dot W(s)\le e^{-\lambda_1 s}\Phi,\forall s\in T_{us}$. Denote by $t_1^{us},\ldots,t_{N_\sigma^{us}}^{us}$, and $t_1^s,\ldots,t_{N_\sigma^s}^s$ the switching instants on the interval T_{us} and T_s respectively.

Case 1: $T_{us}=T_c, T_s=(T-T_c)$.

We first consider the time interval $[T_c,T)$, following the results of theorems 2.6, 4.2 and (4.23), one has

$$\begin{aligned} W(T^-)\le{}& \varpi^{N_\sigma^s}e^{(\lambda_1+\lambda_0)T_c^-}W(T_c^-)\\ &+\sum_{i=1}^{N_\sigma^s}\left(\varpi^{N_\sigma^s-i}\chi_i^s\right)+\sum_{j=1}^{N_\sigma^s}\left(\varpi^{N_\sigma^s-j}G_{t_j^s}^{t_{j+1}^{s-}}(\lambda_0)\right)\end{aligned}\tag{4.24}$$

where $\chi_k^s\triangleq 2e^{\lambda_0 t_k^s}\bar\alpha_2\xi_2^2|e_x(t_k^{s-})|^2$, $t_{N_\sigma+1}^s=T$. Similarly to the iterative procedure in Theorem 2.6, we can obtain

$$\begin{aligned} W(T_c^-)\le{}& \varpi^{N_\sigma^{us}}W(0)+\sum_{i=1}^{N_\sigma^{us}}\left(\varpi^{N_\sigma^{us}-i}\chi_i^{us}\right)\\ &+\sum_{j=0}^{N_\sigma^{us}}\left(\varpi^{N_\sigma^{us}-j}G_{t_j^{us}}^{t_{j+1}^{us-}}(-\lambda_1)\right)\end{aligned}\tag{4.25}$$

where $\chi_k^{us}\triangleq 2e^{-\lambda_1 t_k^{us}}\bar\alpha_2\xi_2^2|e_x(t_k^{us-})|^2$.

Combining (4.24) and (4.25) leads to

$$\begin{aligned} W(T^-)\le{}& \varpi^{N_\sigma^s+N_\sigma^{us}}e^{(\lambda_1+\lambda_0)T_c^-}W(0)+e^{(\lambda_1+\lambda_0)T_c^-}\sum_{i=1}^{N_\sigma^{us}}\left(\varpi^{N_\sigma^{us}+N_\sigma^s-i}\chi_i^{us}\right)\\ &+e^{(\lambda_1+\lambda_0)T_c^-}\sum_{j=0}^{N_\sigma^{us}}\left(\varpi^{N_\sigma^{us}+N_\sigma^s-j}G_{t_j^{us}}^{t_{j+1}^{us-}}(-\lambda_1)\right)\\ &+\sum_{i=1}^{N_\sigma^s}\left(\varpi^{N_\sigma^s-i}\chi_i^s\right)+\sum_{j=1}^{N_\sigma^s}\left(\varpi^{N_\sigma^s-j}G_{t_j^s}^{t_{j+1}^{s-}}(\lambda_0)\right)\end{aligned}\tag{4.26}$$

From the condition (4.21), choose a number $\lambda<\lambda_0-\frac{\ln\varpi}{\tau_s}$, one has the following inequalities

$$\begin{aligned} \varpi^{N_\sigma^s+N_\sigma^{us}}e^{(\lambda_1+\lambda_0)T_c^-} &\le \varpi^{2N_0}e^{\tau_s(\lambda_0-\lambda)(\frac{T-T_c}{\tau_s})}\varpi^{\frac{T_c}{\tau_{us}}}e^{(\lambda_1+\lambda_0)T_c}\\ &\le \varpi^{2N_0}e^{(\lambda_0-\lambda)T}e^{\ln\varpi\frac{T_c}{\tau_{us}}+(\lambda_1+\lambda_0)T_c}\\ &\le \varpi^{2N_0}\Delta(\tau_{us},T_c)e^{(\lambda_0-\lambda)T}\end{aligned}\tag{4.27}$$

where $\Delta(\tau_{us},T_c)\triangleq e^{\ln\varpi\frac{T_c}{\tau_{us}}+(\lambda_1+\lambda_0)T_c}$ is a positive number.

$$e^{(\lambda_1+\lambda_0)T_c^-}\varpi^{N_\sigma^{us}+N_\sigma^s-i}\chi_i^{us}\leq 2\varpi^{2N_0}\Delta(\tau_{us},T_c)\bar{\alpha}_2\xi_2^2|e_x(t_i^-)|^2e^{\lambda_0 T} \tag{4.28}$$

$$e^{(\lambda_1+\lambda_0)T_c^-}\varpi^{N_\sigma^{us}+N_\sigma^s-j}G_{t_j^{us}}^{t_{j+1}^{us-}}(-\lambda_1)\leq\varpi^{2N_0}\Delta(\tau_{us},T_c)e^{(\lambda_0-\lambda)T}G_{t_j^{us}}^{t_{j+1}^{us-}}(\lambda) \tag{4.29}$$

Substituting (4.27)-(4.29) into (4.26), together with the results of theorems 2.6 and 4.1, yields

$$\bar{\alpha}_1|x(T)|^2\leq\varpi^{2N_0}\Delta(\tau_{us},T_c)e^{-\lambda T}(\bar{\alpha}_2|x(0)|^2+G_0^T(\lambda))+\bar{\gamma}_4(\|e_x(t_i^-)\|_{[t_1,t_{N_\sigma}]} \tag{4.30}$$

where the function $\bar{\gamma}_4\in\mathscr{K}_\infty$. The ISS result follows from Theorem 2.6.

Case 2: $T_s=T_c, T_{us}=T-T_c$.

Similar to (4.26), we can obtain

$$\begin{aligned}W(T^-)\leq{}&\varpi^{N_\sigma^s+N_\sigma^{us}}e^{-(\lambda_1+\lambda_0)T_c^-}W(0)+e^{-(\lambda_1+\lambda_0)T_c^-}\sum_{i=1}^{N_\sigma^s}\left(\varpi^{N_\sigma^{us}+N_\sigma^s-i}\chi_i\right)\\&+e^{(\lambda_1+\lambda_0)T_c^-}\sum_{j=0}^{N_\sigma^s}\left(\varpi^{N_\sigma^{us}+N_\sigma^s-j}G_{t_j^s}^{t_{j+1}^{s-}}(\lambda_0)\right)\\&+\sum_{i=1}^{N_\sigma^{us}}\left(\varpi^{N_\sigma^{us}-i}\chi_i^{us}\right)+\sum_{j=1}^{N_\sigma^{us}}\left(\varpi^{N_\sigma^{us}-j}G_{t_j^{us}}^{t_{j+1}^{us-}}(-\lambda_1)\right)\end{aligned} \tag{4.31}$$

From the condition (4.22), choose a number λ satisfying $\lambda<\min\left\{\lambda_0-\frac{\ln\varpi}{\tau_s},\lambda_0-\ln\varpi\frac{T_{us}}{\tau_{us}\cdot\tau_s}-\frac{\lambda_1 T_{us}}{\tau_s}\right\}$. The following inequalities can be obtained

$$\begin{aligned}\varpi^{N_\sigma^s+N_\sigma^{us}}e^{-(\lambda_1+\lambda_0)T_c^-}&\leq\varpi^{2N_0+1}e^{\tau_s\lambda_0(\frac{T_c}{\tau_s}-1)}\varpi^{\frac{T_c}{\tau_{us}}}e^{-(\lambda_1+\lambda_0)T_c}\\&\leq\varpi^{2N_0+1}e^{\lambda_0 T_s+\ln\varpi\frac{T_{us}}{\tau_{us}}-(\lambda_1+\lambda_0)T_s}\\&\leq\varpi^{2N_0+1}e^{-\lambda_1 T}e^{-\lambda\tau_s}\end{aligned} \tag{4.32}$$

Since $\lambda_\cdot>0$, there always exists a $\lambda^*>0$ such that $\lambda^*T=\lambda\tau_s$,

$$e^{-(\lambda_1+\lambda_0)T_c^-}\varpi^{N_\sigma^{us}-i}\chi_i^{us}\leq 2\varpi^{2N_0}\bar{\alpha}_2\xi_2^2|e_x(t_i^-)|^2e^{\lambda_0\tau_s}e^{-\lambda_1 T} \tag{4.33}$$

$$e^{-(\lambda_1+\lambda_0)T_c^-}\varpi^{N_\sigma^{us}-j}G_{t_j^{us}}^{t_{j+1}^{us-}}(-\lambda_1)\leq\varpi^{2N_0}e^{\lambda_0\tau_s}e^{-\lambda^*T}e^{-\lambda_1 T}G_{t_j^{us}}^{t_{j+1}^{us-}}(\lambda) \tag{4.34}$$

Substituting (4.32)-(4.34) into (4.31), together with the results of theorems 2.6, 4.1 and Case 1, yields

$$\bar{\alpha}_1|x(T)|^2 \le \varpi^{2N_0+1}e^{-\lambda^* T}\bar{\alpha}_2|x(0)|^2 + \bar{\gamma}_5(\|e_x\|_{[0,T)}) + \bar{\gamma}_6(\|e_\theta\|_{[0,T)}) \quad (4.35)$$

where the functions $\bar{\gamma}_5, \bar{\gamma}_6 \in \mathcal{K}_\infty$. This completes the proof. □

Theorem 4.3 relaxes the condition that all modes are required to be ISS, the overall HIS in the presence of faults can still be ISS with partial ISS modes. This result is very useful for stabilization of HIS and non-impulsive hybrid sytems with unstable modes due to faults.

Example 4.1: An example borrowed from [69] is given to illustrate the theoretical results. Consider a HIS with two modes as

$$\text{mode 1: } \begin{cases} \dot{x}_1 = \frac{1}{8}x_1 - x_2 \\ \dot{x}_2 = x_1 + \frac{1}{8}x_2 + (\sin^2 x_1 + \sin x_1)\theta_1 + u_1 \\ y = x_1 - x_2 \end{cases}$$

$$\text{mode 2: } \begin{cases} \dot{x}_1 = -4x_1 + x_2 \\ \dot{x}_2 = x_1 - 3x_2 + (\sin^2 x_1)\theta_2 + u_2 \\ y = x_1 - x_2 \end{cases}$$

$$f_{1,2}: \begin{cases} x_1 = \frac{2}{3}x_1 + \theta^d_{1,2}(x) \\ x_2 = \frac{1}{3}x_1 + \frac{2}{3}x_2 \end{cases}, \quad f_{2,1}: \begin{cases} x_1 = x_1 + \theta^d_{2,1}(x) \\ x_2 = \frac{1}{2}x_1 + x_2 \end{cases}$$

where $\theta_{H1} = \frac{1}{8}$, $\theta_{H2} = 1$, the bounds of faulty parameters are assumed to be $\bar{\theta}_{f1} = \frac{1}{8}$, $\bar{\theta}_{f2} = 1$, and $\bar{\theta}_1 = \frac{1}{4}$, $\bar{\theta}_2 = 2$. It can be seen that $E_1 = E_2 = [0\ \ 1]^\top$, $L_1 = 3$, $L_2 = 2$, $\bar{g}_1 = 2$, $\bar{g}_2 = 1$.

As for mode 1, the matrix K_1 and Q_1 are chosen as $Q_1 = \begin{bmatrix} 0.9993 & -0.5788 \\ -0.5788 & 1.9412 \end{bmatrix}$ and $K_1 = \begin{bmatrix} -1 \\ -5 \end{bmatrix}$, we can obtain $P_1 = \begin{bmatrix} 1.3564 & -0.3376 \\ -0.3376 & 0.3376 \end{bmatrix}$ and $R_1 = -0.3376$. Note that Assumption 4.1 holds, which implies that the FD observer works well.

On the other hand, by choosing $W_1 = I_{2\times 2}$, we obtain the matrix H_1 from (4.7) as $H_1 = \begin{bmatrix} 2.0048 & -0.5003 \\ -0.5003 & 1.0646 \end{bmatrix}$. Simple calculation leads to that $\eta_1 = 0.6366$ in (4.8), it can be checked that $-\lambda_{\min}(W_1) + \eta_1|H_1| = 0.5136$, which means mode 1 is not ISS.

As for mode 2, K_2 and Q_2 are chosen as $Q_2 = \begin{bmatrix} 10.8180 & -0.7843 \\ -0.7843 & 1.0912 \end{bmatrix}$, $K_2 = \begin{bmatrix} -1 \\ -5 \end{bmatrix}$, one has $R_2 = -0.0341$ and $P_2 = \begin{bmatrix} 1.7348 & -0.0341 \\ -0.0341 & 0.0682 \end{bmatrix}$. Assumption 4.1 also holds.

By choosing $W_2 = I_{2\times 2}$, we obtain $H_2 = \begin{bmatrix} 0.1350 & 0.0417 \\ 0.0417 & 0.1708 \end{bmatrix}$, and $\eta_2 = 2.8497$, it follows that $-\lambda_{\min}(W_2) + \eta_2|H_1| = -0.3572$, which implies mode 2 is ISS w.r.t. e_x, e_θ.

From above calculations, we get $\bar{\alpha}_1 = 0.1076$, $\bar{\alpha}_2 = 2.2212$ in (4.41), $\lambda_0 = 0.3572$, $\lambda_1 = 0.5136$ in (4.19).

Now consider the impulsive dynamics, assume $\theta_{1,2}^d = \frac{1}{3}x_1$, $\theta_{2,1}^d = x_1$, we have $\xi_1 = 1.8028$ in (4.17), it follows that $\ln\varpi = 4.8992$.

Now we illustrate the results of Theorem 4.3, we consider two cases: the HIS is initialized at mode 1 then switches to mode 2, and the converse. For the former case, the HIS is ended at ISS mode 2, from the condition (4.21), if the dwell time of mode 2 is larger than $\frac{\ln\varpi}{\lambda_0} = 13.6876s$, then HIS is ISS w.r.t. e_x, e_θ. For the latter case, the HIS is ended at non-ISS mode 1, provided that the dwell time of mode 1 is $10s$, i.e., $T_{us} = 10s$, from the condition (4.22), if the dwell time of mode 2 is larger than $28.0751s$, then HIS is still ISS w.r.t. e_x, e_θ.

4.2 Stochastic Switching Case

In this section, we address the stability issue of a class of stochastic HS called switching diffusion processes (SDP) where each mode is represented by a stochastic differential equation, the mode switching is governed by a Markov process [98], [81]. This work is motivated by the fact that SDP often models stochastic systems with faults, since SDP model can represent the fault process in different state spaces such that the consideration of FD and FTC is natural [81, 133].

The main idea is to transfer the FTC problem of a stochastic system into the stability problem of a SDP. It will be shown that the fault tolerability of a stochastic system relies on the trade-off among the fault occurrence transition rate, the frequency of switching, and the decreasing rate of Lyapunov functions along the solution of the SDP.

4.2.1 Preliminaries

The SDP takes the form

$$dx(t) = f_{\sigma(t)}(x(t),u(t))dt + g_{\sigma(t)}(x(t),u(t))dW(t) \tag{4.36}$$

where the state $x \in \Re^n$, the input $u \in \Re^m$. W is an r-dimensional standard Brownian motion. Both f_σ and g_σ satisfy the Lipschitz and the linear growth conditions which guarantee that each mode has a unique solution for any initial state.

Denote $\mathbf{P}(\cdot)$ as the probability, whereas $\mathbf{E}[\cdot]$ represents the expectation. Let $(\Omega,\mathscr{F},\mathbf{P})$ be a complete probability space of the fault occurrence, the switching function $\sigma(t)$ is a right-continuous Markov chain on the probability space taking values in a finite state space $Q = \{1,2,...,N\}$ with generator matrix $\Gamma = (\rho_{ij})_{N\times N}$ given by

$$\mathbf{P}\{\sigma(t+\Delta) = j|\sigma(t) = i\} = \begin{cases} \rho_{ij}\Delta + o(\Delta) & \text{if } i \neq j \\ 1+\rho_{ii}\Delta + o(\Delta) & \text{if } i = j \end{cases} \tag{4.37}$$

where $0 \leq \rho_{ij} < 1$ represents the fault occurrence rate from mode i to mode j if $i \neq j$, and $\rho_{ii} = -\sum_{j\neq i}\rho_{ij}$. $\Delta > 0$ is the infinitesimal transition time interval and $o(\Delta)$ is composed of infinitesimal terms of order higher than that of $\Delta > 0$. We assume that the Markov chain σ is independent of the Brownian motion W.

For any given $V_q(x) : \Re^n \to \Re_+ \in \mathscr{C}^2$ associated with the mode q of the system (4.36), we define the differential operator as

$$\mathscr{L}V_q(x) = \frac{\partial V_q(x)}{\partial x} f_q(x,u) + \frac{1}{2}\mathrm{Tr}[g_q^\top(x,u)\frac{\partial^2 V_q(x)}{\partial x^2} g_q(x,u)] \tag{4.38}$$

We also define the following generator

$$\overline{\mathscr{L}}V_q(x) = \mathscr{L}V_q(x) + \sum_{j=1}^{N} \rho_{qj} V_j(x) \tag{4.39}$$

According to the generalized Itô formula [112], one has

$$\mathbf{E}[V_{\sigma(t_2)}(x(t_2))] = \mathbf{E}[V_{\sigma(t_1)}(x(t_1))] + \mathbf{E}\left[\int_{t_1}^{t_2} \overline{\mathscr{L}}V_{\sigma(t)}(x(t))dt\right] \tag{4.40}$$

for any stopping times t_1, t_2 as long as the involved integrals exist and are finite. In the following, we assume that the integrals in (4.40) always exist and are finite for any $0 \leq t_1 \leq t_2 < \infty$.

Definition 4.3. *The system (4.36) is said to be* input-to-state stable *(ISS) w.r.t the input u if there exist functions $\beta \in \mathscr{K}\mathscr{L}$, $\alpha, \gamma \in \mathscr{K}_\infty$ such that for any initial $x(0)$, we have*

$$\mathbf{E}[\alpha(|x(t)|)] \leq \beta(|x(0)|, t) + \gamma(\|u\|_{[0,t)}), \quad \forall t \geq 0$$

The difference of Definition 4.3 from usual ISS formula (Definition 4.1) is the introduction of the expectation. It has been proven in [75] that the following ISS property of the single stochastic system holds.

Lemma 4.3. *The system $dx = f(x,u)dt + g(x,u)dW$ is ISS w.r.t. u, if there exist α_1, α_2, α_3, $\gamma_1 \in \mathscr{K}_\infty$, and a smooth function $V(x) \in \mathscr{C}^2(\Re^n;\Re_+)$ such that $\alpha_1(|x|) \leq V(x) \leq \alpha_2(|x|)$, $\mathscr{L}V(x) \leq -\alpha_3(|x|) + \gamma_1(|u|)$.*

In the following, a series of sufficient conditions of fault tolerance are derived such that the SDP can be stabilized in the sense of ISS in general cases:

1) where each individual mode is ISS, i.e. the stochastic system is ISS separately in the healthy situation and in the faulty situations.
2) where some modes are ISS, while others are not ISS. This comes from the fact that some modes representing the faulty situations may be not ISS.
3) where no mode is ISS. This is the worst case where the stochastic system is not ISS separately whatever it is healthy or not.

4.2.2 Fault Tolerance Analysis

We shall first establish the general fault tolerability conditions.

Theorem 4.4. *The SDP (4.36) is ISS w.r.t. u if there exist* $\alpha_1, \alpha_2, \chi \in \mathcal{K}_\infty$, $\omega > 0$ *and smooth functions* $V_k \in \mathscr{C}^2(\Re^n; \Re_+)$ $k \in Q$ *such that* $\forall q \in Q$

$$\alpha_1(|x|) \leq V_q(x) \leq \alpha_2(|x|) \tag{4.41}$$

$$\overline{\mathscr{L}}V_q(x) \leq -\omega V_q(x) + \chi(|u|) \tag{4.42}$$

where the generator $\overline{\mathscr{L}}$ *is defined in (4.38)-(4.39).*

Proof: Since any t is supposed to be a stop time, applying the generalized Itô formula derives

$$\mathbf{E}[V_{\sigma(t)}(x(t))] = \mathbf{E}[V_{\sigma(0)}(x(0))] + \mathbf{E}\Big[\int_0^t \Big(\omega V_{\sigma(s)}(x(s)) + \overline{\mathscr{L}}V_{\sigma(s)}(x(s))\Big)ds\Big]$$

We further have

$$\begin{aligned}
\mathbf{E}[e^{\omega t}V_{\sigma(t)}(x(t))] &= \mathbf{E}[V_{\sigma(0)}(x(0))] + \mathbf{E}\Big[\int_0^t e^{\omega s}\Big(\omega V_{\sigma(s)}(x(s)) + \overline{\mathscr{L}}V_{\sigma(s)}(x(s))\Big)ds\Big] \\
&\leq \alpha_2(|x(0)|) + \mathbf{E}\Big[\int_0^t e^{\omega s}\Big(\omega V_{\sigma(s)}(x(s)) - \omega V_{\sigma(s)}(x(s)) + \chi(|u|)\Big)ds\Big] \\
&\leq \alpha_2(|x(0)|) + \mathbf{E}\Big[\int_0^t e^{\omega s}\Big(\chi(|u|)\Big)ds\Big] \\
&\leq \alpha_2(|x(0)|) + \frac{1}{\omega}(e^{\omega t} - 1) \sup_{\tau \in [0,t)} \{\chi(|u|)\}
\end{aligned}$$

Consequently, we obtain

$$\mathbf{E}[\alpha_1(|x(t)|)] \leq e^{-\omega t}\alpha_2(|x(0)|) + \frac{1}{\omega} \sup_{\tau \in [0,t)} \{\chi(|u|)$$

This completes the proof. □

It is interesting to analyze the condition (4.42). It follows from the definition of $\overline{\mathscr{L}}$ in (4.39) that if there is only one mode in the system or there is a common $V(x)$ for all modes, then $\overline{\mathscr{L}}V_q(x) = \mathscr{L}V_q(x)$. In these two cases, Theorem 4.4 and Lemma 4.3 are equivalent. Thus we obtain the property

- If there is a common ISS-Lyapunov function for the normal and all faulty modes, then ISS of each individual mode implies ISS of the overall stochastic system.

This property is very useful in the practical situation, if we find that the healthy and faulty mode share the same ISS-Lyapunov function, then what we need to do is just to preserve ISS of each individual mode without consideration for the transient behavior.

We continue to observe (4.39), it is clear that

$$\sum_{j=1}^{N} \rho_{qj}V_j(x) = \sum_{j \neq q} \rho_{qj}V_j(x) - |\rho_{qq}|V_q(x)$$

we further conclude from (4.42) that

- If $\sum_{j\neq q}\rho_{qj}V_j(x) \geq |\rho_{qq}|V_q(x)$, then ISS of SDP implies the ISS of mode q.
- If $\sum_{j\neq q}\rho_{qj}V_j(x) < |\rho_{qq}|V_q(x)$, then ISS of SDP can be achieved even mode q is non ISS.
- If $\sum_{j\neq q}\rho_{qj}V_j(x) < |\rho_{qq}|V_q(x)$, $\forall q \in Q$, then ISS of SDP can be achieved without any ISS mode.

The above three properties reflect to some extents the effect of fault occurrence transition rate on the fault tolerance of the stochastic system. However, these properties are not easy to be verified since $V_q(x)$ is not unique.

Theorem 4.4 implicitly quantifies the trade-off between frequency of switching/dwell time and rate of decrease of the Lyapunov function. In order to analyze more precisely the relations among these factors to achieve the ISS, we will adopt the method in determined hybrid systems in the following discussions. The three cases where all modes are ISS, only some modes are ISS and no mode is ISS, will be successively studied.

Consider the SDP (4.36) where each mode is ISS. More formally, suppose that there exist α_1, α_2, $\chi \in \mathcal{K}_\infty$, $\lambda_0 > 0$, $\mu > 1$ and smooth functions $V_k \in \mathcal{C}^2(\Re^n;\Re_+)$ $k \in Q$, such that $\forall p,q \in Q$

$$\alpha_1(|x|) \leq V_q(x) \leq \alpha_2(|x|) \tag{4.43}$$
$$\mathcal{L}V_q(x) \leq -\lambda_0 V_q(x) + \chi(|u|) \tag{4.44}$$
$$V_p(x) \leq \mu V_q(x) \tag{4.45}$$

Theorem 4.5. *The SDP (4.36) satisfying (4.43)-(4.45) is ISS w.r.t. u if*

$$\mu < \frac{\tilde{\lambda}}{\bar{\lambda}}, \quad \text{with } \bar{\lambda} \triangleq \max\{|\rho_{ii}| | i \in Q\}, \tilde{\lambda} \triangleq \max\{\rho_{ij} | i,j \in Q\} \tag{4.46}$$

where μ is defined in (4.45), ρ_{ij} represents the fault occurrence rate defined in (4.37).

In order to prove Theorem 4.5, the following lemma is needed.

Lemma 4.4. *[17] Suppose that σ is a Markov chain satisfying (4.37). It holds that $\forall t \geq 0$, $\forall k \in \mathbb{N}$*

$$\mathbf{P}(N_\sigma(t) = k) \leq \frac{e^{-\bar{\lambda}t}(\tilde{\lambda}t)^k}{k!}$$

where $\tilde{\lambda}$ and $\bar{\lambda}$ are defined in (4.46). $N_\sigma(t)$ denotes the number of switchings of σ over the interval $[0,t)$ as defined in Definition 2.4.

Proof of Theorem 4.5: Let $T > 0$ be an arbitrary time. Denote by $\tau_1, \ldots, \tau_{N_\sigma(T,0)}$ the switching instants on the interval $[0,T)$, where $N_\sigma(T,0)$ is defined in (4.49). Denote $G_a^b(\lambda) = \int_a^b e^{-\lambda(b-s)}\chi(|u|)ds$. Since $\frac{d}{dt}\mathbf{E}[V(x)] = \mathbf{E}[\mathcal{L}V(x)]$, it follows from (4.44) and (4.45) that for $t \in [\tau_{i+1}, \tau_{i+2})$

$$\begin{aligned}
\mathbf{E}[V_{\sigma(\tau_{i+1})}(x(t))] &\leq \mathbf{E}[V_{\sigma(\tau_{i+1})}(x(\tau_{i+1}))]e^{-\lambda_0(t-\tau_{i+1})} + \mathbf{E}[G^t_{\tau_{i+1}}(\lambda_0)] \\
&\leq \mathbf{E}[\mu V_{\sigma(\tau_i)}(x(\tau_{i+1}))e^{-\lambda_0(t-\tau_{i+1})}] + \mathbf{E}[G^t_{\tau_{i+1}}(\lambda_0)] \\
&\leq \mathbf{E}[\mu V_{\sigma(\tau_i)}(x(\tau_i))e^{-\lambda_0(t-\tau_i)}] \\
&\qquad + \mathbf{E}[\mu e^{-\lambda_0(t-\tau_{i+1})}G^{\tau_{i+1}}_{\tau_i}(\lambda_0) + G^t_{\tau_{i+1}}(\lambda_0)] \\
&\leq \mathbf{E}[\mu^2 V_{\sigma(\tau_{i-1})}(x(\tau_{i-1}))e^{-\lambda_0(t-\tau_{i-1})}] + \mathbf{E}[\mu^2 e^{-\lambda_0(t-\tau_i)}G^{\tau_i}_{\tau_{i-1}}(\lambda_0)] \\
&\qquad + \mathbf{E}[\mu e^{-\lambda_0(t-\tau_{i+1})}G^{\tau_{i+1}}_{\tau_i}(\lambda_0) + G^t_{\tau_{i+1}}(\lambda_0)] \\
&\qquad\qquad \vdots \\
&\qquad\qquad \vdots
\end{aligned}$$

Denote $N_\sigma \triangleq N_\sigma(T,0)$, following the above iterative procedure, we finally obtain

$$\begin{aligned}
\mathbf{E}[V_{\sigma(T)}x(T)] &\leq \mathbf{E}[\mu^{N_\sigma}]e^{-\lambda_0 T}V_{\sigma(0)}(x(0)) + \mathbf{E}[G^T_{\tau_{N_\sigma}}(\lambda_0)] \\
&\qquad + \mathbf{E}\Big[\sum_{j=0}^{N_\sigma-1} \mu^{N_\sigma-j}e^{-\lambda_0(T-\tau_{j+1})}G^{\tau_{j+1}}_{\tau_j}(\lambda_0)\Big]
\end{aligned} \tag{4.47}$$

Based on Lemma 4.4, one has

$$\mathbf{E}[\mu^{N_\sigma}] = \sum_{k=0}^{\infty} \mu^k \mathbf{P}(N_\sigma = k) \leq \sum_{k=0}^{\infty} \mu^k \frac{e^{-\bar{\lambda}T}(\bar{\lambda}T)^k}{k!} = e^{(\mu\bar{\lambda}-\bar{\lambda})T} \tag{4.48}$$

Under the condition (4.46), substituting (4.48) into (4.47) leads to

$$\mathbf{E}[V_{\sigma(T)}x(T)] \leq e^{-\lambda_0 T}V_{\sigma(0)}(x(0)) + \mathbf{E}\Big[\int_0^T \chi(|u|)ds\Big]$$

The result follows directly from the proof of Theorem 4.4. □

Roughly speaking, if each mode is ISS, and the fault occurrence transition rate $\max\{\rho_{ij}\}$ is large enough, then the ISS of the stochastic system is guaranteed. It can be seen from the proof of Theorem 4.5 that, the condition (4.46) can be removed if there is only one mode in the system or there is a common $V(x)$ for all modes. This is consistent with Theorem 4.4.

Theorem 4.5 completely depends on the fault occurrence transition rate without consideration of the frequency of switching and the decreasing rate of ISS-Lyapunov functions. We shall relax the condition (4.46) by introducing the concept of *stochastic average dwell time* defined as follows.

Definition 4.4. *If there exist a series positive numbers τ_k $\forall k \in \mathbb{N}\cup\{0\}$ such that*

$$k \leq N_0 + \frac{T-t}{\tau_k}, \quad \forall T \geq t \geq 0 \tag{4.49}$$

where $N_0 > 0$ *denotes the chattering bound, then* $\Theta = \{\tau_k, k \in \mathbb{N} \cup \{0\}\}$ *is called the set of* stochastic average dwell time *(sADT) of* σ *over* $[t,T)$.

Definition 4.4 extends the ADT (see Definition 2.4 in Chapter 2.2) to the stochastic case, which means that for each possible switching number k, there may exist some switchings separated by less than τ_k, but the average dwell period among switchings is not less than τ_k.

Theorem 4.6. *The SDP (4.36) satisfying (4.43)-(4.45) is ISS w.r.t.* u *if* $\forall k \in \mathbb{N} \cup \{0\}$

$$\mu < \min\left\{\frac{\lambda_0 + \tilde{\lambda}}{\bar{\lambda}}, e^{\lambda_0 \tau_k}\right\} \tag{4.50}$$

$$\bar{\lambda} \leq \tilde{\lambda} \tag{4.51}$$

where $\tau_k \in \Theta$, Θ *is the set of* sADT *defined in Definition 4.4,* $\tilde{\lambda}$ *and* $\bar{\lambda}$ *are defined in (4.46).*

Proof: Following the proof of Theorem 4.5, inequality (4.47) is rewritten as

$$\begin{aligned} \mathbf{E}[V_{\sigma(T)}x(T)] &\leq \mathbf{E}[\mu^{N_\sigma}]e^{-\lambda_0 T}V_{\sigma(0)}(x(0)) \\ &+ \sum_{j=0}^{N_\sigma - 1} \mathbf{E}\left[\mu^{N_\sigma - j}e^{-\lambda_0(T-\tau_{j+1})}G_{\tau_j}^{\tau_{j+1}}(\lambda_0)\right] + \mathbf{E}[G_{\tau_{N_\sigma}}^T(\lambda_0)] \end{aligned} \tag{4.52}$$

It follows from (4.50), (4.51) and Lemma 4.4 that

$$\mathbf{E}[\mu^{N_\sigma}] \leq e^{(\mu\bar{\lambda} - \tilde{\lambda})T} \leq e^{\lambda_0 T} \tag{4.53}$$

$$\begin{aligned} \mathbf{E}\left[\mu^{N_\sigma - j}e^{-\lambda_0(T-\tau_{j+1})}\right] &= \sum_{k=0}^{\infty} \mathbf{P}(N_\sigma = k)\left(\mu^{k-j}e^{-\lambda_0(T-\tau_{j+1})}\right) \\ &\leq \sum_{k=0}^{\infty} \mathbf{P}(N_\sigma = k)\left(\mu^{N_0 + \frac{T}{\tau_k} - j + 1 - 1}e^{-\lambda_0(T-\tau_{j+1})}\right) \\ &\leq \sum_{k=0}^{\infty} \mathbf{P}(N_\sigma = k)\left(\mu^{1+N_0}e^{\tau_k\lambda_0(\frac{T}{\tau_k} - j - 1)}e^{-\lambda_0(T-\tau_{j+1})}\right) \\ &\leq \sum_{k=0}^{\infty} \mathbf{P}(N_\sigma = k)\left(\mu^{1+N_0}e^{\lambda_0(T-\tau_{j+1})}e^{-\lambda_0(T-\tau_{j+1})}\right) \\ &\leq \mu^{1+N_0}\sum_{k=0}^{\infty} \frac{e^{-\tilde{\lambda}T}(\tilde{\lambda}T)^k}{k!} \leq \mu^{1+N_0} \end{aligned} \tag{4.54}$$

Substituting (4.53), (4.54) into (4.52), we have

$$\mathbf{E}[V_{\sigma(T)}x(T)] \leq V_{\sigma(0)}(x(0)) + \mu^{1+N_0}\mathbf{E}\left[\int_0^T \chi(|u|)ds\right]$$

The result follows. □

Theorem 4.6 shows that, if the ADT is large enough, i.e., the switching is slow averagely, the stochastic system does not change too frequently among healthy and faulty modes, then a less restrictive condition on transition rates is required than (4.46) to achieve the ISS of the SDP.

Remark 4.4. *Generally, the condition (4.50) can not be used to verify* a priori *whether the system is ISS. Since the switching of SDP (i.e. the instant of fault occurrence) is random, $N_\sigma(T,t)$ is not determined at each time T, a series of ADT have to be provided to include all possible switching numbers. However, inequality (4.50) is very useful to check on-line the ISS of the system in the current situation.*

Now we consider the case that some modes of SDP are ISS while others may be not. This work is motivated by the fact that some modes that represent the faulty situations are often not ISS.

Define two subsets of Q as $Q = Q_s \cup Q_{us}$, where Q_s (Q_{us}) denotes the set of modes that are (not) ISS.

Consequently, the inequality (4.44) is modified as

$$\mathscr{L}V_q(x) \le -\lambda_0 V_q(x) + \chi(|u|), \ \ \forall q \in Q_s \tag{4.55}$$

$$\mathscr{L}V_q(x) \le \lambda_1 V_q(x) + \chi(|u|), \ \ \forall q \in Q_{us} \tag{4.56}$$

where $\lambda_1 > 0$. In this case, the continuous flow in mode $q \in Q_{us}$ can potentially destroy ISS.

Similarly to Chapter 4.1, divide the time interval $[t,T) = T_s \cup T_{us}$, where T_s (T_{us}) denotes the dwell period of ISS (non-ISS) modes in $[t,T)$. Then we define the *double sADT*, which generalizes Definition 4.4 and provides two ADT scales for the SDP with both ISS and non-ISS modes.

Definition 4.5. *Let $N^s_\sigma(T,t) = \varepsilon_1$, $\varepsilon_1 \in \mathbb{N} \cup \{0\}$ ($N^{us}_\sigma(T,t) = \varepsilon_2$, $\varepsilon_2 \in \mathbb{N} \cup \{0\}$) denote the number of switchings of σ during the period T_s (T_{us}). If there exists a series of positive numbers $\tau^s_{\varepsilon_1}$ $\forall \varepsilon_1 \in \mathbb{N} \cup \{0\}$, and $\tau^{us}_{\varepsilon_2}$ $\forall \varepsilon_2 \in \mathbb{N} \cup \{0\}$ such that*

$$\varepsilon_1 \le N_0 + \frac{T_s}{\tau^s_{\varepsilon_1}}, \quad \varepsilon_2 \le N_0 + \frac{T_{us}}{\tau^{us}_{\varepsilon_2}}, \quad \forall T \ge t \ge 0 \tag{4.57}$$

where $N_0 > 0$, then $\Theta_d = \Theta_s \cup \Theta_{us}$ is called the set of double stochastic average dwell time *of σ over $[t,T)$, where $\Theta_s = \{\tau^s_{\varepsilon_1}, \varepsilon_1 \in \mathbb{N} \cup \{0\}\}$, and $\Theta_{us} = \{\tau^s_{\varepsilon_2}, \varepsilon_2 \in \mathbb{N} \cup \{0\}\}$.*

Consider the time interval $[0,T)$ for arbitrary time $T > 0$. For the sake of simplicity, in the following, we divide $[0,T) = [0,T_c^-) \cup [T_c,T)$ and focus on two cases: **Case 1:** $T_{us} = T_c^-, T_s = T - T_c$, i.e., non-ISS modes work in $[0,T_c^-)$ and ISS ones work in $[T_c,T)$. **Case 2:** $T_s = T_c^-, T_{us} = T - T_c$, i.e., ISS modes work in $[0,T_c^-)$ and non-ISS ones work in $[T_c,T)$.

Theorem 4.7. *The SDP (4.36) satisfying (4.43),(4.45),(4.55)-(4.56) is ISS w.r.t. u if $\forall \varepsilon_1, \varepsilon_2 \in \mathbb{N} \cup \{0\}$*

$$\begin{cases} \mu < \min\left\{\frac{\lambda_0+\tilde{\lambda}}{\tilde{\lambda}}, e^{\lambda_0 \tau^s_{\varepsilon_1}}\right\} \\ \bar{\lambda} \le \tilde{\lambda} \end{cases} \textit{for } \textbf{\textit{Case 1}} \tag{4.58}$$

$$\begin{cases} \mu < \min\left\{\frac{\lambda_0+\tilde{\lambda}}{\tilde{\lambda}}, e^{\lambda_0 \tau^s_{\varepsilon_1}}, e^{\frac{(\lambda_0 \tau^s_{\varepsilon_1} - \lambda_1 T_{us})\tau^{us}_{\varepsilon_2}}{T_{us}}}\right\} \\ \bar{\lambda} \le \tilde{\lambda} \end{cases} \textit{for } \textbf{\textit{Case 2}} \tag{4.59}$$

where $\tilde{\lambda}$ *and* $\bar{\lambda}$ *are defined in (4.46).*

Proof of Theorem 4.7: Denote by $\tau_1^{us}, \ldots, \tau^{us}_{N^{us}_\sigma}$, and $\tau^s_1, \ldots, \tau^s_{N^s_\sigma}$ the switching instants on the interval T_{us} and T_s respectively.

Case 1: $T_{us} = [0, T_c^-), T_s = [T_c, T)$.

During the time interval $[T_c, T)$, following the proof of Theorem 4.5, one has

$$\begin{aligned} \mathbf{E}[V_{\sigma(T)}x(T)] &\le \mathbf{E}[\mu^{N^s_\sigma} e^{-\lambda_0(T-T_c)} V_{\sigma(T_c^-)}(x(T_c^-))] \\ &+ \mathbf{E}\Big[\sum_{j=0}^{N^s_\sigma-1} \mu^{N^s_\sigma - j} e^{-\lambda_0(T-\tau^s_{j+1})} G^{\tau^s_{j+1}}_{\tau^s_j}(\lambda_0)\Big] + \mathbf{E}[G^T_{\tau^s_{N^s_\sigma}}(\lambda_0)] \end{aligned} \tag{4.60}$$

Similarly, during the time interval $[0, T_c^-)$, we obtain

$$\begin{aligned} \mathbf{E}[V_{\sigma(T_c^-)}x(T_c^-)] \le \mathbf{E}[\mu^{N^{us}_\sigma} e^{\lambda_1 T_c} V_{\sigma(0)}(x(0))] + \mathbf{E}[G^{T_c}_{\tau^{us}_{N^{us}_\sigma}}(-\lambda_1)] \\ + \mathbf{E}\Big[\sum_{j=0}^{N^{us}_\sigma-1} \mu^{N^{us}_\sigma - j} e^{\lambda_1(T_c - \tau^{us}_{j+1})} G^{\tau^{us}_{j+1}}_{\tau^{us}_j}(-\lambda_1)\Big] \end{aligned} \tag{4.61}$$

Combining (4.60) and (4.61) leads to

$$\begin{aligned} \mathbf{E}[V_{\sigma(T)}x(T)] \le\; & \mathbf{E}[\mu^{N^s_\sigma + N^{us}_\sigma} e^{-\lambda_0 T + (\lambda_0+\lambda_1)T_c} V_{\sigma(0)}(x(0))] \\ & + \mathbf{E}\Big[\sum_{j=0}^{N^{us}_\sigma-1} \mu^{N^s_\sigma + N^{us}_\sigma - j} e^{-\lambda_0 T + (\lambda_0+\lambda_1)T_c - \lambda_1 \tau^{us}_{j+1}} G^{\tau^{us}_{j+1}}_{\tau^{us}_j}(-\lambda_1)\Big] \\ & + \mathbf{E}[\mu^{N^s_\sigma} e^{-\lambda_0(T-T_c)} G^{T_c}_{\tau^{us}_{N^{us}_\sigma}}(-\lambda_1)] \\ + \mathbf{E}\Big[\sum_{j=0}^{N^s_\sigma-1} \mu^{N^s_\sigma - j} e^{-\lambda_0(T-\tau^s_{j+1})} G^{\tau^s_{j+1}}_{\tau^s_j}(\lambda_0)\Big] & + \mathbf{E}[G^T_{\tau^s_{N^s_\sigma}}(\lambda_0)] \end{aligned} \tag{4.62}$$

Under condition (4.58), it follows that $\mathbf{E}[\mu^{N^s_\sigma + N^{us}_\sigma}] \le e^{\lambda_0 T}$, and

$$\begin{aligned} \mathbf{E}[\mu^{N^s_\sigma}] &= \sum_{\varepsilon_1=0}^{\infty} \mu^{\varepsilon_1} \mathbf{P}(N^s_\sigma(T, T_c) = \varepsilon_1) \\ &\le \sum_{\varepsilon_1=0}^{\infty} \mu^{\varepsilon_1} \frac{e^{-\tilde{\lambda}(T-T_c)}(\bar{\lambda}(T-T_c))^{\varepsilon_1}}{\varepsilon_1 !} = e^{(\mu\bar{\lambda}-\tilde{\lambda})(T-T_c)} \le e^{\lambda_0(T-T_c)} \end{aligned}$$

On the other hand, $G_{\tau_j^{us}}^{\tau_{j+1}^{us}}(-\lambda_1) \leq e^{(\lambda_1+\lambda_0)\tau_{j+1}^{us}} G_{\tau_j^{us}}^{\tau_{j+1}^{us}}(\lambda_0)$. Consequently, we get

$$\begin{aligned}
\mathbf{E}[V_{\sigma(T)}x(T)] \leq\ & e^{(\lambda_0+\lambda_1)T_c} V_{\sigma(0)}(x(0)) + e^{(\lambda_0+\lambda_1)T_c} \mathbf{E}\Big[\sum_{j=0}^{N_\sigma^{us}-1} G_{\tau_j^{us}}^{\tau_{j+1}^{us}}(\lambda_0)\Big] \\
& + e^{(\lambda_0+\lambda_1)T_c} \mathbf{E}\Big[G_{\tau_{N_\sigma^{us}}^{us}}^{T_c}(\lambda_0)\Big] + \mu^{1+N_0} \mathbf{E}\Big[\sum_{j=0}^{N_\sigma^{s}-1} G_{\tau_j^{s}}^{\tau_{j+1}^{s}}(\lambda_0)\Big] \\
& + \mathbf{E}[G_{\tau_{N_\sigma^{s}}^{s}}^{T}(\lambda_0)]
\end{aligned}$$

Define $\Upsilon \triangleq \max\{e^{(\lambda_0+\lambda_1)T_c}, \mu^{1+N_0}\}$, we further have

$$\mathbf{E}[V_{\sigma(T)}x(T)] \leq \Upsilon V_{\sigma(0)}(x(0)) + \Upsilon \mathbf{E}\Big[\int_0^T \chi(|u|)ds\Big]$$

The ISS result follows.

Case 2: $T_s = [0, T_c^-), T_{us} = [T_c, T)$.

Similarly to (4.62), we can obtain

$$\begin{aligned}
\mathbf{E}[V_{\sigma(T)}x(T)] \leq\ & \mathbf{E}[\mu^{N_\sigma^s+N_\sigma^{us}} e^{\lambda_1 T-(\lambda_0+\lambda_1)T_c} V_{\sigma(0)}(x(0))] \\
& + \mathbf{E}\Big[\sum_{j=0}^{N_\sigma^s-1} \mu^{N_\sigma^s+N_\sigma^{us}-j} e^{\lambda_1 T-(\lambda_0+\lambda_1)T_c+\lambda_0\tau_{j+1}^s} G_{\tau_j^s}^{\tau_{j+1}^s}(\lambda_0)\Big] \\
& + \mathbf{E}[\mu^{N_\sigma^{us}} e^{\lambda_1(T-T_c)} G_{\tau_{N_\sigma^s}^s}^{T_c}(\lambda_0)] + \mathbf{E}[G_{\tau_{N_\sigma^{us}}^{us}}^{T}(-\lambda_1)] \\
& + \mathbf{E}\Big[\sum_{j=0}^{N_\sigma^{us}-1} \mu^{N_\sigma^{us}-j} e^{\lambda_1(T-\tau_{j+1}^{us})} G_{\tau_j^{us}}^{\tau_{j+1}^{us}}(-\lambda_1)\Big]
\end{aligned} \tag{4.63}$$

From the condition (4.59), the following inequalities can be obtained

$$\begin{aligned}
\mu^{\varepsilon_1+\varepsilon_2} e^{\lambda_1 T-(\lambda_0+\lambda_1)T_c} & \leq \mu^{2N_0+1} e^{\tau_{\varepsilon_1}^s \lambda_0(\frac{T_c}{\tau_{\varepsilon_1}^s}-1)} \mu^{\frac{T-T_c}{\tau_{\varepsilon_2}^{us}}} e^{\lambda_1 T-(\lambda_0+\lambda_1)T_c} \\
& \leq \mu^{2N_0+1} e^{-\tau_{\varepsilon_1}^s \lambda_0 + \ln\mu \frac{T_{us}}{\tau_{\varepsilon_2}^{us}} + \lambda_1 T_{us}} \\
& \leq \mu^{2N_0+1}
\end{aligned} \tag{4.64}$$

$$\begin{aligned}
\mu^{\varepsilon_1+\varepsilon_2-j} e^{\lambda_1 T-(\lambda_0+\lambda_1)T_c+\lambda_0\tau_{j+1}^s} & \leq \mu^{2N_0+1} e^{\tau_{\varepsilon_1}^s \lambda_0(\frac{T_c}{\tau_{\varepsilon_1}^s}-j-1)} \mu^{\frac{T-T_c}{\tau_{\varepsilon_2}^{us}}} e^{\lambda_1 T-(\lambda_0+\lambda_1)T_c+\lambda_0\tau_{j+1}^s} \\
& \leq \mu^{2N_0+1} e^{-\tau_{\varepsilon_1}^s \lambda_0 + \ln\mu \frac{T_{us}}{\tau_{\varepsilon_2}^{us}} + \lambda_1 T_{us}} \\
& \leq \mu^{2N_0+1}
\end{aligned} \tag{4.65}$$

Since $\lambda_0 > 0$, there always exists a $\lambda^* > 0$ such that $\lambda^* T_c = \lambda_0 \tau^s_{\varepsilon_1}$. It holds that

$$\mu^{\varepsilon_2} e^{\lambda_1 (T-T_c)} = e^{\lambda^* T_c} e^{-\lambda_0 \tau^s_{\varepsilon_1}} \mu^{\varepsilon_2} e^{\lambda_1 (T-T_c)} \leq e^{\lambda^* T_c} \mu^{N_0} \tag{4.66}$$

$$\mu^{\varepsilon_2 - j} e^{\lambda_1 (T-\tau^{us}_{j+1})} \leq \mu^{\varepsilon_2} e^{\lambda_1 T} e^{-\lambda_1 \tau^{us}_{j+1}} \leq e^{(\lambda^*+\lambda_1)T_c} \mu^{N_0} e^{-\lambda_1 \tau^{us}_{j+1}} \tag{4.67}$$

Let us come back to inequality (4.63), under (4.64)-(4.67), we further have

$$\begin{aligned}
\mathbf{E}[V_{\sigma(T)} x(T)] \leq{} & \mu^{2N_0+1} \sum_{\varepsilon_1=0}^{\infty} \sum_{\varepsilon_2=0}^{\infty} \mathbf{P}(N^s_\sigma = \varepsilon_1, N^{us}_\sigma = \varepsilon_2) \\
& \cdot \Bigg(V_{\sigma(0)}(x(0)) + e^{(\lambda^*+\lambda)T_c} \sum_{j=0}^{N^{us}_\sigma - 1} \int_{\tau^{us}_j}^{\tau^{us}_{j+1}} \chi(|u|) ds \\
& + e^{\lambda^* T_c} G^{T_c}_{\tau^s_{N^s_\sigma}}(\lambda_0) + \sum_{j=0}^{N^s_\sigma - 1} G^{\tau^s_{j+1}}_{\tau^s_j}(\lambda_0) + e^{(\lambda^*+\lambda_1)T_c} \int_{\tau^{us}_{N^{us}_\sigma}}^{T} \chi(|u|) ds \Bigg) \\
\leq{} & \mu^{2N_0+1} V_{\sigma(0)}(x(0)) + \mu^{2N_0+1} e^{(\lambda^*+\lambda_1)T_c} \int_0^T \chi(|u|) ds
\end{aligned}$$

This completes the proof. □

Theorem 4.7 shows explicitly the balance of dwell periods between ISS modes and non ISS ones that is needed for ISS of overall SDP.

Remark 4.5. *It can be seen that even the stochastic system is not separately ISS in faulty situations, the overall system process may be still ISS. This means that it is not necessary to design the FTC law to guarantee the stability of each system mode as in [81], less control effort is required. Compared with the general FTC methods [10], our results imply that we do not always have to stabilize the system in the post-fault situation.*

Finally, let us consider the worst case where no mode is ISS. The inequality (4.44) is changed into

$$\mathscr{L} V_q(x) \leq \lambda_1 V_q(x) + \chi(|u|), \quad \forall q \in Q \tag{4.68}$$

where $\lambda_1 > 0$. We have the following result:

Theorem 4.8. *The SDP (4.36) satisfying (4.43),(4.45) and (4.68) is ISS w.r.t. u if*

$$\mu \bar{\lambda} + \lambda_1 < \tilde{\lambda} \tag{4.69}$$

where $\tilde{\lambda}$ and $\bar{\lambda}$ are defined in (4.46), λ_1 is given in (4.68).

Proof: Following the similar procedure in the proof of Theorem 4.5 yields

$$\mathbf{E}[V_{\sigma(T)} x(T)] \leq \mathbf{E}[\mu^{N_\sigma} e^{\lambda_1 T} V_{\sigma(0)}(x(0))] + \mathbf{E}\Big[\sum_{j=0}^{N_\sigma - 1} \mu^{N_\sigma - j} e^{\lambda_1 (T-\tau_{j+1})} G^{\tau_{j+1}}_{\tau_j} + G^t_{\tau_{N_\sigma}} \Big] \tag{4.70}$$

Under condition (4.69), it holds that

$$\mathbf{E}[\mu^{N_\sigma}] \leq e^{(\mu\bar{\lambda}-\tilde{\lambda})T} \leq e^{-\lambda_1 T} \tag{4.71}$$

Equality (4.71), together with (4.70), leads to the result. □

Theorem 4.8 shows that if the fault occurrence transition rate $\max\{\rho_{ij}\}$ is larger than that of any previous cases (All ISS modes, partial ISS modes), the ISS of SDP is achieved without any ISS mode. This result implies that, under the condition (4.69), we do not need to design the stabilizing controller even the stochastic system is not stable separately in the healthy and faulty situations.

Example 4.2: A fault-prone manufacturing system originated from [37] is a good example to illustrate our results. Consider a machine producing a single commodity, the SDP model takes the form

$$dx(t) = (f_{\sigma(t)}(x(t)) - d)dt + g(x)dW(t) \tag{4.72}$$

where the state $x(t) \in \Re$ denotes the inventory, $d \geq 0$ is a constant representing the demand rate, which is regarded as the input. W is a one-dimensional Brownian motion independent of $\sigma(t)$. $f_\sigma(x)$ is the state feedback control policy which is the production effort. The term $g(x)dW$ is often interpreted as "sales return", "inventory spoilage", or "sudden demand fluctuations".

Two modes are considered, $\sigma(t) = 1$ or 2, depending on whether the manufacturing system is in the functional state or the actuator faulty situation respectively. $\sigma(t)$ is modeled as a Markov chain with generator $-\rho_{11} = \rho_{12} > 0$ and $-\rho_{22} = \rho_{21} > 0$. This means $\bar{\lambda} = \tilde{\lambda}$.

In mode 1, $f_1(x) = -x$, $g(x) = \frac{1}{2}x$, this means that the backlogged demand is required in the healthy situation. In mode 2, the actuator fault occurs due to the abnormal behavior of the machine's production scheme. Here our objective is to check whether the overall system process is ISS w.r.t. the demand rate d in spite of the faults.

Two faulty cases are considered.

Faulty case 1: $f_2(x) = -2x$.

Choosing $V_1(x) = V_2(x) = x^2$ leads to

$$\overline{\mathscr{L}}V_1(x) = \mathscr{L}V_1(x) = -2x^2 - 2xd + \frac{1}{4}x^2$$

$$\overline{\mathscr{L}}V_2(x) = \mathscr{L}V_2(x) = -4x^2 - 4xd + \frac{1}{4}x^2$$

The condition of Theorem 4.4 is satisfied, so the SDP is ISS. On the other hand, both two modes are ISS and share the same ISS-Lyapunov function, Theorem 4.5 could also be used to verify the ISS property. The condition of Theorem 4.6 also holds, in this case, we do not have to reconfigure the controller after the fault occurs, and the frequency of the fault occurrence also has no effect on the ISS of the system. In the

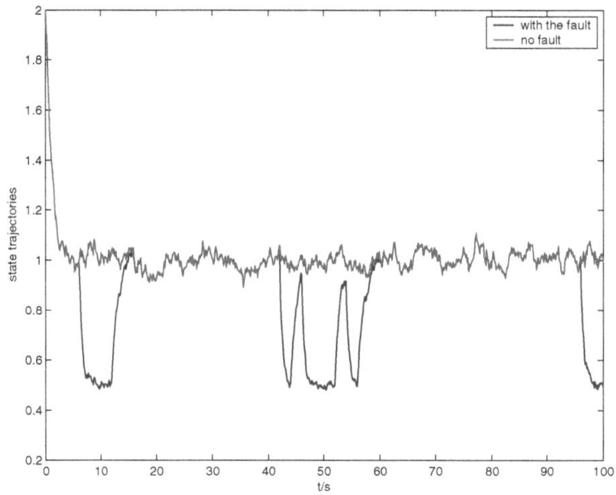

Fig. 4.1 State trajectories in faulty case 1

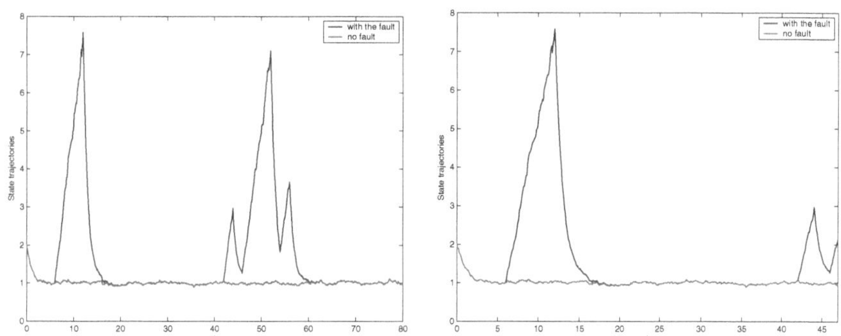

(a) State trajectories (ended at mode 1) (b) State trajectories (ended at mode 2)

Fig. 4.2 State trajectories in faulty case 2

simulation, suppose that $-\rho_{11} = \rho_{12} = 0.5$ and $-\rho_{22} = \rho_{21} = 0.8$, $d = 1$. Fig. 4.1 illustrates the state trajectory, from which we can see that the system is always ISS with respect to the demand rate d in spite of the fault.

Faulty case 2: $f_2(x) = 2x$.

We can get

$$\overline{\mathscr{L}}V_2(x) = \mathscr{L}V_2(x) = 4x^2 - 4xd + \frac{1}{4}x^2$$

Theorem 4.4 is unavailable now. It can be seen that the faulty mode may become non-ISS. Choose $\lambda_0 = 1.5$, $\lambda_1 = 4.5$ that satisfy (4.55) and (4.56). It follows from Theorem 4.7 that, if the system is ended at mode 1, i.e. the machine finishes the work normally, then ISS is achieved as shown in Fig. 4.2(a). If the machine stops at the faulty mode (mode 2), it is obtained from (4.59) that if $\tau^s \geq 3T_{us}$, i.e. the dwell period of the healthy mode is large enough and the time that the machine works in faulty mode is small enough, then ISS of the overall system process is still guaranteed. Suppose that the system stops at $t = 47s$, Fig. 4.2(b) shows the trajectories in this case, the system is stable in the sense of ISS.

4.3 Conclusion

The main contribution of this chapter is the generalization of ISS theory to hybrid impulsive systems and stochastic hybrid systems. For a hybrid impulsive system, it has been shown that ISS is achieved in spite of partial non-ISS modes. Whereas for a stochastic hybrid system, ISS is maintained even no mode is ISS. The obtained results are useful for stabilization of HS with unstable faulty modes.

Chapter 5
Hybrid Systems with Discrete Specifications

In chapters 2-4, FTC design methods for several different classes of HS have been discussed from the stability point of view (i.e., continuous states are globally convergent whatever mode is activated). These methods are based on the continuous system theory, and limited for more general discrete faults, especially, when certain discrete specifications are required.

This chapter considers HS with certain discrete specifications, i.e., it has to follow some specifications imposed on the discrete part of the system. A discrete fault would violate these specifications. As for such fault, one natural idea is to reconfigure the discrete part after faults occur, which can be achieved from discrete event system (DES) point of view. However, compared with pure DES, continuous system behaviors must be taken into account in HS. Two major discrete event system models, namely *finite state machine* and *Petri net* are used respectively.

5.1 Qualitative FTC Based on Finite State Machine

In this section, we consider a class of HS with certain discrete specifications, i.e., it has to follow the desired switching sequence and finally reach the target mode. A discrete fault would change the sequence and violate these specifications. Fault tolerability properties of such HS is analyzed in a qualitative manner.

5.1.1 Problem Formulation

As for the considered discrete fault, one natural idea is to reconfigure the switching sequence after faults occur to maintain the specification, which could be achieved by the discrete event system (DES) supervisory control theory [101]. However, compared with pure DES, continuous dynamics have to be considered for HS, the reachability must be checked after reconfiguration of the sequence. It has been shown that checking reachability for very simple class of HS is a difficult work, and the accurate mathematic model of the system must be known. In fact, the hybrid models of physical environment in real world are usually too large and complicated. How to

H. Yang et al.: Fault Tolerant Control Design for Hybrid Systems, LNCIS 397, pp. 113–148.
springerlink.com

link the continuous and discrete parts for the purpose of fault tolerance analysis is one challenge that is to be faced in this work.

Abstraction is a technique to reduce the complexity of the system design, while preserving some of its relevant behaviors, so that the simplified system is more accessible to analysis tools [2]. Such method leads to a lower computation level than that for the original system. Qualitative abstraction (QA) originates from the qualitative theory that has been shown to be effective tools to analyze system behavior in the absence of complete knowledge [64]. Several results have been reported about QA for HS, e.g., [5, 121], most of these works focus on the linear HS. In [8], a qualitative description of the nonlinear systems' behavior is proposed while HS are not considered.

Above results of QA inspire us to link qualitatively the continuous and discrete parts of HS. The novelty of this work is that a new clue to solve the FTC problem of HS is provided, that is in a qualitative manner and from discrete event system (DES) point of view.

A hierarchical model is developed to describe the HS as in [140]. Such model not only represents the discrete-event dynamics that is appropriate to find the supervisor, but also provides absolute temporal information. Moreover, the discrete and continuous parts are linked qualitatively such that the reachability and fault tolerability properties of HS can be analyzed.

The proposed model consists of four parts from bottom to top: hybrid automaton, qualitative abstraction, discrete abstraction, and supervisor as in Fig.5.1. The hybrid automaton models the original HS; QA is a finite state machine which captures the information of the time derivatives and the positions of continuous states, and describes the qualitative behavior of HS based on the incomplete system's knowledge. QA is a link between continuous and discrete part of HS, the reachability can be analyzed effectively in this level; Discrete abstraction is also a finite state machine which represents the discrete modes and the switchings among them, while the behavior of continuous modes is removed. Fault tolerance is discussed in this level using DES supervisory control theory [101]; Supervisor determines whether the controllable switching between modes is activated or not, and reconfigures the switching sequence after faults occur.

It will be shown that under this model, it is easy to check if the switching sequence design based on discrete abstraction is available for original HS, the reachability and fault tolerability properties of HS can be analyzed systematically. The main contributions of this work are twofold:

1. A qualitative description is derived for a class of HS. Such qualitative model links well the continuous and discrete parts of HS. Reachability can be analyzed in a qualitative manner.
2. Fault tolerance of HS is discussed from DES point of view, which is effective for HS with the desired discrete specification. The intelligent supervisor is less conservative than those robust ones.

It should be pointed out that this model is similar to that in [78] and [79]. However, in [78], the Petri net is used to model the abstraction of the system. Our work

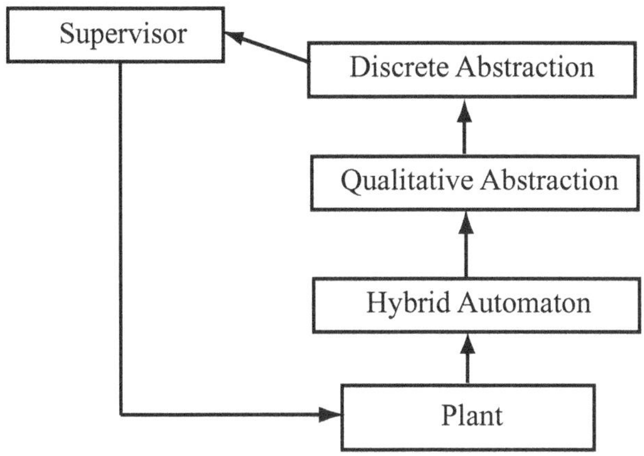

Fig. 5.1 The hierarchical FTC model

utilizes the finite state automata which is more suitable for fault tolerance analysis of discrete fault. In [79], the objective is to handle a kind of continuous systems with the discrete inputs. The qualitative abstraction in [79] relies on the behavior theory which is also quite different from ours.

The HS is modeled by a hybrid automaton $\mathscr{H}$ as Definition 1.1 where the outputs and continuous faults are not involved. The trajectories of $\mathscr{H}$ that start from some initial state $(q_0, x_0) \in Init$ consist of a sequence of continuous flows and discrete transitions. When the discrete state q is maintained, the continuous state x evolves according to the differential equation $\dot{x} = f^q(x)$, as long as $x \in Inv(q)$. After x reaches the guard set, the system would switch into next mode under discrete controller v.

As mentioned before, under certain discrete specifications, the HS has to follow a marked switching sequence and reach the target mode to complete the task as shown in Fig.5.2 (which describes a system with 2 continuous states and 4 modes).

The considered *discrete faults* F_d affect the discrete transitions $E : V \times F_d \to Q \times Q$, that forces the system to switch into a mode which is not the prescribed

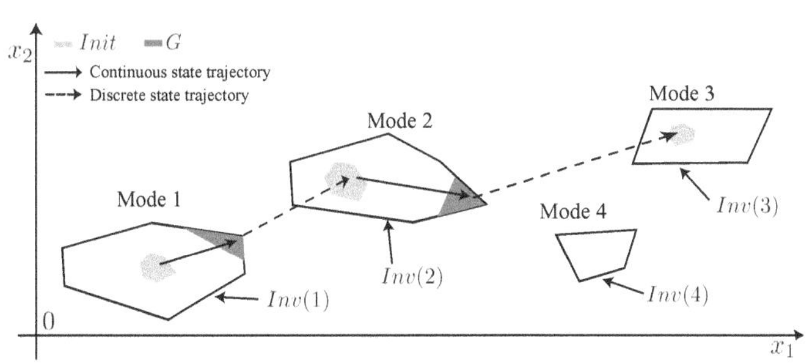

Fig. 5.2 Illustration of the system trajectory

successor. such faulty switching is not affected by the discrete controller V, this implies that discrete faults are *uncontrollable switchings* that change the prescribed sequence.

The fault tolerance problem can be described as: *make the HS reach the target mode according to discrete specifications in spite of discrete faults.* We will analyze detailed procedures of the hierarchical model (Fig.5.1) in the following sections.

5.1.2 Qualitative Abstraction for Nonlinear System

The purpose of qualitative behavior analysis is to build a bridge that connects the continuous and discrete parts of HS. This section derives a qualitative description for HS. For the sake of notational simplicity, the superscript q that denotes mode q is omitted in this section.

We first define *sign* as the symbol of the scalar, i.e., for a scalar α, $sign(\alpha) = -1$, if $\alpha < 0$; $sign(\alpha) = 0$, if $\alpha = 0$; $sign(\alpha) = 1$, if $\alpha > 0$. Define the set $S \triangleq \{\varsigma = [\varsigma_1, ..., \varsigma_n]^T; \varsigma_k \in \{-1, 1\}\}$. It is clear that S contains 2^n elements. Denote ς^k as the kth element of S. The matrix $diag(\beta)$ is the diagonal matrix having the vector β on its main diagonal.

Consider the nonlinear system

$$\dot{x} = f(x) \tag{5.1}$$

where $x = [x_1, ..., x_n]^T \in \Re^n$, $f(x) = [f_1(x), ..., f_n(x)]^T$ is a smooth function. It is not assumed to have the complete knowledge of $f(x)$, only two qualitative properties are required: $sign(f_i(x))$ and $sign(\frac{\partial f_i}{\partial x_j})$, $\forall i, j \in \{1, 2, ..., n\}$.

Consider a point x^*, which divides the domain $\Re^n$ of x into 2^n regions, denoted as $\Omega_q \triangleq \{x \in \Re^n | diag(\varsigma^q)(x - x^*) > \bar{0}\}$, for $\varsigma^q \in S$. $\bar{0}$ is a null vector of dimension n. Each region Ω_q has n neighbors that share a hyperplane with Ω_q. It is clear that one of continuous states x_j equals x_j^* which is on this hyperplane, for $j \in \{1, 2, ..., n\}$. Denote $M_{(pq,j)}$ as the hyperplane between two neighboring regions Ω_p, Ω_q, and $x_j = x_j^*$. Also denote $sign(f_i(x))_{\Omega_q}$ as the sign of $f_i(x)$ in Ω_q. When x goes from one region to another, we say that a *continuous transition* occurs.

Definition 5.1. *The nonlinear system (5.1) is said to be* point monotonous *if there exists a point x^* such that*
1) $\forall i \in \{1, 2, ..., n\}$, $\forall \varsigma^q \in S$, $sign(f_i(x))_{\Omega_q}$ is fixed, i.e. the value of $sign(f_i(x))_{\Omega_q}$ is unique.
2) there is a fixed $k \in \{1, 2, ..., n\}$, s.t. $sign(f_k(x))_{\Omega_q} \neq sign(f_k(x))_{\Omega_p}$, and $\forall j \in \{1, 2, ..., n\}, j \neq k$, $sign(f_j(x))_{\Omega_q} = sign(f_j(x))_{\Omega_p}$, where Ω_p, Ω_q are two neighboring regions.

The following Lemma gives the method to check which f_i changes the symbol between neighboring regions.

Lemma 5.1. *Consider a point monotonous nonlinear system (5.1) and two neighboring regions Ω_p, Ω_q. If $\forall x \in M_{(pq,s)}$, for $s, i \in \{1, 2, ..., n\}$, we have $f_i(x) = 0$ and*

$$\sum_{j\in\{1,2,...,n\},j\neq i} \left(\frac{\partial f_i}{\partial x_j} f_j\right) \neq 0$$

then $sign(f_i(x))_{\Omega_q} \neq sign(f_i(x))_{\Omega_p}$.

Proof: Differentiating equation (5.1) w.r.t the time leads to

$$\ddot{x}_i = \frac{\partial f_i}{\partial x_i}\dot{x}_i + \sum_{j\in\{1,2,...,n\},j\neq i} \frac{\partial f_i}{\partial x_j} f_j \tag{5.2}$$

Note that $\forall x \in M_{(pq,s)}$, $f_i(x) = 0$ implies x_i admits an extremum at $M_{(pq,s)}$. The following two cases are considered.

- For the case that $sign(\sum_{j\in\{1,2,...,n\},j\neq i} \frac{\partial f_i}{\partial x_j} f_j) = 1$, it is obtained from (5.2) that $\ddot{x}_i > 0$ $\forall x \in M_{(pq,s)}$. So $\dot{x}_i$ changes monotonously at $M_{(pq,s)}$. On the other hand, $sign(f_i(x))$ is fixed respectively in Ω_p and Ω_q, we have that $sign(f_i(x))_{\Omega_q} > 0$, and $sign(f_i(x))_{\Omega_p} < 0$, x_i reaches a maximal point or $sign(f_i(x))_{\Omega_q} < 0$ and $sign(f_i(x))_{\Omega_p} > 0$, x_i reaches a minimal point. It follows that $sign(f_i(x))_{\Omega_q} \neq sign(f_i(x))_{\Omega_p}$.
- For the case that $sign(\sum_{j\in\{1,2,...,n\},j\neq i} \frac{\partial f_i}{\partial x_j} f_j) = -1$, it follows that $\ddot{x}_i < 0$ $\forall x \in M_{(pq,s)}$. Thus $\dot{x}_i$ also changes monotonously at $M_{(pq,s)}$, the similar procedure as for case 1 can be done to obtain the results. □

Lemma 5.1 captures the symbolic change of $\dot{x}$ among different regions, this is very useful for continuous transition analysis as in Lemma 5.2.

Lemma 5.2. *Consider a point monotonous nonlinear system (5.1), Ω_p, Ω_q are two neighboring regions sharing $M_{(pq,j)}$, and $sign(f_i(x))_{\Omega_q} \neq sign(f_i(x))_{\Omega_p}$.*

-If $sign(x_j - x_j^*)_{\Omega_p} = -sign(f_j)_{\Omega_p}$, then crossing of $M_{(pq,j)}$ is possible from Ω_p to Ω_q.

-If $sign(x_j - x_j^*)_{\Omega_q} = -sign(f_j)_{\Omega_q}$, then crossing of $M_{(pq,j)}$ is possible from Ω_q to Ω_p.

Proof: For the case that $sign(x_j - x_j^*)_{\Omega_p} = -sign(f_j)_{\Omega_p}$, without loss of generality, assume that in Ω_p, $sign(x_j - x_j^*) = 1$, and $sign(f_j) = -1$, then x_j would converge to x_j^* and reach the $M_{(pq,j)}$ from Ω_p, then go to Ω_q. If in Ω_p, $sign(x_j - x_j^*) = -1$, and $sign(f_j) = 1$, then x_j would also converge to x_j^* and reach the $M_{(pq,j)}$. The proof of the transition from Ω_q to Ω_p is the same as above, which is omitted. □

Based on the above analysis, we define the *qualitative states* of point monotonous system (5.1) as $\vartheta : x \to \mathbb{M} = \{1,2,...,2^n\}$, which is a finite set of variables interpreted over 2^n regions of system (5.1), e.g., $\forall x \in \Omega_q$, $\vartheta(x) = q$, which is denoted as ϑ_q.

Example 5.1: Consider the Lotka-Volterra system describing the relation between a population of preys (x_1) and a population of predators (x_2):

$$\begin{cases} \dot{x}_1 = ax_1 - bx_1x_2 \\ \dot{x}_2 = -cx_2 + dx_1x_2 \end{cases}$$

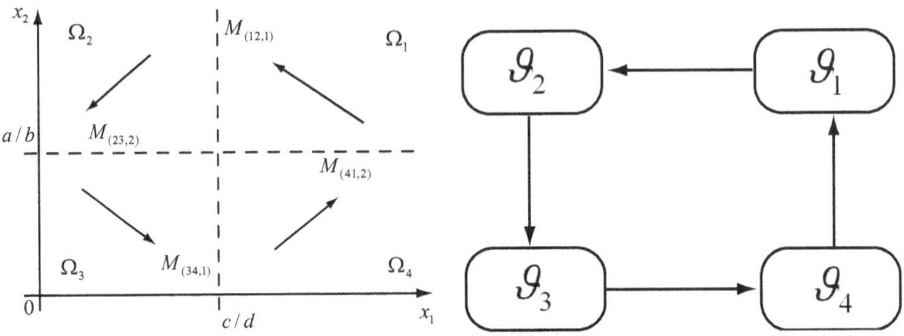

(a) vector field in each region (b) qualitative state graph

Fig. 5.3 qualitative behavior of the system

where a, b, c and d are positive. The system is point monotonous with $x^* = [c/d \;\; a/b]^T$. Fig. 5.3(a) shows its four regions divided by x^*: Ω_1, Ω_2, Ω_3 and Ω_4. $sign(f_i)$ is also fixed in each region as in Fig.5.3: $sign(f_1)_{\Omega_1} < 0$, $sign(f_1)_{\Omega_2} < 0$, $sign(f_1)_{\Omega_3} > 0$, $sign(f_1)_{\Omega_4} > 0$. $sign(f_2)_{\Omega_1} > 0$, $sign(f_2)_{\Omega_2} < 0$, $sign(f_2)_{\Omega_3} < 0$, $sign(f_2)_{\Omega_4} > 0$. Based on Lemma 5.2, four continuous transitions can be obtained: from Ω_1 to Ω_2, from Ω_2 to Ω_3, from Ω_3 to Ω_4, from Ω_4 to Ω_1. Fig. 5.3(b) shows the graph of qualitative states.

5.1.3 Qualitative Abstraction of Hybrid Systems

The above qualitative description can be extended naturally to HS with all modes satisfying the point monotony property. For HS, two transitions have to be considered:

-Continuous transitions (in each mode): The rule for constructing continuous transitions in each mode is the same as in Lemmas 5.1, 5.2. Denote E_c as the continuous transition set of HS.

-Discrete transitions (between modes): As in Definition 5.1, when x in mode i reaches the guard set G, and $(i,i') \in E$, then the system can be switched from mode i to i' under v.

The region of mode i is denoted as $\Omega^i = [\Omega_1^i, ..., \Omega_{2^n}^i]^T$, where

$$\Omega_q^i \triangleq \{x \in \Re^n | (diag(\varsigma^q)(x - x^{*(i)}) > \bar{0}) \cap Inv(i), \varsigma^q \in S\} \tag{5.3}$$

where $x^{*(i)}$ is related to the point monotony property of mode i. From (5.3), it follows that some regions Ω_q^i may be empty if $(diag(\varsigma^q)(x - x^*) > \bar{0}) \cap Inv(i) = \emptyset$. The qualitative states of mode i are denoted as $\vartheta^i = [\vartheta_1^i, ..., \vartheta_{\iota(i)}^i]^T$, where $0 < \iota(i) \leq 2^n$, ϑ_q^i is related to $x \in \Omega_q^i \neq \emptyset$.

Definition 5.2. *The region Ω_q^i is said to be* determined *if only discrete transitions or only one continuous transition may occur from Ω_q^i.*

The "determined region" is a special case of the "good region" defined in [3], which ensures the uniqueness of the continuous transition from each region.

The QA of HS is constructed as a finite state machine $\mathscr{QA} = (\hat{Q}, \hat{q}_0, \hat{\Sigma}, T)$, where $\hat{Q} = \bigcup_{i \in Q} \vartheta^i$ is the state set, $\hat{Q}_0 = \bigcup_{\forall (x,i) \in Init, x \in \Omega_q^i \neq \emptyset} \vartheta_q^i$ is the initial state, $\hat{\Sigma} = E \bigcup E_c$ is the transition set $T : \hat{\Sigma} \times \hat{Q} \rightarrow \hat{Q}$ is the activated transition[1]. Divide $T = T_c \cup T_d$, where T_c is the set of continuous transitions, T_d the discrete one. The number of operations required to build the QA depends on the number of qualitative states that the HS generates. The computational complexity is $\mathscr{O}(\sum_{i=1}^{N} |\iota(i)|)$.

Compared with the abstraction in the usual sense [2], [121], $\mathscr{QA}$ corresponds to the qualitative behavior of HS, since the continuous state in each region of HS has a unique related state in $\mathscr{QA}$. The following theorem proves the qualitative reachability equivalence between $\mathscr{QA}$ and the original system.

Theorem 5.1. *Consider a HS where each mode is point monotonous and all the regions are determined. For any $x_1 \in \Omega_q^i$, $x_2 \in \Omega_p^j$, if there exists a solution $x(t)$ of HS and t_1, t_2, s.t. $0 \leq t_1 \leq t_2$, $x(t_1) = x_1$, $x(t_2) = x_2$, then $\mathscr{QA}$ has a solution $\hat{q}(t) \in \hat{Q}$ s.t. $\hat{q}(t_1) = \vartheta_q^i$, and $\hat{q}(t_2) = \vartheta_p^j$.*

Proof: Consider two points $x_1 \in \Omega_q^i$, $x_2 \in \Omega_p^j$, s.t. $x(t_1) = x_1, x(t_2) = x_2$ for $0 \leq t_1 \leq t_2$. From the structure of $\mathscr{QA}$, it follows that for every transition of HS, there is a unique related transition in $\mathscr{QA}$. If $t_1 = t_2$, then $\hat{q}(t)$ is a trivial solution satisfying the theorem. If $t_1 < t_2$, denote $\hat{q}^0, \hat{q}^1, \ldots \hat{q}^m$ as the state sequence of $\mathscr{QA}$ in the interval $[t_1, t_2]$. If x_1 and x_2 are within the same region of the same mode, i.e. $i = j$, $p = q$, then no transition occurs, m=0, $\hat{q}^0 = \hat{q}(t_1) = \hat{q}(t_2) = \vartheta^i$. If x_1 and x_2 are within the different regions of the same mode, i.e., $i = j$, $p \neq q$, continuous transitions must occur from $x(t_1)$ to $x(t_2)$, so $m > 0$, and all $\hat{q}^0, \ldots, \hat{q}^m \in \vartheta^i$, with $\hat{q}^0 = \hat{q}(t_1) = \vartheta_q^i$, and $\hat{q}^m = \hat{q}(t_2) = \vartheta_p^i$. If x_1 and x_2 are within the different regions of the different modes, i.e., $i \neq j$, $p \neq q$, then both discrete and continuous transitions occur, we have $\hat{q}^0 = \hat{q}(t_1) = \vartheta_q^i$, and $\hat{q}^m = \hat{q}(t_2) = \vartheta_p^j$. □

The converse version of Theorem 5.1 is not true, i.e., for $\hat{q}(t_1) = \vartheta_q^i$, and $\hat{q}(t_2) = \vartheta_p^j$ $x_1 \in \Omega_q^i$, $x_2 \in \Omega_p^j$, it may not hold that the solution $x(t)$ of HS satisfies $x(t_1) = x_1$, $x(t_2) = x_2$, but $x(t_1) \in \Omega_q^i$, $x(t_2) \in \Omega_p^j$ as in the following corollary.

Corollary 5.1. *Consider a HS where each mode is point monotonous and all the regions are determined. For any $x_1 \in \Omega_q^i$, $x_2 \in \Omega_p^j$, if there exists t_1, t_2, s.t. $0 \leq t_1 \leq t_2$, $\hat{q}(t_1) = \vartheta_q^i$, and $\hat{q}(t_2) = \vartheta_p^j$, then there exists a solution $x(t)$ s.t. $x(t_1) \in \Omega_q^i$, $x(t_2) \in \Omega_p^j$.*

Proof: Consider the state sequence $\hat{q}^0, \hat{q}^2, \ldots \hat{q}^m$ of $\mathscr{QA}$ in the interval $[t_1, t_2]$, Similarly to the proof of Theorem 5.1, if m=0, then both x_1, x_2 are within the same region of the same mode, which leads to $x(t_1), x(t_2) \in \Omega_q^i$. In the sequel, suppose $m > 0$,

[1] In some literatures of DES, Σ is also called "Event", and T is called "transition".

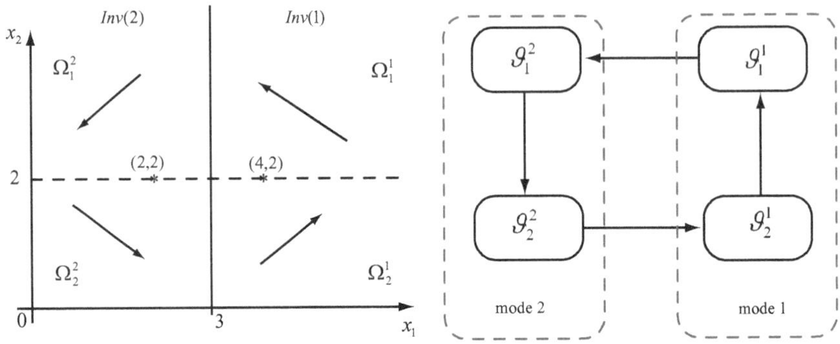

(a) vector field in each region (b) qualitative state graph

Fig. 5.4 qualitative behavior of the HS

we have $\hat{q}(t_1) = \vartheta_q^i$, $\hat{q}(t_2) = \vartheta_p^j$. From the definition of ϑ_p^j, there must be a solution $x(t)$ s.t. $x(t_1) \in \Omega_q^i$, $x(t_2) \in \Omega_p^j$. □

Example 5.2: Consider a HS with two modes as

$$\text{mode 1: } \begin{cases} \dot{x}_1 = 3x_1 - 1.5x_1x_2 \\ \dot{x}_2 = -2x_2 + x_1x_2 \end{cases}, \quad \text{mode 2: } \begin{cases} \dot{x}_1 = 2x_1 - x_1x_2 \\ \dot{x}_2 = -4x_2 + x_1x_2 \end{cases}$$

where $Inv(1) = \{x_1 \geq 3\}$, $Inv(2) = \{x_1 \leq 3\}$. The guard set $G(1,2) = \{x \in \Re^2, x_1 \leq 3\}$, $G(2,1) = \{x \in \Re^2, x_1 \geq 3\}$, and x is continuous everywhere. Both two modes are point monotonous with $x^{*(1)} = (2,2)$, $x^{*(2)} = (4,2)$. Consequently, four regions can be divided for each mode. However, only two regions are used to build the QA as shown in Fig. 5.4(a), since other regions of the mode do not intersect its invariant set. It can be seen that all regions are determined, from the transition set, we have that $x \in \Omega_2^2$ could be reached from $x \in \Omega_1^1$. This is also reflected in the qualitative state graph as in Fig. 5.4(b).

5.1.4 *Discrete Abstraction*

Discrete abstraction (DA) is connected with the supervisor, which can be viewed as a reduction of $\mathscr{QA}$ by removing the qualitative behavior of each mode, such that the DES theory can be applied.

The DA of HS is also constructed as a finite state machine

$$\mathscr{DA} = (Q_d, \Sigma, T_d, Q_{d0}, Q_{dm})$$

where $Q_d = Q$ and $\Sigma = E$ are the same as in Definition 5.1; T_d denotes the activated discrete transition as in $\mathscr{QA}$, $Q_{d0} = \bigcup_{\forall (x,q) \in Init} q$. $Q_{dm} \subseteq Q$ is the set of marked states.

The following theorem shows discrete reachability equivalence between $\mathscr{DA}$, $\mathscr{QA}$ and the original system.

Theorem 5.2. *Consider a HS where each mode is point monotonous and all the regions are determined. For any $x_1 \in Inv(p)$, $x_2 \in Inv(q)$, if there exists a solution $x(t)$ of HS and t_1, t_2, s.t. $0 \leq t_1 \leq t_2$, $x(t_1) = x_1$, $x(t_2) = x_2$, then*
-there is a solution $\hat{q}(t)$ of $\mathscr{QA}$ s.t. $\hat{q}(t_1) \in \vartheta^i$, and $\hat{q}(t_2) \in \vartheta^j$.
-there is a solution $q_d(t)$ of $\mathscr{DA}$ s.t. $q_d(t_1) = p$, and $q_d(t_2) = q$.

Proof: The result can be obtained following the same procedure in proof of Theorem 5.1. □

The following definition gives the relations between the transitions of $\mathscr{QA}$ and $\mathscr{DA}$.

Definition 5.3. *A transition sequence $s = e_1 e_2 \cdots e_m$ of $\mathscr{QA}$, $e_i \in \hat{\Sigma}, i = 1, 2 \ldots, m$ for $m > 0$* is consistent with *the transition sequence $\delta = \varepsilon_1 \varepsilon_2 \cdots \varepsilon_u$ of $\mathscr{DA}$, $\varepsilon_i \in \Sigma, i = 1, 2 \ldots, u$ for $u > 0$ if at any t, s.t. $\hat{q}(t) \in \vartheta^i$ under s, then $\hat{q}_d(t) = i$ under δ, where $\hat{q}(t)$ is the solution of $\mathscr{QA}$ along s, and $\hat{q}_d(t)$ is the solution of $\mathscr{DA}$ along δ.*

5.1.5 Fault Tolerance

Let us recall that the *discrete faults* force the system to switch into an unprescribed successive mode, and may violate the discrete specification of the HS.

From sections 5.1.3 and 5.1.4, it can be seen that $\mathscr{QA}$ and $\mathscr{DA}$ are generated by the normal HS off-line, and then works in parallel with HS. $\mathscr{QA}$ receives all the switchings and state information from HS, and triggers its corresponding transitions to keep itself synchronized with HS. Meanwhile $\mathscr{QA}$ sends the information of discrete transitions to $\mathscr{DA}$, such that $\mathscr{DA}$ is also synchronized with HS. Since the set of faulty switchings is not included in normal HS, once the fault occurs, its information may be missing for $\mathscr{QA}$ or sent to $\mathscr{QA}$ by the HS with delays of several discrete steps. This motivates the following definition:

Definition 5.4. *A fault is said to be* diagnosable *w.r.t $\mathscr{QA}$ and $\mathscr{DA}$, if when such faulty switching occurs in HS, the corresponding transitions occur in $\mathscr{QA}$ and $\mathscr{DA}$ before the next discrete switching in HS occurs.*

Definition 5.4 is equivalent to the diagnosability with 0-delay step in [92]. This is possible for HS. Compared with the pure DES system, the continuous dynamics in HS can help to detect the discrete faults rapidly based on the trajectory of continuous states. Such property of HS also allows us not to consider the observability of faults as in [108]. If the full measurement of continuous states is unavailable, some more complicated techniques have to be applied, e.g. the multi-mode identifier in Chapter 2.2.

Once $\mathscr{QA}$ receives the information of faulty switching, and sends it to $\mathscr{DA}$, the transition set Σ_d in $\mathscr{DA}$ would be updated, and partitioned as $\Sigma = \Sigma_n \cup \Sigma_f$, where Σ_n are normal transition sets corresponding to E in $\mathscr{H}$, and Σ_f corresponds to faulty transitions.

In this section, fault tolerance problem is discussed on $\mathscr{DA}$ under the DES framework, the resulting fault tolerant supervisor is applied to the original system as shown in Fig. 5.1.

We first introduce some notations used in the following, the reader is referred to [101] for more detailed notations. Σ^* denotes the set of all finite strings of elements of Σ, including the empty string ε. A subset of Σ^* is called a language over Σ. The prefix closure of $L \subseteq \Sigma^*$ is defined as $\overline{L} := \{u \in \Sigma^* | uu' \in L \text{ for some } u' \in \Sigma^*\}$. The closed behavior of $\mathscr{DA}$ is $L(\mathscr{DA}) := \{s \in \Sigma^* | T_d(s, Q_{d0}) \text{ is defined}\}$, which is the set of transition sequences. The marked behavior of $\mathscr{DA}$ is $L_m(\mathscr{DA}) := \{s \in \Sigma^* | T_d(s, Q_{d0}) \in Q_{dm}\}$ which represents the completed tasks. $A(q_d)$ denotes the set of transitions that are possible at q_d. A supervisor $\mathscr{S} \subseteq \Sigma^* \to 2^{\Sigma}$ specifies the set of transitions for the system's desired specification. The behavior of $\mathscr{DA}$ supervised by $\mathscr{S}$ is denoted by $L(\mathscr{S}/\mathscr{DA})$.

Assume that a supervisor $\mathscr{S}$ has been designed for the healthy system satisfying the specifications. In the following, we will focus on how to update $\mathscr{S}$ for the purpose of fault tolerance. Denote B_{id} and B_a as the ideal behavior and the acceptable behavior respectively, where $B_{id} \subseteq B_a \subseteq L_m(\mathscr{DA})$. Moreover, for $\kappa, \kappa' \in \Sigma^*$, define

$$L(q_d, \omega) := \{\kappa | T_d(\kappa', Q_{d0}) = q_d, \text{and } \kappa' \omega \kappa \in L(\mathscr{DA})\}$$

which is a set of transition sequences generated in $\mathscr{DA}$ after the transition ω occurs at the state q_d. Similarly, define

$$L_{B_{id}}(q_d, \omega) := \{\kappa | T_d(\kappa', Q_{d0}) = q_d, \text{and } \kappa' \omega \kappa \in B_{id}\}$$

$$L_{B_a}(q_d, \omega) := \{\kappa | T_d(\kappa', Q_{d0}) = q_d, \text{and } \kappa' \omega \kappa \in B_a\}$$

The following definition gives conditions for discrete faults to be tolerable.

Definition 5.5. *The transition ω that occurs at q_d is an* absolutely tolerable fault *w.r.t. Q_{dm}, if*

1) *There exists a nonempty $K \subseteq L_{B_a}(q_d, \omega)$ s.t. $\overline{K}\Sigma_f \cap L(q_d, \omega) \subseteq \overline{K}$.*
2) *There exist a transition sequence s from $\hat{q}_i$ in $\mathscr{QA}$ that is consistent with K, where $\hat{q}_i$ is the state of $\mathscr{QA}$ reached due to the fault.*

Condition 1) means that K is controllable w.r.t. $L(q_d, \omega)$ [101], which implies that for a faulty transition ω, the system $\mathscr{DA}$ can still be driven to the marked states. Condition 2) ensures that the discrete switching sequence designed from $\mathscr{DA}$ is possible for HS, i.e., there exists a trajectory of continuous states in HS that is consistent with the discrete sequence.

Definition 5.5 is a little conservative since the worst case that all possible faults occur simultaneously is considered. This is relaxed in the following definition, where the single faulty case is considered.

Definition 5.6. *The transition ω that occurs at q_d is a* tolerable fault *w.r.t. Q_{dm}, if 2) in Definition 5.5 holds, and there exists a nonempty $K \subseteq L_{B_a}(q_d, \omega)$.*

Denote $\Sigma_{tf} \subseteq \Sigma_f$ as the set of tolerable faults. In the following, we restrict our attention to the single faulty case, the results can be extended straightly to multi-faulty case.

Now we design the ideal fault tolerant supervisor.

Definition 5.7. *The transition sequence* $s = e_1 e_2 \cdots e_m \in B_{id}(\mathscr{DA})$, $e_i \in \Sigma, i = 1,2\ldots,m$ *for* $m > 0$ *is called an* ideal tolerable fault marked sequence (ITFMS) *if any transition in* $A(q_{d(i-1)}) - \{e_i\} \in \Sigma \cup \Sigma_{tf}$*, where* $q_{d_{i-1}} := T_d(e_{i-1}, Q_{d0})$ *and* $e_0 := \varepsilon$.

Definition 5.7 means that ITFMS is a transition sequence which can drive the initial state to the marked states within an *ideal behavior* in spite of the interference of tolerable faults.

Define $IT(\mathscr{DA}) := \{t \in L(\mathscr{DA}) | t \text{ is an ITFMS}\}$. For a language L and $s \in \overline{L}$, define $\psi_L(s) := \{\alpha \in \Sigma | s\alpha \in \overline{L}\}$. We have the following theorem.

Theorem 5.3. *Consider a HS where each mode is point monotonous and all the regions are determined. Suppose that* $\mathscr{DA}$ *has a fault transition* $\omega \in \Sigma_f$ *at* q_d*, which is diagnosable w.r.t* $\mathscr{QA}$, $\mathscr{DA}$*, and* $IT(\mathscr{DA}) \neq \emptyset$*. There exists an ideal fault tolerant supervisor* $\mathscr{S}_{id}$ *for HS if*

1) $\exists K \subseteq IT(\mathscr{DA})$ *s.t.* $\forall s \in \overline{K}, \psi_K(s) \subseteq \psi_{IT(\mathscr{DA})}(s)$ *and* $\{s\omega\} \cap \overline{IT(\mathscr{DA})} \subseteq \overline{K}$.
2) There exists a transition sequence s *from* $\hat{q}_i$ *in* $\mathscr{QA}$ *that is consistent with* K*, where* $\hat{q}_i$ *is the state of* $\mathscr{QA}$ *that is reached due to the fault.*

Proof: Consider a supervisor $\mathscr{S}_{id} = \psi_K(s) = \{\alpha \in \Sigma | s\alpha \in \overline{K}\}$, which represents the set of enabled transitions after the string s. Firstly, let $s = \varepsilon$. Since $IT(\mathscr{DA}) \neq \emptyset$, we have $\psi_{IT(\mathscr{DA})}(\varepsilon) \neq \emptyset$, which implies that $\psi_{L(\mathscr{S}_{id}/\mathscr{DA})}(\varepsilon) \neq \emptyset$, and $\psi_{L(\mathscr{S}_{id}/\mathscr{DA})}(\varepsilon) \subseteq \psi_{IT(\mathscr{DA})}(\varepsilon)$. Secondly, let $s \neq \varepsilon$, following the same procedure, it can be proven that $\psi_{L(\mathscr{S}_{id}/\mathscr{DA})}(s) \subseteq \psi_{IT(\mathscr{DA})}(s), \forall \psi_{IT(\mathscr{DA})}(s) \neq \varepsilon$.

Also, since $\{s\omega\} \cap \overline{IT(\mathscr{DA})} \subseteq \overline{K}$, we have $s\omega \in \overline{K}$ and $\omega \in \psi_K(s) = \mathscr{S}_{id}(s)$, which further leads to that $\omega \in \psi_{L(\mathscr{S}_{id}/\mathscr{DA})}(s)$. To this end, it can be obtained that $L(\mathscr{S}_{id}/\mathscr{DA}) = \overline{K}$, which means that the abstraction $\mathscr{DA}$ under $\mathscr{S}_{id}$ generates a nonempty subset of ITFMS's set.

On the other hand, Condition 2) implies that after a fault occurs, there exists a state trajectory of $\mathscr{QA}$ from $\hat{q}_i$ that reaches the $\hat{q}_m$, s.t., $\hat{q}_m \in \vartheta^{Q_{dm}}$. From theorems 5.1 and 5.2, it follows that there exists a continuous state trajectory from the region related to $\hat{q}_i$ in HS that is consistent with the discrete sequence, and will reach the marked mode. □

Theorem 5.3 gives not only the existence condition of the ideal fault tolerant supervisor, but also its construction method. Obtaining $\overline{IT(\mathscr{DA})}$ requires the calculations of $\psi_{IT(\mathscr{DA})}(s)$ which needs $\mathscr{O}(N)$ computation, where N is the number of states in Q.

We denote $\mathscr{S}_{updated} = \mathscr{S}_{id}$, which is the updated version of the original $\mathscr{S}$ after a fault occurs. This means that $\mathscr{S}$ is applied to the normal plant. Once a discrete fault occurs and $\mathscr{S}_{id}$ has been obtained, $\mathscr{S}$ will be self updated into $\mathscr{S}_{updated} = \mathscr{S}_{id}$,

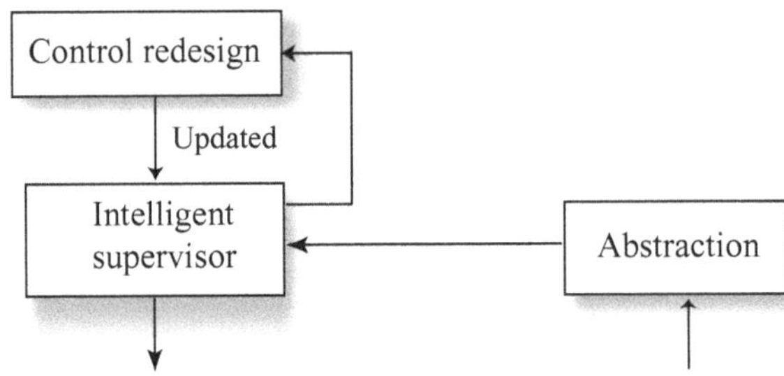

Fig. 5.5 Updating of the supervisor

as shown in Fig. 5.5. In the healthy situation, it can be proven that $L(\mathscr{S}/\mathscr{DA}) = L(\mathscr{S}_{id}/\mathscr{DA})$ following the procedure in [108].

Remark 5.1. *The proposed intelligent supervisor is less conservative than those ones obtained assuming that possible faults occur from the beginning as in [105]. Compared with the pure DES in [105], the reachability problem of continuous systems has to be considered for HS, as the condition 2) in Theorem 5.3.*

When the ideal behavior is not feasible under discrete faults, one natural idea is to let the system work in an acceptable behavior that exceeds but stays close to the ideal behavior. We have the following definition.

Definition 5.8. *The transition sequence $s = e_1e_2\cdots e_m \in B_a(\mathscr{DA})$, $e_i \in \Sigma, i = 1\ldots m$ for $m > 0$ is called an* acceptable tolerable fault marked sequence (ATFMS) *if any transition in $A(q_{d_{i-1}}) - \{e_i\} \in \Sigma \cup \Sigma_{tf}$, where $q_{d_{i-1}} := T_d(e_{i-1}, Q_{d0})$ and $e_0 := \varepsilon$.*

From Definition 5.8, it follows that ATFMS can still drive the initial state to the marked states within an *acceptable* behavior in spite of tolerable faults. Define $AT(\mathscr{DA}) := \{t \in L(\mathscr{DA}) | t$ is an ATFMS$\}$. The following corollary is an extension of Theorem 5.1.

Corollary 5.2. *Consider a HS where each mode is point monotonous and all the regions are determined. Suppose that $\mathscr{DA}$ has a fault transition $\omega \in \Sigma_f$ at q_d which is diagnosable w.r.t $\mathscr{QA}$, $\mathscr{DA}$, and $AT(\mathscr{DA}) \neq \emptyset$. There exists an ideal fault tolerant supervisor $\mathscr{S}_a$ for $\mathscr{H}$ if Condition 2) in Theorem 5.3 holds and*

1) $\exists K \subseteq AT(\mathscr{DA})$ s.t. $\forall s \in \overline{K}, \psi_K(s) \subseteq \psi_{AT(\mathscr{DA})}(s)$ and $\{s\omega\} \cap \overline{AT(\mathscr{DA})} \subseteq \overline{K}$.

Note that the candidate transition sequences $L(\mathscr{S}_a/\mathscr{DA})$ obtained from Corollary 5.2 may not be unique. We propose an optimized choosing method, which makes the system work within an acceptable behavior that is most similar to the ideal one.

Definition 5.9. *[15] A (1-bounded) metric space is a pair $(X_d; d)$ consisting of a nonempty set X_d and a function $d : X_d \times X_d \to [0,1]$ which satisfies the following conditions:*

1) $d(a,b) = 0$ if and only if $a = b$;
2) $d(a,b) = d(b,a)$, $\forall a,b \in X_d$;
3) $d(a,c) \leq d(a,b) + d(b,c)$, $\forall a,b,c \in X_d$.

The distance $d(a,b)$ measures the similarity between a and b. The less the distance is, the more similar the two elements are. We endow Σ^* with the metric induced by d, which measures the distance between transition sequences, Let $s = e_1 e_2 \cdots e_{l(s)}$ and $\rho = \rho_1 \rho_2 \cdots \rho_{l(\rho)}$ be two transition sequences in Σ^*, and $l(s,\rho) := \max\{l(s), l(\rho)\}$. If $l(s) \neq l(\rho)$, e.g. $l(s) < l(\rho)$, set $e_i = \varepsilon\ \forall i > l(s)$. Define

$$d_r(s,\rho) := \sum_{i=1}^{l(s,\rho)} \frac{1}{2^i} d(e_i, \rho_i)$$

set $d(a,\varepsilon) = d(\varepsilon,a) = 1$ for $a \in \Sigma$. It is verified that d_r gives rise to a metric on Σ^*. The following corollary gives an optimal choosing method, which can be verified from Definition 5.9.

Corollary 5.3. *Consider a candidate transition sequence $s \in L(\mathscr{S}_a/\mathscr{DA})$ obtained from Corollary 5.1. If $s = \arg\min\{\max_{\rho \in B_{id}} d_r(s,\rho)\}$, then s is a sequence that is most similar to the ideal one in the sense of* metric space.

Once the optimal s, denoted by s^* is obtained, $\mathscr{S}$ is updated into $\mathscr{S}_{update}$ s.t. $L(\mathscr{S}_{update}/\mathscr{DA}) = s^*$.

Consider a HS where each mode is point monotonous and all the regions are determined, based on previous analysis, a fault tolerance framework for HS can be provided as

(1) Build $\mathscr{H}$ of HS.
(2) Describe the qualitative behavior of each mode based on Section 5.1.2.
(3) Build $\mathscr{QA}$ and $\mathscr{DA}$ of the HS based on sections 5.1.3, 5.1.4.
(4) Apply the supervisor $\mathscr{S}$ to the HS, let the $\mathscr{QA}$ and $\mathscr{DA}$ work in parallel with HS. Once a fault occurs, go to (5). When the task is completed without fault, go to (8).
(5) If 2) in Theorem 5.3 holds, send the information of fault to $\mathscr{QA}$ and $\mathscr{DA}$, go to (6), else go to (8).
(6) If 1) in Theorem 5.3 holds, update $\mathscr{S}$ into $\mathscr{S}_{update} = \mathscr{S}_{id}$, apply $\mathscr{S}_{update}$ to HS until the task is completed, go to (8), else go to (7).
(7) If condition 1) in Corollary 5.2 holds, update $\mathscr{S}$ into $\mathscr{S}_{update}$, apply $\mathscr{S}_{update}$ to HS, until the task is completed, go to (8), else go to (8).
(8) Stop the system.

Example 5.3 (Example 1.2 revisited): Recall Example 1.2 as shown in Fig. 1.3. The system can be modeled as a hybrid automaton, two continuous states x_1 and x_2 represent the positions of the fingertip. The angle θ of the fingertip is assumed to be adjusted according to its position by the robot arm. Discrete states, i.e., modes are defined based on the contact configuration between the hose and the plug, as shown in Fig. 5.6. Eight configurations are considered, each one can be further divided

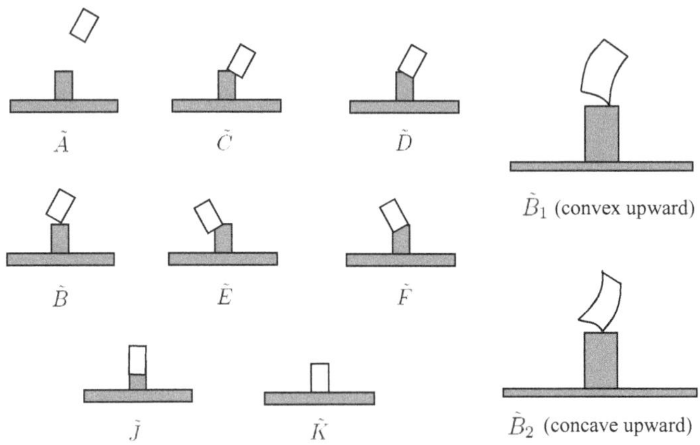

Fig. 5.6 Contact configuration

into two configurations with different form of hose, e.g. $\tilde{B}_1$ (convex upward) and $\tilde{B}_2$ (concave upward). The completed work is to insert the hose onto the plug. There are two operation situations of the fingertip: "PUT" vertically in $\tilde{A}$, $\tilde{J}$, and "SWING" horizontally in other modes. According to the working process, three normal switching sequences are designed: $\tilde{A} \to \tilde{C}_1 \to \tilde{D}_1 \to \tilde{J} \to \tilde{K}, \tilde{A} \to \tilde{B}_1 \to \tilde{E}_2 \to \tilde{F}_2 \to \tilde{J} \to \tilde{K}$, and $\tilde{A} \to \tilde{B}_1 \to \tilde{B}_2 \to \tilde{C}_2 \to \tilde{D}_1 \to \tilde{J} \to \tilde{K}$. A hybrid automaton is constructed as in Fig. 5.7.

Since x in each mode has constant derivatives, all modes are point monotonous with $x^* = [\infty \ \ \infty]^T$, and there is no continuous transition in each mode. It can be checked that each region is determined. The $\mathscr{Q}\mathscr{A}$ is not given, which is the same as $\mathscr{D}\mathscr{A}$. The $\mathscr{D}\mathscr{A}$ of the system is shown in Fig. 5.8, where the $\tilde{K}$ is the target mode, $e_1, \ldots, e_{11}$ are corresponding events. The marked behavior is $L_m(\mathscr{A}) = \{e_1e_2e_3e_4,\ e_5e_6e_7e_8e_4,\ e_5e_9e_{10}e_{11}e_4\}$. The desired specification is that in one working process, modes $\tilde{C}_1$ and $\tilde{D}_1$ must be visited, and $\tilde{D}_1$ is not visited until $\tilde{C}_1$ has been visited. So $B_{id} = e_1e_2e_3e_4$. In the healthy situation, the supervisor $\mathscr{S}$ enables events $\{e_1, e_2, e_3, e_4\}$ while disable others. The initial of the fingertip's position is $(2,5)$. Fig. 5.9 shows the continuous state trajectory, which implies that the work is completed with an ideal behavior.

Now consider two faulty cases of abrupt changes of the fingertip's position, which, as shown in Fig. 5.10, are due to physical faults of the robot arm.

Case 1: The system is switched into $(x_1, x_2) = (0.7, 3)$ of $\tilde{B}_1$ from $(x_1, x_2) = (1.5, 3)$ of $\tilde{C}_1$.

Case 2: The system is switched into $(x_1, x_2) = (-1.5, 3)$ of $\tilde{E}_2$ from $(x_1, x_2) = (1.5, 3)$ of $\tilde{C}_1$.

In Case 1, $\tilde{B}_1$ is activated due to the fault. The conditions of Theorem 5.3 hold. Indeed, there exists a ITFMS satisfying the system specification as in Fig. 5.11(a) $e_1e_{f1}e_9e_{10}e_{11}e_3e_4$. Assume that there is a time delay of $0.2s$ to detect this fault. The ideal fault tolerant supervisor $\mathscr{S}_{id}$ enables events $\{e_1, e_2, e_3, e_4, e_9, e_{10}, e_{11,e_{f1}}\}$, and

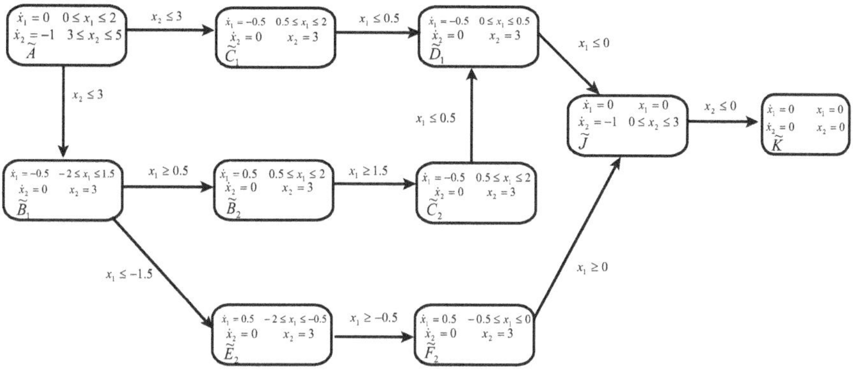

Fig. 5.7 The hybrid automaton of the system

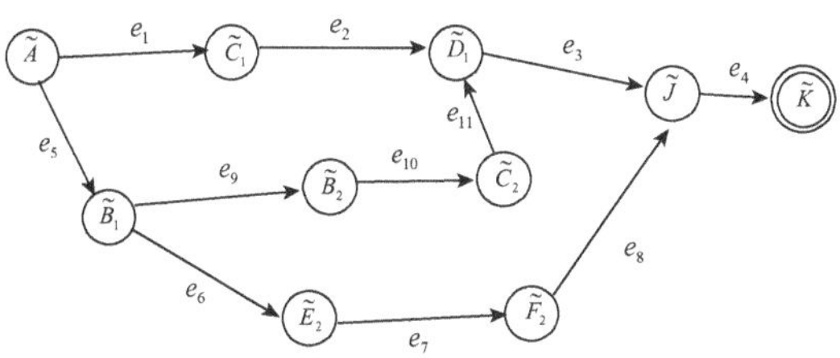

Fig. 5.8 Abstraction of the hybrid automaton

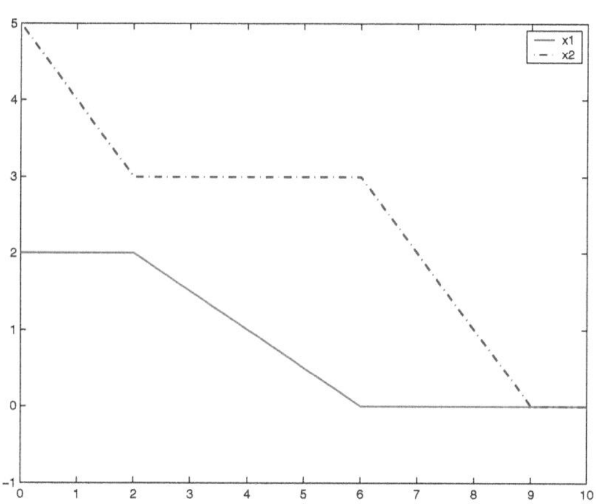

Fig. 5.9 Continuous state trajectory

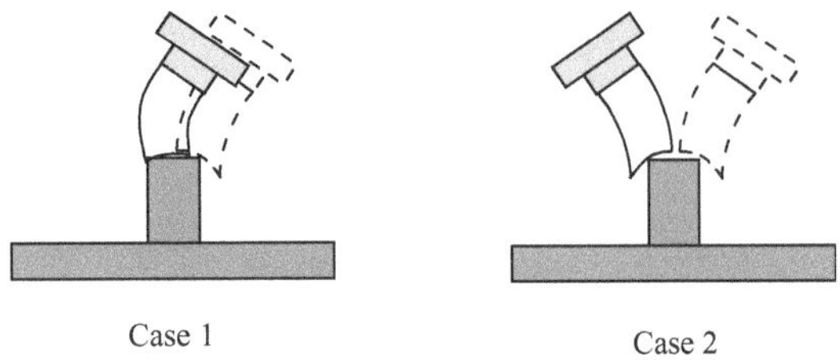

Fig. 5.10 Abrupt change of the fingertip's position

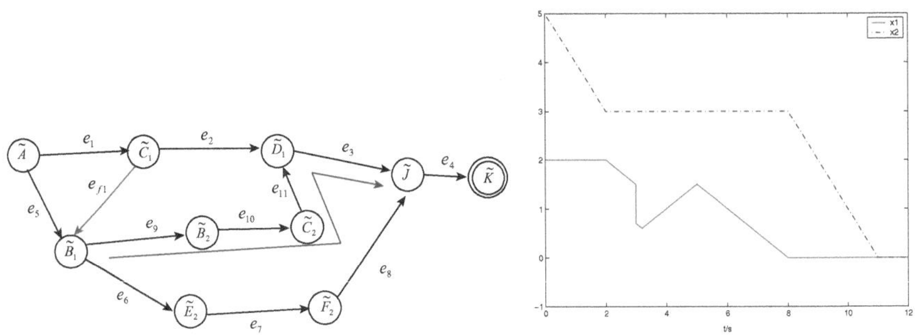

(a) Event sequence in Case 1 (b) Continuous state trajectory in Case 1

Fig. 5.11 FTC performance

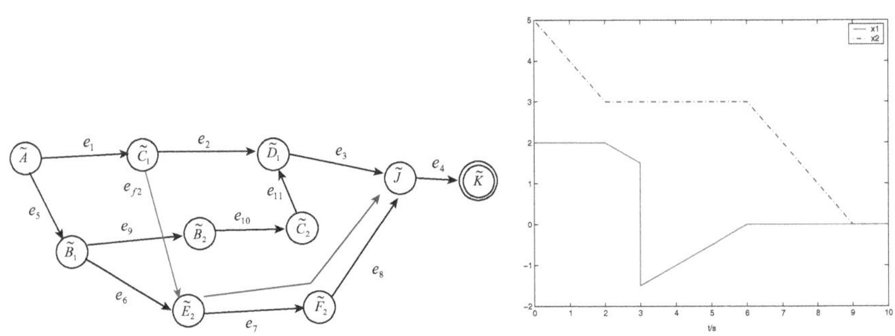

(a) Event sequence in Case 2 (b) Continuous state trajectory in Case 2

Fig. 5.12 FTC performance

set $\mathscr{S}_{update} = \mathscr{S}_{id}$. Fig. 5.11(b) shows the continuous state trajectory, it can be seen that fault tolerance goal is achieved.

In Case 2, the conditions of Theorem 5.3 do not hold, which means that no sequence can satisfy the system specification. However, the conditions of Corollary 5.2 hold. There still exists one and only one ATFMS driving the system to the marked state as in Fig. 5.12(a): $e_1 e_{f2} e_7 e_8 e_4$, which can be considered as the most similar behavior to the ideal one. The acceptable fault tolerant supervisor $\mathscr{S}_a$ enables events $\{e_1, e_2, e_3, e_4, e_7, e_8, e_{f2}\}$, and set $\mathscr{S}_{update} = \mathscr{S}_a$. Fig. 5.12(b) shows the continuous state trajectory, which implies that the fault tolerance goal is achieved with an acceptable behavior.

5.2 FTC via Hybrid Petri Nets

This section proposes a novel FTC scheme for HS modeled by hybrid Petri nets (HPNs). HPNs are widely used for modeling hybrid complex systems [21, 111] e.g., autonomous manufacturing, traffic control, and chemical process. Such model inherits all the advantages of the PNs model such as the ability to capture behaviors including concurrency, synchronization and conflicts.

To the best of our knowledge, until now, only few attempts have been made to FTC for HS modeled by HPNs such as [51], where the FTC goal is to prevent the system from deadlock. However, fruitful results of diagnosis methods for PN can be used as the basis of further FTC researches. In [7], an unfolding based diagnosis approach is provided for asynchronous discrete-event systems. A diagnoser is given based on the concept of basis marking in [41]. An on-line diagnosis method is proposed in [102], where the output information of marking has to be used. The identification scheme developed in [24] relies on full observable events. The method derived in [67] is based on marking variation and causality relationships. In [128], the parity space method is extended to Petri net. In most of these works, it is assumed that either the partial marking is measurable or initial marking is known, such that the current marking just before faults occur can be calculated. Most recently, the marking estimation from event observations with unknown initial marking has been discussed in [39] and [40]. However, these works do not consider faulty behaviors.

The faulty behaviors considered in this work are represented in two forms:

- (F1) Faults produce unobservable and uncontrollable discrete transitions as in [41, 67], which may violate timed-PN's general mutual exclusion constraints (GMEC) that is the basic requirement for system's stability (the former definition of GMEC will be given later), or affect continuous PNs where the optimality should be kept.
- (F2) Faults generate the normal discrete transitions that occur at abnormal time as in [128], which do not violate GMEC, but affect the optimality of continuous PNs.

The faults in continuous PNs, similar to the continuous faults defined in switched system or hybrid automata (See chapters 2-4), will not be addressed in this section.

In this work, we propose a novel hierarchical FTC scheme which consists of two parts: a FTC law in discrete PN and a reconfiguration rule in continuous PNs. The main contributions are as follows:

1. An observer-based FD method is proposed for discrete timed-PN with unknown initial marking and the known initial macromarking (defined later), which estimates the unmeasurable marking in discrete place and meanwhile, diagnoses the fault (F1).

2. Based on the marking estimates, an adaptive FTC scheme is designed for timed-PN with (F1) to maintain the GMEC. The general condition of controller design, that imposes the GMEC is not affected by unobservable transitions, is relaxed.
3. FTC for faults (F2) and (F1) that do not violate GMEC in discrete timed-PN is achieved in continuous PNs by adjusting the firing speed of continuous transitions. Such FTC rule maintains the optimality of the system.
4. Finally, the proposed method is applied to an intelligent transportation system consisting of automated vehicles on a bridge.

5.2.1 Model Setting

We first recall the HPNs formalism. The reader can find a more detailed presentation of HPNs in [21, 111, 30, 31] and PNs in [89]. A HPN structure is the 5-tuple $N = (P, T, Pre, Post, h)$, where P is a set of m places, T is a set of n transitions; The set of places P (resp. transitions T) is split into two subsets: m^d discrete places P^D (resp. n^d discrete transitions T^D) and m^c continuous places P^C (resp. n^c continuous transitions T^C), where $m = m^d + m^c$, $n = n^d + n^c$.

$Pre : P \times T \rightarrow \{\mathbb{R}^+, \forall p_i \in P^C, \text{ or } \mathbb{N}, \forall p_i \in P^D\}$ that assigns a weight to any arc between a transition t_j and its input place p_i, where $\mathbb{R}^+$ denotes the set of positive real numbers, and $\mathbb{N}$ the set of natural numbers. $Post : P \times T \rightarrow \{\mathbb{R}^+, \forall p_i \in P^C, \text{ or } \mathbb{N}, \forall p_i \in P^D\}$ that assigns a weight to any arc between a transition t_j and its output place p_i. The preset and postset of a node $X \in P \cup T$ are denoted ${}^\bullet X$ and $X^\bullet$.

The marking of an HPN is the function $M : P \rightarrow \{\mathbb{R}^+, \forall p_i \in P^C, \text{ or } \mathbb{N}, \forall p_i \in P^D\}$ which assigns a nonnegative integer number of tokens to each discrete places and a nonnegative real number to each continuous place.

The following two transition rules are considered:

Firing of discrete transitions: A discrete transition $t \in T^D$ is enabled, if $M \geq Pre(\cdot, t)$ and may fire yielding $M' = M + C(\cdot, t)$, where $C(p,t) \triangleq Post(p,t) - Pre(p,t)$, $\forall p \in P^D$. Firing of $t_j \in T^D$ lasts d_j time units, where d_j is a nonnegative deterministic number. Denote $M[\omega\rangle M'$ such that the enabled sequence of transitions ω may fire at M yielding M'.

Firing of continuous transitions: A continuous transition $t_j \in T^C$ is enabled if $M(p) \geq Pre(p, t_j), \forall p \in {}^\bullet t_j \cap P^D$, and $M(p) \geq 0, \forall p \in {}^\bullet t_j \cap P^C$. Note that t_j is affected by the discrete places of timed-PN if ${}^\bullet t_j \cap P^D \neq \emptyset$ (this is consistent with our application as shown later). Given two time instants τ and τ', the evolution of the marking is given as $M(p, \tau) = M(p, \tau') + \vartheta(p, t, \tau, \tau')$, $\forall p \in P^C$, where $\vartheta(p, t, \tau, \tau') \triangleq \sum_{t_j \in {}^\bullet p} Post_{p,t_j} \cdot \int_{\tau'}^{\tau} v_{t_j}(s) ds - \sum_{t_k \in p^\bullet} Pre_{p,t_k} \cdot \int_{\tau'}^{\tau} v_{t_k}(s) ds$, v_{t_j} and v_{t_k} denote the firing speeds of t_j and t_k at the time s respectively.

In general, both continuous and discrete transitions may have input and output places that are either continuous or discrete. In this work, we suppose that all discrete input places must also be output places, and vice-versa with arcs of the same weight. The firing of a continuous transition cannot modify the marking of the discrete part (This property will also be illustrated in our application).

Define two sets $P^{DC} = \{p \in P^D | \exists t \in T^C, p \in {}^\bullet t \cap t^\bullet\}$ and $T^{CD} = \{t \in T^C | \exists p \in P^D, t \in p^\bullet \cap {}^\bullet p\}$. A place $p \in P^{DC}$ and a transition $t \in T^{CD}$ are *related* if $p \in {}^\bullet t \cap t^\bullet$

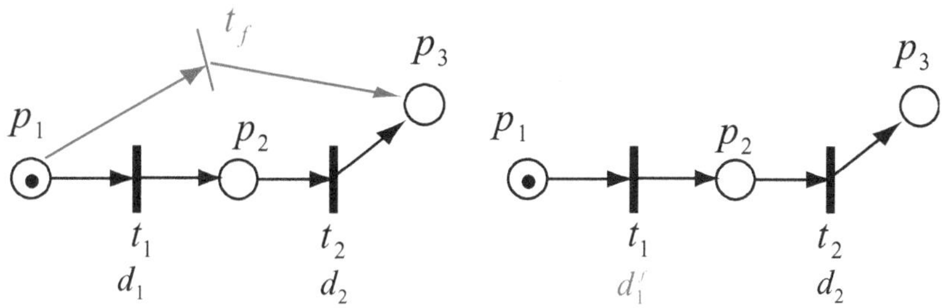

(a) A PN with fault t_f(F1)   (b) A PN with fault t_1 (d_1^f) (F2)

Fig. 5.13 A PN with the fault

($t \in p^\bullet \cap {}^\bullet p$). Note that P^{DC} and T^{CD} describe the relations between discrete part and continuous part of the HPN.

As already described, the fault is defined as two sets T_f and $TIME_f$, where $T^D = T_N \cup T_f$, $T_N \cap T_f = \emptyset$ with T_N the set of normal transitions, T_f the set of faults (F1), that is the set of unobservable and uncontrollable discrete transitions. $TIME_f$ denotes the set of faults (F2), such that the firing of normal transition in T_N lasts abnormal time denoted as d^f. Fig. 5.13 shows a net with these two types of faults.

From a graphical point of view, discrete places are represented by circles, discrete transitions are represented by thick bars (thin bars denote the immediate discrete transitions i.e., $d = 0$) whereas continuous places are represented by double circles and continuous transitions are represented by boxes. Finally, the marking are represented by the dot in places.

Example 5.4 (Example 1.3 revisited): Recall the traffic flow control problem shown in Fig. 1.4. The specification can be described as

P1 *(stable)*: the AVs from different input roads never get into the bridge simultaneously. This is the basic requirement on the initial performance, which must be guaranteed, otherwise the AVs may crash.

P2 *(optimal)*: The AVs flow from different input roads keeps a safe distance with others on the bridge as shown in Fig. 5.14. This is the optimal specification for the safe purpose.

We consider the worst case where all input roads have infinitely long AVs flows. The proposed method can be modified for the better case.

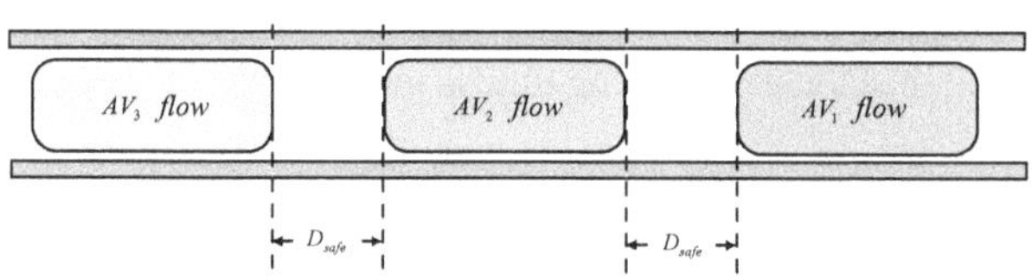

Fig. 5.14 Segment of AVs flow on the bridge

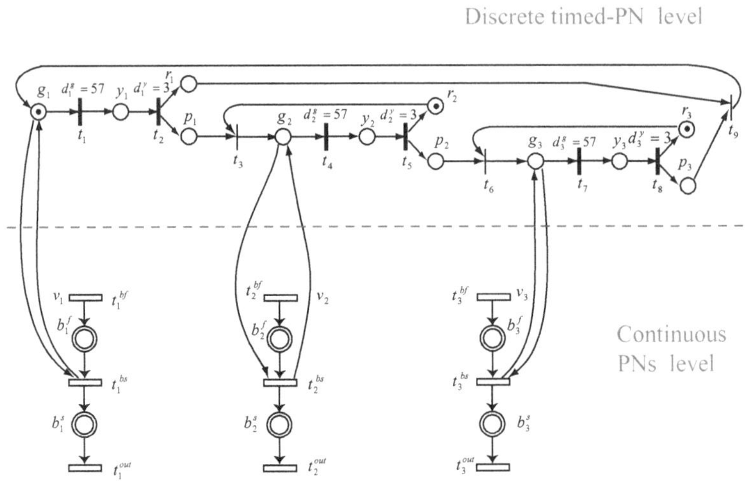

Fig. 5.15 HPN model of AVs transportation of the bridge

The fault considered in this system represents the abnormal behavior of the supervising light, i.e. the logic lights do not work as prescribed. Under such fault, both P1 and P2 may be affected. Note that P2 can be achieved in the presence of local faults in AVs by adjusting the speed of AVs on-line as in [38, 99], where the problem of online-speed adjustment has been intensively investigated, which is not addressed in this work. We assume that, after an AVs flow from one input road accelerates to a speed v_{normal}, it always keeps v_{normal} on the bridge.

The HPN model of the of AVs transporting process of bridge related to Fig.1.4 is shown in Fig.5.15, where a discrete PN illustrates the working process of supervisory lights, and three continuous PNs model the AVs flows from different input roads. A detailed description of places and transitions is given in Table 5.1. Compared with the HPN model in [30, 31], red lights are considered in ours, which is more suitable for fault modeling and FTC design. For more complicated traffic networks as in [4, 25], coloured timed Petri nets can be used to model the vehicle flows.

Note that the first and second parts of the bridge b_i^f and b_i^s are divided as in Fig.1.4, where the length of the first part is equal to $3v_{normal}$, i.e. the distance that AVs pass through in the yellow period. In the healthy case, the initial speed of AVs flows from ith input road is $v_i(\tau) = \min\{a_i(\tau - \tau_i^s), v_{normal}\}$, where a_i is the accelerate speed and τ_i^s is the time when t_i^{bf} starts firing. We assume that $a_i = a, i = 1,2,3$. Since the accelerating time is much less than green period, we further have $D_{safe} = 3v_{normal} + \frac{v_{normal}^2}{2a}$. Moreover, $v_i^{\max}$ represents the maximal speed, and M^b denotes the capacity of $(b_1^f + b_1^s)$, i.e. the full length of the bridge. From physical point of view, b_i^f and b_i^s for $i = 1,2,3$ in Fig. 5.15 represent the same bridge.

Table 5.1 Places and transitions of the HPN in Fig.5.16

Place	Meaning
g_i	green period of AVs flow from i th input road
y_i	yellow period of AVs flow from i th input road
r_i	red period of AVs flow from i th input road
b_i^f	the first part of the bridge from i th input road
b_i^s	the second part of the bridge from i th input road
Transition	**Meaning**
t_i^{bf}	AVs flow get into the first part of the bridge from ith input road
t_i^{bs}	AVs flow get into the second part of the bridge from ith input road
t_i^{out}	AVs flow from ith input road gets out from the bridge

From Fig. 5.15, it is much more clear that the FTC objectives are :

1) (stability) To reconfigure the discrete timed-PN such that at each time, only one green light is activated in the presence of faults (F1).
2) (optimality) To adjust the firing speed of t_i^{bf} such that D_{safe} is kept in the presence of faults (F2).

To make the HPNs mode closer to our application, we impose the following hypothesis throughout Section 5.2.

H1 **(timed-PN).** All $t \in T_N$ are controllable and observable. All $t \in T_f$ are uncontrollable and unobservable. $\forall p \in P^D$, $Pre(p,\cdot) = Post(p,\cdot)$. $M(p)$ is unmeasurable. The initial marking is unknown while the initial macromarking is known.

H2 **(continuous PN).** All $t \in T^{CD}$ are measurable with alterable firing speeds. All $t \in T^C \setminus T^{CD}$ are measurable with fixed firing speeds. $\forall p \in P^C$, $M(p)$ is unmeasurable. $\forall X \in P^C \cup T^C$, both ${}^\bullet X \cap (P^C \cup T^C)$ and $X^\bullet \cap (P^C \cup T^C)$ are singleton.

H3 **(interconnection).** For $p \in P^{DC}$ and $t \in T^{CD}$ that are related, $Pre(p,t) = Post(p,t)$, such that the firing of a continuous transition cannot modify the marking of discrete places.

5.2.2 FD and Marking Estimation

In this section, we consider the problem of FD and observer design in the level of discrete timed-PN. If there is no additional remark, all the PNs, places and transitions discussed in this section are related to the discrete PN.

Denote TS as the set of transition sequences, and TS^o the set of observable transition sequences. Similarly to finite state machine formulation in [107], let $\mathbb{P} : TS \to TS^o$ denote a projection operator that "erases" the unobservable transitions in a transition sequence. The inverse projection operator is defined as

$$\mathbb{P}^{-1}(y) = \{s \in TS : \mathbb{P}(s) = y\}$$

The diagnosability definition of finite state machine in [107] is extended to PN as follows:

Definition 5.10. *A PN is* diagnosable *with respect to $t \in T_f$, if $\exists n \in \mathbb{N}$, and an observable transition sequence ω, such that $\|\omega\| \geq n \Rightarrow t \in \mathbb{P}^{-1}(\mathbb{P}(\psi(t)\omega))$, where $\psi(t)$ denotes the observable sequence that is ended at t, $\|\omega\|$ is the length of the sequence ω.*

The above definition of diagnosability means the following: Let $\psi(t)$ be any transition sequence that is ended at a fault $t \in T_f$, and let ω be any sufficiently long continuation of $\psi(t)$. $t \in \mathbb{P}^{-1}(\psi(t)\omega)$ means that every transition sequence, that produces the same record of observable transitions as the sequence $\psi(t)\omega$, should contain a fault in it. This implies that along every continuation ω of $\psi(t)$, one can detect the occurrence of a fault t with a finite delay (n steps).

Before giving the conditions of diagnosability, the following definitions are also introduced.

Definition 5.11. *Given a PN N, and a subset $T' \subseteq T$ of its transitions, we define the* T'-induced subnet *of N as the new net $N' = (P, T', Pre', Post')$ where Pre' and $Post'$ are the restriction of $Pre, Post$ to T'.*

The net N' can also be obtained from N by removing all transitions in $T \setminus T'$.

Definition 5.12. *An induced subnet of an unobservable transitions subset of a PN is* acyclic *if no oriented cycle of sequences in this PN occurs that contains only unobservable transitions in this subset.*

Definition 5.13. *A PN is* forward conflict *(FC) if there exist two transitions which have at least one common input place. A PN is* backward conflict *(BC) if there exist two transitions which have at least one common output place. A PN is* absolutely conflict *(AC) if it is both FC and BC.*

We also say that a PN is forward (resp. backward) conflict free (FCF (resp. BCF)) if it is not forward (resp. backward) conflict.

Lemma 5.3. *A PN is* diagnosable *with respect to $t \in T_f$, if*

1) T_f-induced subnet is acyclic.
2) T_f-induced subnet N_f is not AC.
3) the initial marking $M_0(p_b) = M_0(p_a) = 0$, where $p_b \in^{\bullet} t$, $p_a \in t^{\bullet}$.
4) ${}^{\bullet}p_a \setminus t$ do not fire before $p_b^{\bullet} \setminus t$ or $p_a^{\bullet}$ fire.
5) After one transition from ${}^{\bullet}p_b$ fired, ${}^{\bullet}p_b$ do not fire again before $p_b^{\bullet}$ fire.

Proof: From the graph point of view, there exist two transition sets ${}^{\bullet}p_b$ and $p_a^{\bullet}$ before and after t. Condition 3) implies that a transition $t_b \in^{\bullet} p_b$ must fire before t since $M_0(p_b) = 0$. Condition 1) means that the occurrence of a fault must be interconnected with the firing of normal transitions. Under the condition 2), three cases are considered as follows:

Case 1: The N_f is FCF and BCF. Since $Pre(p,\cdot) = Post(p,\cdot)$ from H1 and $M_0(p_b) = 0$, Condition 5) implies that if $\exists p \in p_b^\bullet \setminus t$ fires, then t must not occur. On the other hand, $M_0(p_a) = 0$, Condition 4) means that before we determined whether the fault occurs or not, ${}^\bullet p_a \setminus t$ do not fire, i.e., $M_0(p_a)$ do not change due to the firing of ${}^\bullet p_a \setminus t$. Thus t can be diagnosed once $t_a \in p_a^\bullet$ fires.

Case 2: The N_f is FC and BCF. Several faults share one same input place. The fault t may not be identified from t_b, while a smaller region than T_f in which the fault belongs to can be determined. The property of BCF ensures that t can still be diagnosed once t_a fires.

Case 3: The N_f is BC and FCF. Since each fault has one different input place, the fault that may occur can be distinguished from t_b. Although several faults share one same output place, t can be diagnosed once t_a fires. □

Remark 5.2. *The conditions in Lemma 5.3 depend only on the observable transitions, no information of marking is needed [139], while the marking information is required in [102, 67, 128]. The diagnosis procedure derived in [41] also relies on the observable transitions while the initial marking has to be known.*

Remark 5.3. *Lemma 5.3 guarantees that at most one fault from one input place really occurs. Under Conditions 1)-5), the unique fault identification is always achieved. Such diagnosability property is also available when several faults from different places occur simultaneously. Checking 5) may require some external information of macromarking (as shown in our application), if 5) is removed, multiple faults may occur from one input place simultaneously, more restrictive conditions need to be imposed.*

The purpose of the observer design for discrete timed-PN is to provide the marking estimates in discrete places in the presence of faults. The partial information of the initial marking in discrete places is available in the form of macromarking defined as follows.

Definition 5.14. *Assume that the set of places P^D can be written as the union of $r+1$ subsets: $P^D = P_0 \cup P_1 \cup \ldots \cup P_r$ such that $P_0 \cap P_j = \emptyset$, $\forall j > 0$. The number of tokens contained in $P_j (j > 0)$ is known to be b_j, while the number of tokens in P_0 is unknown. For each P_j, let v_j be its characteristic vector, i.e., $v_j(p) = 1$ if $p \in P_j$, else $v_j(p) = 0$. Let $V = [v_1, \ldots, v_r]$ and $b = [b_1, \ldots, b_r]$. The* macromarking *is defined as the set $\mathscr{V}(V,b) = \{M \in \mathbb{N}^{m^d} | V^T M = b\}$.*

The following definition describes *consistent markings* as in [39].

Definition 5.15. *After the transition sequence ω has been observed, we define the set of ω-*consistent markings *$\mathscr{C}(\omega) = \{M | \exists M \in \mathbb{N}^{m^d}, M'[\omega\rangle M\}$ as the set of all markings in which the system may be given the observed behavior and the initial macromarking.*

Denote $\chi(\mathbf{t})$ as the set of all transition sequences that $\mathbf{t}$ may follow, with $\mathbf{t} = \{t_1, t_2, ..., t_q\} \in T_f$.

Assumption 5.1. *If $\omega_i \in \chi(\mathbf{t})$, then all the faults in* **t** *share the same input place.*

Assumption 5.1 means that once the faults from one input place occurred and have not been diagnosed, the faults from other input places do not occur. That is to say we just take into account the possible faults from one input place before they are diagnosed.

Based on the conditions in Lemma 5.3 and Assumption 5.1, the following algorithm provides the marking estimates in the form of consistent markings iteratively in spite of faults.

Algorithm 5.1. *marking estimation with event observation, initial macromarking and faults*

1. *Let the initial estimate $M^e_{\omega_0}(p) = 0$, the initial complementary estimates $M^c_{\omega_0} = M^e_{\omega_0}$.*
2. *Let the initial bound $B_{\omega_0} = b - V^T M^e_{\omega_0}$, the initial complementary bound $B^c_{\omega_0} = B_{\omega_0}$.*
3. *Let $i = 1$.*
4. *Wait until $t_{\alpha i}$ fires.*
   If for $i \geq 2$, $t_{\alpha i} \in t_j^{\bullet\bullet}$ where $t_j \in T_f$, then
   $M^e_{\omega_{i-1}} = M^{cj}_{\omega_i}$, $B_{\omega_{i-1}} = B^{cj}_{\omega_i}$, go to 6.
   end if.
5. *If for $i \geq 2$, $\omega_i \in \chi(\mathbf{t})$ then*
   Let $M'_{\omega_i}(p) = \max\{M^e_{\omega_{i-1}}(p), Pre(p, t_{\alpha i})\}$,
   Let $M^e_{\omega_i} = M'_{\omega_i} + C(\cdot, t_{\alpha i})$, $B_{\omega_i} = B_{\omega_{i-1}} - V^T \cdot (M'_{\omega_i} - M^e_{\omega_{i-1}})$.
   Let $M^{cj'}_{\omega_{i+1}}(p) = \max\{M^{cj}_{\omega_i}(p), Pre(p, t_{\alpha i})\}$,
   Let $M^{cj}_{\omega_{i+1}} = M^{cj'}_{\omega_{i+1}} + C(\cdot, t_{\alpha i})$, $B^{cj}_{\omega_{i+1}} = B^{cj}_{\omega_i} - V^T \cdot (M^{cj'}_{\omega_{i+1}} - M^{cj}_{\omega_i})$, go to 9.
   end if.
6. *Let $M'_{\omega_i}(p) = \max\{M^e_{\omega_{i-1}}(p), Pre(p, t_{\alpha i})\}$.*
7. *Let $M^e_{\omega_i} = M^c_{\omega_i} = M'_{\omega_i} + C(\cdot, t_{\alpha i})$, $B_{\omega_i} = B^{cj}_{\omega_i} = B_{\omega_{i-1}} - V^T \cdot (M'_{\omega_i} - M^e_{\omega_{i-1}})$.*
8. *If $\exists \bar{p} \in t^{\bullet}_{\alpha i}$, and $t_1, \ldots, t_q \in T_f$, such that $\bar{p} \in {}^{\bullet} t_j, (1 \leq j \leq q)$ then*
   For $1 \leq j \leq q$
   Let $M^{cj'}_{\omega_{i+1}}(\bar{p}) = \max\{M^e_{\omega_i}(\bar{p}), Pre(\bar{p}, t_j)\}$.
   Let $M^{cj}_{\omega_{i+1}} = M^{ci'}_{\omega_{i+1}} + C(\cdot, t_j)$, $B^{cj}_{\omega_{i+1}} = B^{cj}_{\omega_i} - V^T \cdot (M^{ci'}_{\omega_{i+1}} - M^{cj}_{\omega_i})$
   End for.
   Let $M^c_{\omega_i} = \bigcup M^{cj}_{\omega_i}$, $B^c_{\omega_i} = \bigcup B^{cj}_{\omega_i}$.
   end if.
9. *Let $i = i + 1$, go to 4.* ■

Algorithm 5.1 extends the algorithm in [39] to the faulty case. Its novelty is the utilization of complementary estimates. The main idea behind Algorithm 5.1, as shown in Fig. 5.16, is that, when we predict that a fault may occur at next transition (steps 8 and 5), we consider all the possible markings that may be reached under this fault, which are recorded in the complementary marking estimate M^c_ω. When we determine that the fault has occurred (Step 4), M^c_ω will be used to update the marking estimate M^e_ω. Otherwise, M^e_ω will not be changed (steps 6 and 7).

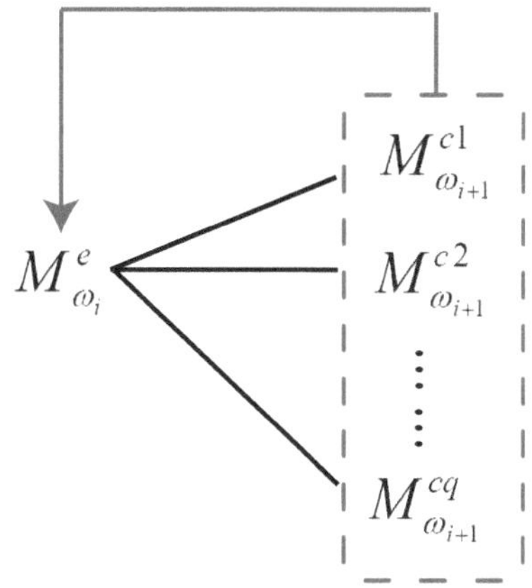

Fig. 5.16 Marking estimation of Algorithm 5.1

Now we analyze the computational complexity. Once a transition is observed, Algorithm 5.1 not only updates M^e_ω, B_ω as the algorithm in [39], but also updates complementary estimates M^c_ω and B^c_ω. The number of operations required depends on how many times the *for* cycle in Step 8 is executed. Both the number of M^c_ω and B^c_ω are q, the complexity is $\mathscr{O}(3 \times |q|)$.

Remark 5.4. *Algorithm 5.1 can also be extended to the case of faults from multiple input places. Suppose that there are N input places such that the faults from these places may occur simultaneously, or before a fault is diagnosed, the faults from other $N-1$ input places may occur. Assume that the ith place may fire q_i possible faults, $1 \le i \le N$. In this case, to consider all possible faults, $(\prod_{i\in[1,N]}(q_i+1)-1)$ complementary marking estimates have to be used, the complexity becomes $\mathscr{O}\Big(3 \times (|\prod_{i\in[1,N]} q_j+1|-1)\Big)$. In the following discussion, Assumption 5.1 always holds, i.e., we only consider the case of faults from the single input place.*

The set of consistent markings provided by Algorithm 5.1 is as follows.

Theorem 5.4. *Supposed that Assumption 5.1 and Conditions 1)-5) in Lemma 5.3 hold. Consider a PN with initial macromarking $\mathscr{V}(V,b)$, an observed transition sequence ω_i, the fault transition $t \in T_f$, and $M^e_{\omega_i}$, B_{ω_i} $M^c_{\omega_{i+1}}$, $B^c_{\omega_{i+1}}$ computed by Algorithm 5.1. The set of ω_i-consistent markings is*

$$\mathscr{C}(\omega_i|V,b) = \begin{cases} \mathscr{C}_1 & \text{if } \omega_i \notin \chi(\mathbf{t}) \\ \mathscr{C}_1 \cup \mathscr{C}_2, & \text{if } \omega_i \in \chi(\mathbf{t}) \end{cases} \tag{5.4}$$

where

$$\mathscr{C}_1 \triangleq \left\{M \in N^{m^d} | V^T M = V^T M^e_{\omega_i} + B_{\omega_i}, M \ge M^e_{\omega_i}\right\}$$

$$\mathscr{C}_2 \triangleq \left\{M \in N^{m^d} | V^T M = V^T M^{cj}_{\omega_{i+1}} + B^{cj}_{\omega_{i+1}}, M \ge \min_j \{M^{cj}_{\omega_{i+1}}\}\right\}$$

Proof: For the case $\omega_i \notin \chi(\mathbf{t})$, i.e., no fault occurs, the proof is similar to [39], and thus is omitted.

For the case $\omega_i \in \chi(\mathbf{t})$, we first consider that the T_f-subnet is FCF, i.e., only one possible fault t may occur after ω_i. In this case, ω_i-consistent markings $\mathscr{C}(\omega_i|V,b)$ should include the marking that may be reached under $\omega_i t$. This can be provided by $M^c_{\omega_i}$ and $B^c_{\omega_i}$ as follows.

Steps 6 and 7 in Algorithm 5.1 ensure $M^e_{\omega_i} = M^c_{\omega_i}$ and $B_{\omega_i} = B^c_{\omega_i}$ before t occurs. Let us show that $\mathscr{C}(\omega_i|V,b) = \{M \in N^{m^d} | V^T M = V^T M^c_{\omega_i} + B^c_{\omega_i}, M \geq M^e_{\omega_i}\} \Rightarrow \mathscr{C}(\omega_i t|V,b) = \{M \in N^{m^d} | V^T M = V^T M^c_{\omega_{i+1}} + B^c_{\omega_{i+1}}, M \geq M^e_{\omega_{i+1}}\}$.

In fact,

$$\begin{aligned}\mathscr{C}(\omega_i t|V,b) &= \{M \in N^{m^d} | \exists M' \in \mathscr{C}(\omega_i|V,b), M' \geq Pre(\cdot,t), M = M' + C(\cdot,t)\} \\ &= \{M \in N^{m^d} | \exists M', V^T M' = V^T M^c_{\omega_i} + B^c_{\omega_i}, \\ &\qquad M' \geq M^e_{\omega_i}, M' \geq Pre(\cdot,t), M = M' + C(\cdot,t)\}\end{aligned}$$

which together with the step 8 of Algorithm 5.1, leads to $M' \geq M^{c\prime}_{\omega_i}$. We further have from the step 8 that $V^T M^c_{\omega_i} + B^c_{\omega_i} = V^T M^{c\prime}_{\omega_{i+1}} + B^c_{\omega_{i+1}}$. Therefore,

$$\begin{aligned}\mathscr{C}(\omega_i t|V,b) &= \{M \in N^{m^d} | \exists M', V^T M' = V^T M^{c\prime}_{\omega_{i+1}} + B^c_{\omega_{i+1}}, \\ &\qquad M' \geq M^e_{\omega_i}, M = M' + C(\cdot,t)\} \\ &= \{M \in N^{m^d} | V^T M = V^T M^c_{\omega_{i+1}} + B^c_{\omega_{i+1}}, M \geq M^e_{\omega_{i+1}}\}\end{aligned}$$

For the case that the T_f-subnet is FC, it can be seen from the analysis above that $\mathscr{C}(\omega|V,b)$ defined in (5.5) includes all markings that may be reached by any fault t_j. Once we determined whether the fault occurs or not from Lemma 5.3, $\mathscr{C}(\omega|V,b)$ will be updated as in Algorithm 5.1, which always gives the set of all markings in which the system may be given the observed behavior. This completes the proof. □

Some properties about the observer of Algorithm 5.1 can also be discussed similar to [39].

Proposition 5.1. *Let ω_i be an observed transition sequence. Under Assumption 5.1 and Conditions 1)-5), the estimate computed by Algorithm 5.1 is a lower bound of actual marking. i.e., $\forall i, \min\{M^e_{\omega_i}, \min_j\{M^{cj}_{\omega_{i+1}}\}\} \leq M_{\omega_i}$.*

Proof: If $\omega_i \notin \chi(\mathbf{t})$, it holds that $M^e_{\omega_i} \leq M_{\omega_i}$, the proof is similar to [39], and thus is omitted.

If $\omega_i \in \chi(\mathbf{t})$, we consider two cases:

Case 1: $t_j \in \mathbf{t}$ really occurs. Before t_j is diagnosed, according to Step 8 of Algorithm 5.1, we can prove that $\min_j\{M^{cj}_{\omega_{i+1}}\} \leq M_{\omega_i}$ using the result of [39]. Once t_j has been diagnosed, from Step 4 of Algorithm 5.1, we have $M^e_{\omega_i} = M^{cj}_{\omega_{i+1}}$, which further leads to $M^e_{\omega_i} = M^{cj}_{\omega_{i+1}} \leq M_{\omega_i}$ from [39].

Case 2: $t_j \in \mathbf{t}$ does not occur. Algorithm 5.1 guarantees that $M^e_{\omega_i}$ is not affected by the fault if it does not occur. Thus we obtain $M^e_{\omega_i} \leq M_{\omega_i}$. □

Denote two positive numbers ε_1 and ε_2 such that, the sequence ω_{ε_1} may be followed by the fault, and the sequence ω_{ε_2} determines whether the fault occurs or not. The following proposition gives the estimating convergence property.

Proposition 5.2. *Given M_{ω_i} and $M^e_{\omega_i}$, under Assumption 5.1 and Conditions 1)-5), the estimation error $e(M_{\omega_i}, M^e_{\omega_i}) = \sum_{p\in P^D}(M_{\omega_i}(p) - M^e_{\omega_i}(p))$ has the following property:*

$$\begin{cases} e(M_{\omega_i}, M^e_{\omega_i}) \geq e(M_{\omega_{i+1}}, M^e_{\omega_{i+1}}) \text{ for } 0 \leq i \leq \varepsilon_1 - 1 \text{ and } \varepsilon_2 \leq i \\ e(M_{\omega_{\varepsilon_1}}, M^e_{\omega_{\varepsilon_1}}) \geq e(M_{\omega_{\varepsilon_2}}, M^e_{\omega_{\varepsilon_2}}) \end{cases} \tag{5.5}$$

Proof: For $0 \leq i \leq \varepsilon_1 - 1$ and $\varepsilon_2 \leq i$, no fault may occur or we have determined whether the fault occurs or not, following the same procedure as in [39], we can prove that $e(M_{\omega_i}, M^e_{\omega_i}) \geq e(M_{\omega_{i+1}}, M^e_{\omega_{i+1}})$, i.e., the estimation error is monotonically nonincreasing. During the interval between the sequences ω_{ε_1} and ω_{ε_2}, Algorithm 5.1 leads to that $M^e_{\omega_i}$ is not affected by the fault before we determine whether the fault occurs or not. If the fault t_j occurs and has been diagnosed, $M^e_{\omega_i}$ is set to the same as $M^{cj}_{\omega_{i+1}}$. Note that $M^{cj}_{\omega_{i+1}}$ is updated according to t_j, thus $e(M_{\omega_{\varepsilon_1}}, M^e_{\omega_{\varepsilon_1}}) \geq e(M_{\omega_{\varepsilon_2}}, M^e_{\omega_{\varepsilon_2}})$. □

Algorithm 5.1 just relies on the observation of discrete transitions. Thanks to the structure of HPNs defined in Section 5.2.1, the following algorithm shows that the information of continuous transitions can help to estimate the marking in discrete places.

Algorithm 5.2. *marking estimation with additional observation of continuous transitions*

1. *Let the initial estimate $M^e_{\omega_0}(p) = 0$*
   If $\exists \bar{t} \in T^{CD}$ related to $\bar{p} \in P^{DC}$ is firing, then
   Let $M^e_{\omega_0}(\bar{p}) = Pre(\bar{p}, \bar{t})$
   end if.
   Let the initial complementary estimates $M^c_{\omega_0} = M^e_{\omega_0}$.
2. *Let the initial bound $B_{\omega_0} = b - V^T M^e_{\omega_0}$, the initial complementary bound $B^c_{\omega_0} = B_{\omega_0}$.*
3. *Let $i = 1$.*
4. *Wait until $t_{\alpha i}$ fires.*
   If for $i \geq 2$, $t_{\alpha i} \in t_j^{\bullet\bullet}$, then
   $M^e_{\omega_{i-1}} = M^{cj}_{\omega_i}$, $B_{\omega_{i-1}} = B^{cj}_{\omega_i}$, go to 6.
   end if.
5. *If for $i \geq 2$, $\omega_i \in \chi(\mathbf{t})$ then*
   Let $M'_{\omega_i}(p) = \max\{M^e_{\omega_{i-1}}(p), Pre(p, t_{\alpha i})\}$, $M^e_{\omega_i} = M'_{\omega_i} + C(\cdot, t_{\alpha i})$,
   Let $M^{cj\prime}_{\omega_{i+1}}(p) = \max\{M^{cj}_{\omega_i}(p), Pre(p, t_{\alpha i})\}$, $M^{cj}_{\omega_{i+1}} = M^{cj\prime}_{\omega_{i+1}} + C(\cdot, t_{\alpha i})$,

If $\exists \bar{t} \in T^{CD}$ *related to* $\bar{p} \in P^{DC}$ *fires, then*
Let $M^e_{\omega_i}(\bar{p}) = \max\{M^e_{\omega_i}(\bar{p}), Pre(\bar{p},\bar{t})\}$, $M^{cj'}_{\omega_{i+1}}(\bar{p}) = \max\{M^{cj'}_{\omega_{i+1}}(\bar{p}), Pre(\bar{p},\bar{t})\}$,
end if.
Let $B_{\omega_i} = B_{\omega_{i-1}} - V^T \cdot (M'_{\omega_i} - M^e_{\omega_{i-1}})$, $B^{cj}_{\omega_{i+1}} = B^{cj}_{\omega_i} - V^T \cdot (M^{cj'}_{\omega_{i+1}} - M^{cj}_{\omega_i})$, *go to 9.*
end if.
6. *Let* $M'_{\omega_i}(p) = \max\{M^e_{\omega_{i-1}}(p), Pre(p,t_{\alpha i})\}$.
7. *Let* $M^e_{\omega_i} = M^c_{\omega_i} = M'_{\omega_i} + C(\cdot,t_{\alpha i})$, $B_{\omega_i} = B^{cj}_{\omega_i} = B_{\omega_{i-1}} - V^T \cdot (M'_{\omega_i} - M^e_{\omega_{i-1}})$.
8. *If* $\exists \bar{p} \in t^\bullet_{\alpha i}$, *and* $t_1,\ldots,t_q \in T_f$, *such that* $\bar{p} \in^\bullet t_j, (1 \le j \le q)$ *then*
Let $M^{cj'}_{\omega_{i+1}}(\bar{p}) = \max\{M^e_{\omega_i}(\bar{p}), Pre(\bar{p},t_j)\}$.
Let $M^{cj}_{\omega_{i+1}} = M^{ci'}_{\omega_{i+1}} + C(\cdot,t_j)$, $B^{cj}_{\omega_{i+1}} = B^{cj}_{\omega_i} - V^T \cdot (M^{ci'}_{\omega_{i+1}} - M^{cj}_{\omega_i})$
Let $M^c_{\omega_i} = \bigcup M^{cj}_{\omega_i}$, $B^c_{\omega_i} = \bigcup B^{cj}_{\omega_i}$.
end if.
9. *Let* $i = i+1$, *go to 4.* ■

Algorithm 5.2 provides the set of ω_i-consistent markings in the same form as (5.5), which, however, is more accurate than, or at least as accurate as that computed by Algorithm 5.1, since in Algorithm 5.2, the observation of continuous transitions may increase M_ω which is closer to the actual marking.

5.2.3 FTC Design

We first give the definition of generalized mutual exclusion constraints (GMEC) for discrete timed-PN that had been considered in [40, 85, 53].

Definition 5.16. *Given an integer matrix* $L = [l_1 \ldots l_s]$ *with* $l_j \in \mathbb{Z}^{m^d}$ *and an integer vector* $k = [k_1,\ldots,k_s]$ *with* $k_j \in \mathbb{Z}$, *a* GMEC *of the discrete timed-PN* (L,k) *defines the set of legal markings* $\mathcal{L} = \{M \in \mathbb{N}^{m^d} | L^T \cdot M \le k\}$.

For the two FTC objectives of our application described in Section 5.2.1, i.e. stability and optimality, we consider three sets of markings:

A set of *forbidden markings* $\mathcal{F} = \{M \in \mathbb{N}^{m^d} | M \notin \mathcal{L}\}$.
A set of *ideal markings* $\mathcal{L}_i$ that is optimal for system's normal operation.
A set of *unideal markings* $\mathcal{L}_u$ that is non-optimal for system's normal operation.

It is clear that $\mathcal{L}_i \cap \mathcal{L}_u = \emptyset$, $\mathcal{F} \subseteq \mathcal{L}_u$. Forbidden markings violate $\mathcal{L}$, which must be prevented from being reached (e.g., in the AVs transportation process, no more than one green light can be activated simultaneously). The FTC for such forbidden markings is designed at timed-PN level. Unideal markings may affect the performance of continuous PNs, the related FTC problem will be considered at continuous PNs level.

Before an adaptive FTC scheme is designed for time-PN, the following assumption is given.

Assumption 5.2. *The initial actual marking in discrete places* $M_0 \in \mathcal{L}$.

Note that Assumption 5.2 is quite general, if the initial situation violates the GMEC, the system would be destroyed at the beginning.

Algorithm 5.3. *Computation of the PN based FTC law using observer*

1. *Given the observed ω_i, solve for each $j(1 \leq j \leq s)$ the IPP*

$$\begin{cases} \max L_j^T \cdot M \\ s.t. \\ M \in \mathscr{C}(\omega_i|V,b) \\ M \in \mathscr{L} \end{cases} \tag{5.6}$$

   and let h_j be its optimal solution.

2. *Update the FTC controller with*

$$\begin{cases} C_{cj} = -L_j C \\ M_{cj} = k_j - h_j \end{cases} \tag{5.7}$$

   where C_{cj} and M_{cj} denote the row j of the incidence matrix and the element j of markings of the controller.

3. *Let $i = i+1$, go to 1.* ■

Remark 5.5. *Compared with the logical control design in [40] and [50], the control law (5.7) is based on place invariants [85], which is updated based on the consistent markings of the observer at each time when a normal discrete transition fires, and disables some controllable discrete transitions such that $\mathscr{F}$ is never reached, and does not require separate computation as in [50].*

Theorem 5.5. *Supposed that Assumptions 5.1, 5.2 and Conditions 1)-5) in Lemma 5.3 hold. The controller (5.7) guarantees that $\mathscr{F}$ is never reached in spite of fault $t \in p^\bullet$, if*

$$M_{\omega_i t_j} \in \mathscr{L}, \forall t_j \in p^\bullet \tag{5.8}$$

Proof: Since $M_0 \in \mathscr{L}$ from Assumption 5.2, and the fault does not occur as the first transition from Lemma 5.3, based on the result in [85], the controller (5.7) ensures that $M_{\omega_1} \in \mathscr{L}$.

As for $i \geq 2$, assume that t may follow ω_i, condition (5.8) guarantees that once a fault from input place p occurs, the GMEC is still not violated. On the other hand, under Assumption 5.1, only the faults from one input place is considered before it is determined to occur or not. So the controller (5.7) only disables the controllable normal transition rather than the fault transitions at each step. From Theorem 5.4, $\mathscr{C}(\omega_i|V,b)$ includes all markings that may be reached by possible faults after observed ω_i, which together with the result in [85], leads to that $\mathscr{F}$ is never reached in spite of faults. □

Remark 5.6. *The condition (5.8) is less restrictive than the general condition in most literature (see for instance [85],[53]), where $L \cdot C(\cdot, t_{uo}) = 0$, i.e., the unobservable transition $t_{uo} \in T_{uo}$ can not change the markings in places that are related to the GMEC. Our method can be applied even if $L \cdot C(\cdot, t) \neq 0$ for $t \in T_f$ as shown in the application.*

If T_f-subset is FC, i.e. some faults share the same input discrete place, then $\mathscr{C}(\omega_i|V,b)$ has to include more possible markings, which would lead to more restrictive controller. The following result can help to analyze the permissiveness of the controller.

Proposition 5.3. *Suppose that the conditions in Theorem 5.5 hold. Let* $\mathscr{C}(\omega_i)_{FC}$*,* $\mathscr{C}(\omega_i)_{FCF}$ *be two sets of* ω_i*-consistent markings under FC and FC free* T_f*-subsets respectively, and the same observable subset. The controller (5.7) based on* $\mathscr{C}(\omega_i)_{FCF}$ *is at least as permissive as that based on* $\mathscr{C}(\omega_i)_{FC}$*.*

Proof: For all ω_i, Theorem 5.4 implies that $\mathscr{C}(\omega_i)_{FCF} \subseteq \mathscr{C}(\omega_i)_{FC}$, it follows that $h_{jFCF} \leq h_{jFC}$, where h_{jFCF} and h_{jFC} denote the solutions of Algorithm 5.3 with $\mathscr{C}(\omega_i)_{FCF}$ and $\mathscr{C}(\omega_i)_{FC}$ respectively, which, together with (5.7), leads to $M_{cjFCF} \leq M_{cjFC}$ i.e., the marking in control places under $\mathscr{C}(\omega_i)_{FCF}$ is equal to or less than that under $\mathscr{C}(\omega_i)_{FC}$. Based on the result in [85], it holds that more controllable transitions may be disabled under $\mathscr{C}(\omega_i)_{FC}$. This completes the proof. □

Remark 5.7. *The observer-based controller may be more restrictive than that obtained when the actual marking is known, which may lead to a deadlock, under such case, the concept of* Siphon *can be used to prevent the PN from the deadlock as in [51],[40] and [53].*

Even the GMEC $\mathscr{L}$ is satisfied, the *unideal* markings may affect the optimality of the continuous PN, e.g., in our transportation system, this corresponds to the case where lights do not switch following the prescribed sequence, such that before the flow from one input road completely passes through the first part of the bridge, the flow from another input road gets in, this makes the distance between vehicles less than D_{safe}. We discuss the FTC problem at the level of continuous PN in this section.

The reconfiguration of continuous PN is achieved by adjusting the firing speed of transitions $t \in T^{CD}$ as shown in the following algorithm. Two time instants τ_i^s and τ_i^e denote respectively, when the transition t_i starts firing and ends firing.

Algorithm 5.4. *Reconfiguration of continuous PN using observer*

1. *Given the current time instant* τ_0.
   If the transition $t_1 \in T^{CD}$ *starts firing at* τ_0 *then*
   Find the transition $t_2 \subset T^{CD}$ *that fired most recently, capture the time information* $(\tau_2^{\bullet\bullet})^s$, $(\tau_2^{\bullet\bullet})^e$, *go to 2*
   else, go to 5.
   end if.
2. *If* $(\tau_2^{\bullet\bullet})^e \leq \tau_0$ *then*
   Go to 3,
   else go to 4,
   end if.
3. *If the equations*

$$\begin{cases} \frac{1}{2}a \cdot (t')^2 + at't'' = M^b - D_{safe} \\ t' + t'' = \frac{M^b - v_2 \cdot d_i^f}{v_2} \end{cases}$$

have the positive solutions t' *and* t'', *then*
Set the firing speed $v_1(\tau) = \min\{a(\tau - \tau_0), at', v_1^{\max}\}$,
else set $v_1(\tau) = \min\{a(\tau - \tau_0), v_1^{\max}\}$
end if, go to 5.

4. *Let* $v_1 = 0$ *until the firing of* $t_2^{\bullet\bullet}$ *ends. Then let* $v_1(\tau) = \min\{a(\tau - (\tau_2^{\bullet\bullet})^e), v_2\}$ *after the time* $\tau = (\tau_2^{\bullet\bullet})^e$.
5. *Go to 1.* ■

Algorithm 5.4 can always be applied without consideration whether there is a fault or not. For the faulty status that makes the distance larger, Step 3 accelerates the firing speed v_1 such that the distance converges to D_{safe}. On the other hand, for the status that shortens such distance, Step 4 sets v_1 to zero until $t_2^{\bullet\bullet}$ end firing. These two schemes guarantee the optimality of continuous PN related to our application.

Example 5.4 (continued): Now we apply the proposed method to the intelligent transportation process of AVs on the bridge. Let us come back to the HPN model in Fig. 5.15. It can be obtained that $\mathcal{L} = \{M \in \mathbb{N}^{12} | M(g_1) + M(g_2) + M(g_3) \leq 1\}$, i.e., only one green light can be activated at one time. $\mathcal{L}_i = \{M \in \mathbb{N}^{12} | M(g_i) + M(r_j) + M(r_h) = 3, M(y_p) + M(r_q) + M(r_m) = 3, i \neq j \neq h, p \neq q \neq m$, with the green light sequence $g_1 \rightarrow g_2 \rightarrow g_3 \rightarrow g_1$, and $d_i^g = 57s, d_i^y = 3s, d_i^r = 120s\}$, this means that if one green light or one yellow light is activated, the other two should be red lights. We also suppose that if more than one green light can be activated simultaneously, the green light that satisfies the ideal marking set is chosen to avoid the conflict. In the simulation, the firing speed is $v_{normal} = 8m/s$, the acceleration of each AVs flow at beginning is $a = 2m/s^2$, the length of the bridge is $4855m$.

Let us first consider the faulty-free case to show the performance of observer-based controller. The macromarking is

$$\begin{cases} M(g_1) + M(y_1) + M(r_1) = 1 \\ M(g_2) + M(y_2) + M(r_2) = 1 \\ M(g_3) + M(y_3) + M(r_3) = 1 \end{cases} \tag{5.9}$$

The initial marking is

$$M(g_1)M(y_1)M(r_1)M(p_1)M(g_2)M(y_2)M(r_2)M(p_2)M(g_3)M(y_3)M(r_3)M(p_3) = (100000100010)$$

which is unknown. The system is initialized when t_1^{pf} fires, i.e., the AVs flow from the first input roads is getting into the bridge. The firing of t_1^{pf} can help to estimate the marking. Fig. 5.17 shows the evolution of the estimation based on Algorithm 5.2, which shows that the estimate is the low bound of actual marking, and equal to the actual marking after t_6 fires, which verifies propositions 5.1 and 5.2. The Fig. 5.18 shows the controller designed from Algorithm 5.3. In the healthy case, the marking always belongs to $\mathcal{L}_i$. Fig. 5.19 illustrates the AVs flows on the bridge, where the accelerating behavior is not reflected. We can see that the AVs flows from three input roads keep the prescribed distance $D_{safe} = 40m$ with each other.

estimates / actual marking

(100000000000/100000100010)

↓ t_1

(010000000000/010000100010)

↓ t_2

(001100000000/001100100010)

↓ t_3

(001010000000/001010000010)

⋮

↓ t_6

(001000101000/001000101000)

Fig. 5.17 Marking estimation

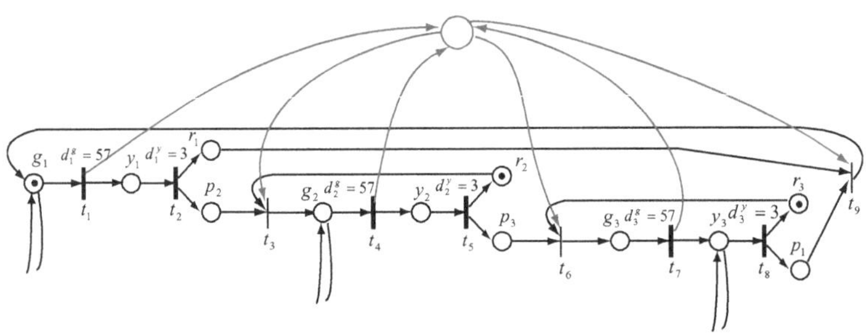

Fig. 5.18 Timed-PN in the healthy case

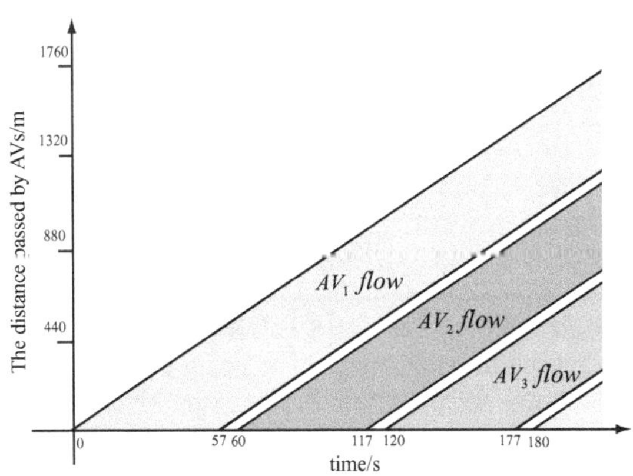

Fig. 5.19 The AVs flow on the bridge in the healthy case

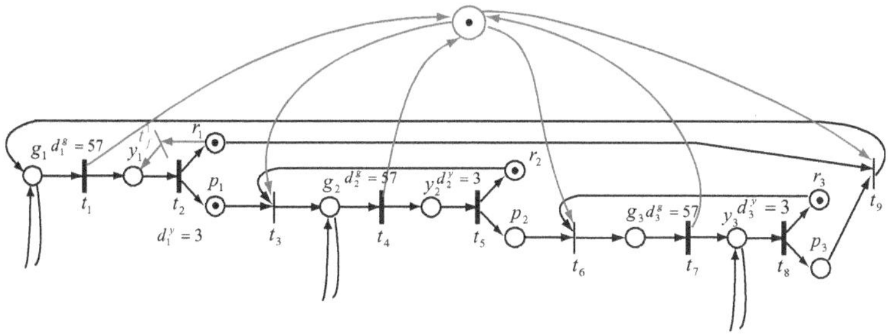

Fig. 5.20 Timed-PN in the faulty case 1

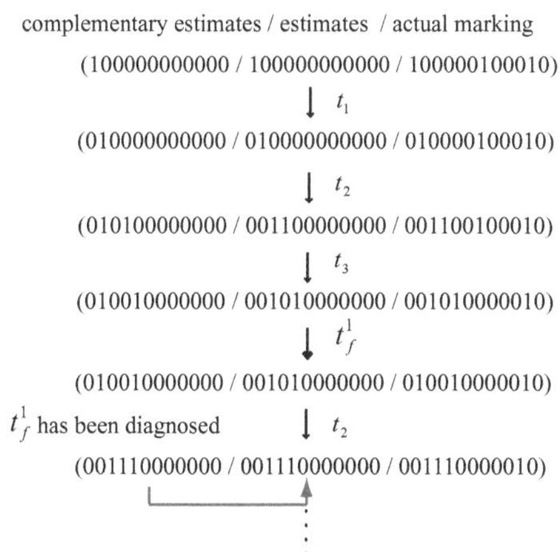

Fig. 5.21 Marking estimation in the faulty case 1

The following 4 faulty cases are simulated:

Case 1: $\exists t_f^1 \in T_f$ as shown in Fig. 5.20. In this case, after t_2 fired, more consistent markings have to be provided. Note that Assumption 5.1 is satisfied, since after t_2 fired, t_2 is impossible to fire again before t_9 or t_f^1 fires. If t_f^1 really occurs, it can be diagnosed once t_2 fires as shown in Lemma 5.3. Fig. 5.21 shows the marking estimation which illustrates Algorithms 5.1, 5.2, one complementary marking estimate is required. After t_2 fires, the marking estimate is updated by the complementary estimate. Propositions 5.1 and 5.2 can also be verified, indeed, $\min\{M^e_{\omega_i}, M^c_{\omega_{i+1}}\} \leq M_{\omega_i}$, for $1 \leq i \leq 5$, and $e(M_{\omega_i}, M^e_{\omega_i}) \geq e(M_{\omega_{i+1}}, M^e_{\omega_{i+1}})$ for $0 \leq i \leq 2$, $e(M_{\omega_3}, M^e_{\omega_3}) = e(M_{\omega_5}, M^e_{\omega_5})$. If t_9 fires before t_2, then it is determined that t_f^1 does not occur. The fault tolerant controller after t_2 fired is also given in Fig. 5.20, which ensures that t_f^1 does not violate the GMEC. Even t_f^1 and t_3 fire simultaneously, GMEC is still maintained.

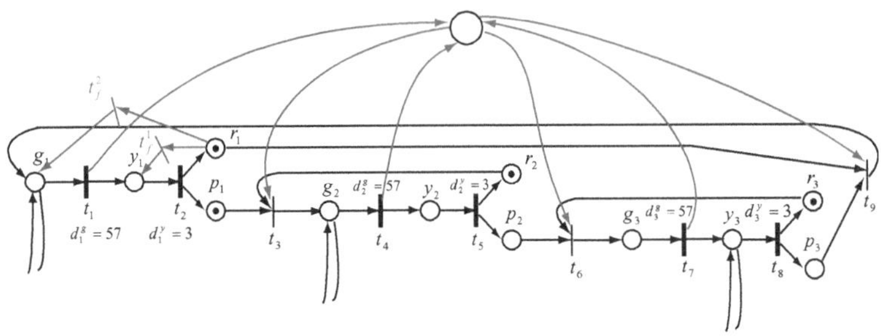

Fig. 5.22 Timed-PN in the faulty case 2

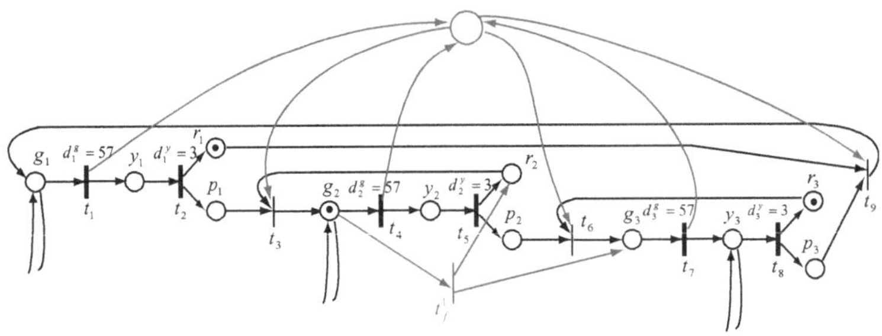

Fig. 5.23 Timed-PN in the faulty case 3

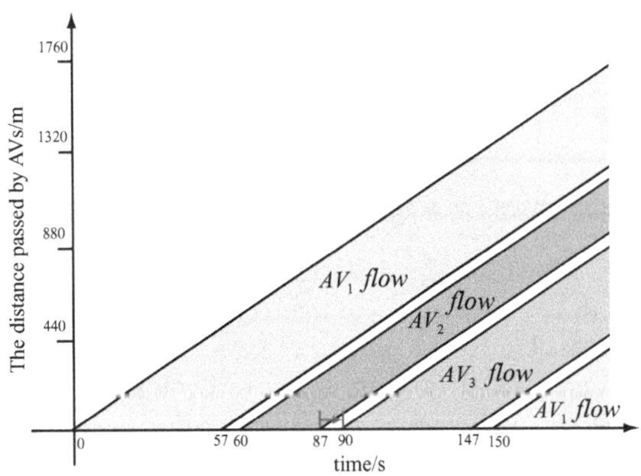

Fig. 5.24 The AVs flow on the bridge in the faulty case 3

Case 2: $\exists t_f^1, t_f^2 \in T_f$ as shown in Fig. 5.23. Note that $LC(\cdot, t_f^2) \neq 0$, which violates the condition in [85] and [53]. Two complementary marking estimates are required. The T_f-subnet is FC since t_f^1 and t_f^2 share the same input place r_1. However, the controller after t_2 fired, shown in Fig. 5.22 is less permissive than that in

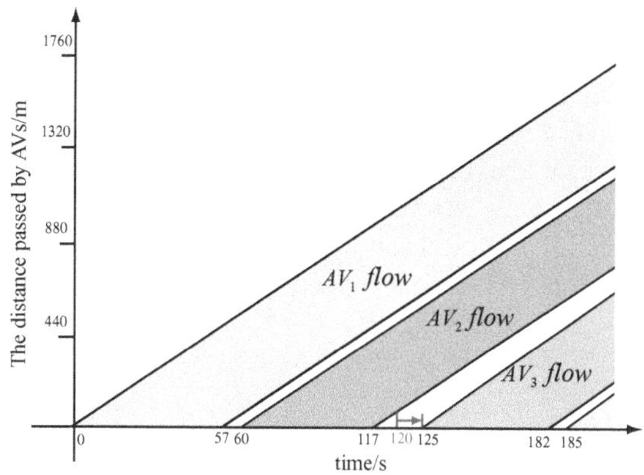

Fig. 5.25 The AVs flow on the bridge in the faulty case 4

Case 1. Due to possible fault t_f^2 which may activate g_1, the controller must disable t_3, i.e., the green light g_2 can not be activated. This verifies the Proposition 5.3. In fact, under t_f^2, the system gets deadlock unless t_f^2 really occurs.

Case 3: $\exists t_f^1 \in T_f$ which occurs at 87s, as in Fig. 5.23, the controller after t_3 fired, shown in Fig. 5.25 is the same as in Case 1, since $\mathscr{L}$ will not be violated. However, the performance of the continuous PN is affected. According to Algorithm 5.4, set the firing speed of t_3^{bf} to zero until the time t_2^{bs} stops firing. Fig. 5.24 shows the AVs flows on the bridge, from which we can see that the prescribed distance is still kept.

Case 4: $\exists t_f^1 \in TIME_f$ such that $d_2^y = 8$, i.e., the firing of t_5 lasts 8s. Such fault also affects the continuous PN. After the first part of the bridge becomes empty, the AVs flow from the 3th input road still stops and does not get into the bridge. According to Algorithm 5.4, we could accelerate the firing speed of t_3^{bf} as $at' = 8.066m/s$ to accommodate this fault, the AVs flows are presented in Fig. 5.25, where the distance between AV_2 and AV_3 converges to D_{safe}, and finally equals D_{safe} when AV_1 completely leaves the bridge.

5.3 Conclusion

In this Chapter, we have provided a new clue to investigate FTC problem of hybrid systems, that is from discrete event point of view. It has been shown that whatever the finite state machine or Petri net mode is used, discrete faults can be accommodated effectively, The continuous system theories described in chapters 2-4 are limited to deal with such kind of faults.

In Section 5.1, the continuous faults can also be considered qualitatively in the QA of the proposed hierarchical model. The sign of the vector field would change due to continuous faults, fault tolerance analysis could be done by rebuilding the continuous transition sequence, and checking whether the designed discrete switching sequence is available for the reconstructed system.

In Section 5.2, the FTC design that deals with faults in both discrete and continuous PNs is still an open problem. In this case, the continuous system theory could be extended under the continuous Petri net framework, which combines the proposed results in this chapter could provide a solution to such problem.

Chapter 6
Hybrid Control Approach in FTC Design

The potential faults in a system often range over a very large region. A single controller (even an adaptive one) is hard to stabilize all faulty situations effectively. However, hybrid control approach can significantly improve the FTC performances including robustness, the speed of response, and optimality, etc. In this chapter, we apply the results of HS proposed in Section 2 to hybrid control design in the FTC system. Three supervisory FTC algorithms are developed. Finally, A four-wheel-steering and four-wheel-driving electric vehicle in LAGIS laboratory is particularly focused on whose actuator faults are analyzed systematically and the hybrid fault tolerant tracking control approach is applied.

6.1 Supervisory FTC via Hybrid System Approaches

Hybrid control seeks to achieve system's performance objectives by switching between members of an a priori specified family of feedback controllers. One of the motivations of HS research arises from the hybrid control problem. HS could present different control configurations. Commutation from one configuration to another one is described using discrete event system model as claimed in [117, 93, 94, 131]. Thus the controlled system becomes hybrid due to the switching control.

The potential faults in a system often range over a very large region. A single controller (even an adaptive one) is hard to stabilize all faulty situations effectively. The supervisory FTC approach assumes that the plant model belongs to a pre-specified set of models, including the nominal situation and all possible faulty situations, and that there exists a finite family of candidate FTC laws controllers such that the faulty system is stabilized when controlled by at least one of those candidate controllers.

Consider the general nonlinear system

$$\dot{x}(t) = G(x(t), u(t), f(t)) \tag{6.1}$$

$$y(t) = H(x(t), f(t)) \tag{6.2}$$

with measurable states $x \in X \subset \Re^n$, inputs $u \in U \subset \Re^p$, outputs $y \in Y \subset \Re^m$. Process and/or actuator and/or sensor faults are represented by the function

H. Yang et al.: Fault Tolerant Control Design for Hybrid Systems, LNCIS 397, pp. 149–179.
springerlink.com

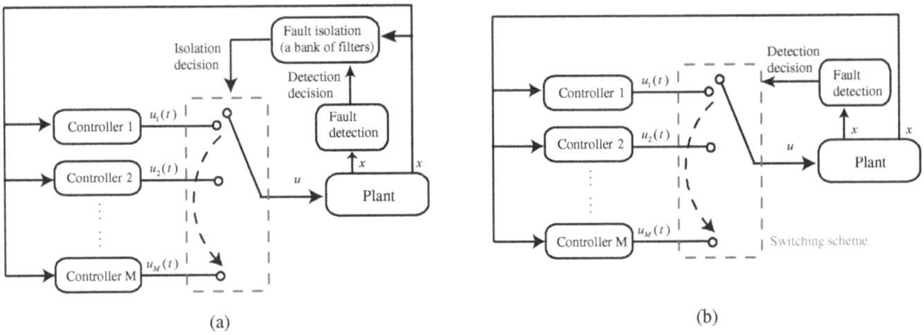

Fig. 6.1 FTC framework

$f \in \mathscr{F} = \bigcup_{i \in Q=\{1,\ldots,M\}} \mathscr{F}_i \subset \Re^q$ and $\mathscr{F}_i$ is the set of fault vectors that are associated with fault mode number i and fault free operation is fault mode $\mathscr{F}_M = \{0\}$. Both G and H are smooth functions.

The classical supervisory FTC approach follows three steps [117]: 1) Detect the occurrence of faults; 2) Identify the current fault mode; 3) Switch to the related controller as shown by Fig. 6.1(a). This scheme obviously introduces a FDI delay to identify the current fault mode. During this delay, the faulty system is controlled using an inappropriate controller, which may result in an unstable behavior.

In our proposed schemes in this section, a sequence of controllers are switched, until the appropriate one is found (Fig. 2(b)). A delay in selecting the correct controller (selection delay) still exists, but no isolation algorithm is required (only fault detection is needed), which makes the scheme simpler and more easily verifiable. Moreover, this selection delay can be controlled, and conditions for the state to remain bounded during this delay can be exhibited.

The novelty of the proposed approaches in this section is twofold:

1) The states are ensured to be bounded during the FDI delay and the functionality of the system is preserved throughout the FDI/FTC process.
2) Unlike the multiple model FDI /FTC method [153, 12] or supervisory control technique [149], we do not need a series of models or filters to work concurrently with the plant in order to identify the current situation. The proposed methods only rely on a simple switching scheme among candidate controllers.

6.1.1 FTC via Overall Regulation

We first apply the overall regulation theories in Section 2.2 to supervisory FTC design. Consider the system (6.1)-(6.2) with the fault and regulated error defined as

$$\dot{f}_i(t) = S_i(f_i(t)),\ \forall i \in Q,\ \forall t \geq t_f,\ \text{with } f_i(t) = 0\ \ \forall t \in [0, t_f) \tag{6.3}$$

$$e(t) = y(t) - y_r(x(t)) \tag{6.4}$$

Supposed that S_i is neurally stable. The initial fault value $f_i(t_f)$ are assumed to be known as a constant $f_{(in)i}$.

Assumption 6.1. *There exists a family of controllers $u_i = \alpha_i(x, f_i)$ for $f_i \in \mathscr{F}_i$, $i \in Q$ solving the fault tolerant regulation problem (FTRP) for system (6.1)-(6.4).*

Assumption 6.1 means that the FTRP of the plant with each fault is solvable under a candidate controller.

Now we consider fault detection problem. Recall the materials in Section 2.2. It can be seen from Theorem 2.3 and Assumption 6.1 that under the FTC law u_i, the system (6.1)-(6.3) with f_i has a center manifold $x = \pi_i(f_i)$ [55]. We further obtain that the equilibrium $(x, f_i) = (0,0)$ of system (6.1) and (6.3) is stable and this center manifold is locally attractive, i.e.,

$$|x(t) - \pi_i(f_i(t))| \leq B_i e^{-a_i(t-t_{ik})}|x(t_{ik}) - \pi_i(f_i(t_{ik}))| \quad \text{for } B_i, a_i > 0 \tag{6.5}$$

where t_{ik} denotes the time at which controller $u_i(t)$ is applied for the kth time.

The following assumption means that all modes are discernable.

Assumption 6.2. *Inequality (6.5) does not hold if the system (6.1)-(6.3) is controlled by u_j, $\forall j \in Q \setminus \{i\}$.*

Consider a time window where the control law u_i and the fault f_i are in adequacy, therefore (6.5) holds, and a simple fault detection law is given

$$|x(t) - \pi_i(f_i(t))| > B_i e^{-a_i(t-t_{ik})}|x(t_{ik}) - \pi_i(f_i(t_{ik}))| \Longrightarrow \text{detection} \tag{6.6}$$

Proposition 6.1. *Under assumptions 6.1 and 6.2, the fault detection law (6.6) is implementable.*

Proof: Note that the state is measurable. Without loss of generality, suppose that there is no fault at the beginning of the system process. The healthy system (6.1)-(6.3) with $i = 0$ is controlled by $u = \alpha_0(x, 0)$. According to Assumption 6.1 and (6.5), we have

$$|x(t)| \leq B_0 e^{-a_0 t}|x(0)|, \quad t \geq 0 \tag{6.7}$$

Once a fault occurs at $t = t_f$, Assumption 6.2 ensures that (6.7) is violated. The following inequality holds

$$|x(t_{fd})| > B_0 e^{-a_0 t_{fd}}|x(t_{fd})| \tag{6.8}$$

where $t_{fd} \geq t_f$, thus the fault can be detected using the detection law (6.6) at $t = t_{fd}$. Note that x is still bounded at $t = t_{fd}$.

Next consider $t \geq t_{ik}$ at which the system (6.1)-(6.3) has the fault f_i and is controlled by $u = \alpha_i(x, f_i)$ (the accurate value of f_i can be approximated via the proposed supervisory FTC scheme as shown later). Inequality (6.5) holds for $t \geq t_{ik}$. Once a fault occurs at $t = t_f \geq t_{ik}$, we can also find a $t_{fd} \geq t_f$ such that (6.5) is violated for $t \geq t_{fd}$, which implies that the fault can be detected using (6.6) at $t = t_{fd}$. This completes the proof. □

The fault detection may have a short time delay $t_{fd} - t_f$. Due to the time varying threshold, $t_{fd} - t_f$ is often much shorter than the activating period of the mode.

The effect of this delay is acceptable in the practical situation. In the following discussion, we assume that there is no fault detection delay, i.e., $t_{fd} = t_f$.

We propose a novel fault isolation method based on control switching. Since a series of controllers have been designed *a priori* for the plant with different faults, the fault isolation problem boils down to the problem of finding the correct controller. Such fault isolation approach also integrates the FTC problem, since the correct controller can be directly applied.

Define $\sigma(t) : [0,\infty) \to Q$ as the *switching function* of the candidate controllers, which is assumed to be a piecewise constant function continuous from the right. Denote by $t_0, t_1, t_2, \ldots$ the switching instants of $\sigma(t)$. These notations will also be used in sections 6.1.2-6.1.3. To exhaustively span all controllers, we will pick a non-repeated switching sequence of controllers as in the following definition.

Definition 6.1. *A switching sequence of controllers is said to be* non-repeated *if* $\sigma(t_i) \neq \sigma(t_j)$ *for* $i \geq 0$, $j \geq 0$, *and* $i \neq j$.

Theorem 6.1. *Consider a system (6.1)-(6.4), and a family of controllers* u_i *satisfying assumptions 6.1, 6.2. Suppose that a fault* $f \in \mathscr{F}_\iota$, $\iota \in Q$ *occurs and is detected simultaneously at* $t = t_f$ *via the threshold (6.6), then there exists a control switching scheme such that the FTRP of system (6.1)-(6.4) is solvable* $\forall t \geq t_f$.

Proof: Choose a constant $\beta > 1$. The switching law is designed as:

Algorithm 6.1. *Switching law of the controllers*

1. *Denote* $t_0 = t_f$; *Let* $s = 0$; *Define* $Q^\star \triangleq Q - \{\sigma(t_f)\}$; *Set* $\sigma(t_0) = i^\star$ *where*

$$i^\star = \arg\min_{i \in Q^\star} \left(y(t_0) - y_r(\pi_i(\hat{f}_i(t_0))) \right) \tag{6.9}$$

   with $\hat{f}_i$ *the* fictitious fault *generated from the system* $\dot{\hat{f}}_i = \hat{S}_i(\hat{f}_i)$ *with the function* $\hat{S}_i(\cdot) = S_i(\cdot)$, *the initial* $\hat{f}_i(t_0) = f_{(in)i}$.
2. *Choose* t_{1+s} *such that*

$$|x(t_{1+s}) - \pi_{i^\star}(\hat{f}_{i^\star}(t_{1+s}))| \leq \sqrt[M-1]{\beta}\, |x(t_s) - \pi_{i^\star}(\hat{f}_{i^\star}(t_s))| \tag{6.10}$$

   If $|x(t_{1+s}) - \pi_{i^\star}(\hat{f}_{i^\star}(t_{1+s}))| \leq B_{i^\star} e^{-a_{i^\star}(t-t_s)} |x(t_s) - \pi_{i^\star}(f_{i^\star}(t_s))|$
   then apply the controller $u_{\sigma(t_s)}(t)$ $\forall t \geq t_{1+s}$; *Stop the switching.*
   else, go to 3.
3. *Let* $Q^\star = Q^\star - \{\sigma(t_s)\}$; *Set* $\sigma(t_{1+s}) = i^\star$ *where*

$$i^\star = \arg\min_{i \in Q^\star} \left(y(t_{1+s}^-) - y_r(\pi_i(\hat{f}_i(t_{1+s}^-))) \right) \tag{6.11}$$

   Apply the controller $u_{\sigma(t_{1+s})}(t)$ *at* $t = t_{1+s}$; *Let* $s = s + 1$; *Go to 2.* ■

We shall prove that Algorithm 6.1 solves the FTRP.

Note that the performance driven switching sequence obtained from (6.9) and (6.11) is non-repeated, since at each switching instant, the next controller is selected from the set $Q^\star$ where the uncorrect controller activated before has been removed (Step 3). Thus at most $M-1$ switchings occur before the controller $u_\iota(t)$ related to $f \in \mathscr{F}_\iota$ is applied. We consider the worst situation that $\sigma(t_{M-2}) = \iota$. The results in other situations are obtained straightly.

Because the function $\hat{S}_i(\cdot) = S_i(\cdot)$, the initial $\hat{f}_i(t_0) = f_{(in)i}$, and the fault detection delay is not considered, there must be one fictitious fault signals $\hat{f}_i$ which is the same as the real fault signal f_i. Note that $\beta > 1$ and control mode $\sigma(t_0)$ is faulty, according to Assumption 6.2, we can choose $t_1 > t_0$ such that (6.10) holds with $s = 0$.

Since $\sigma(t_{M-2}) = \iota$, it holds that $\hat{f}_{\sigma(t_{M-2})} = f_{\sigma(t_{M-2})}$. By induction, we can obtain for $t \geq t_{M-1}$

$$\begin{aligned} &|x(t) - \pi_{\sigma(t_{M-2})}(f_{\sigma(t_{M-2})}(t))| \\ &\quad \leq \beta B e^{-a(t-t_{M-2})} |x(t_0) - \pi_{\sigma(t_0)}(\hat{f}_{\sigma(t_0)}(t_0))| \\ &\quad + B e^{-a(t-t_{M-2})} \sum_{s=1}^{M-1} \Big(\beta^{\frac{s}{M-1}} |\pi_{\sigma(t^u_{M-1-s})}(\hat{f}_{\sigma(t_{M-1-s})}(t_{M-1-s})) \\ &\qquad\qquad\qquad - \pi_{\sigma(t_{M-s})}(\hat{f}_{\sigma(t_{M-s})}(t_{M-s}))| \Big) \end{aligned} \tag{6.12}$$

Inequality (6.12) means that $x - \pi_{\sigma(t_{M-2})}(f_{\sigma(t_{M-2})})$ converges to zero $\forall t \geq t_f$. It follows that $\lim_{t\to 0} e(t) = 0$. □

Remark 6.1. *Under Algorithm 6.1, the switching stops after a finite time. Assumption 6.2 could be loosened as inequality (6.5) still holds under non-relevant controllers. In this case, the FTRP of one faulty plant can be solved by multiple candidate controllers. Less switching numbers are required, and the controller that terminates Algorithm 6.1 maybe not relevant to the current situation. This means that the fault is not isolated accurately. However the FTC goal is still achieved.*

Remark 6.2. *The transient behavior during the controller setting delay largely depends on the value of β. A large β may result in a large overshoot, whereas a small β would make the controllers switch too fast, which may lead to some unexpected phenomena. Section 6.1.3 makes some discussions about this point. Optimal selection of β is still an open problem that deserves further investigation.*

Example 6.1: A DC motor investigated in [116] is employed to illustrate a potential application field of our approach. $x = [\theta_m, \ \omega_m]^T$ is the state, where θ_m, ω_m denote the angular position and velocity of the motor. The system model is:

$$\begin{aligned} \dot{\theta}_m &= \omega_m \\ \dot{\omega}_m &= -\frac{\kappa_e}{J_m}\sin(\theta_m) - \frac{b}{J_m}\omega_m + \frac{c}{J_m}u \\ y &= \theta_m + f_1 \\ e &= y - y_r = y - 2\theta_m = -\theta_m + f_1 \end{aligned} \tag{6.13}$$

where J_m denotes the inertia of the motor. $\kappa_e > 0$ is the elasticity constant. u is the voltage. b and c are the related viscous friction coefficients and the amplifier gain. f_1 denotes the sensor fault.

In the fault-free case, design the controller $u = K(x) = \frac{J_m}{c}\left(\frac{\kappa_e}{J_m}\sin(\theta_m) + \frac{b}{J_m}\omega_m + K_1\theta_m + K_2\omega_m\right)$ such that the matrix $\begin{bmatrix} 0 & 1 \\ K_1 & K_2 \end{bmatrix}$ is Hurwitz. This leads to the asymptotical stability of the origin $x = 0$.

For the sake of clearness, we denote $(\cdot)_{(i)}$ as the parameter of mode i. Three sensor faulty cases are considered as follows which result in a deviation of the output signal from normal:

$$f_{(1)}: \ y = \theta_m + f_1 \tag{6.14}$$
$$f_{(2)}: \ y = \theta_m + 2f_1 \tag{6.15}$$
$$f_{(3)}: \ y = \theta_m + 4f_1 \tag{6.16}$$

where f_1 is generated by the following exosystem

$$\begin{cases} \dot{f}_1 = f_2 \\ \dot{f}_2 = -f_1 \end{cases} \tag{6.17}$$

Choosing a mapping $x = \pi_{(1)}(f) = \begin{bmatrix} \pi_{(1)1}(f) \\ \pi_{(1)2}(f) \end{bmatrix} = \begin{bmatrix} f_1 \\ f_2 \end{bmatrix}$ leads to

$$\begin{aligned} \frac{\partial \pi_{(1)1}(f)}{\partial t} &= \pi_{(1)2}(f) \\ \frac{\partial \pi_{(1)2}(f)}{\partial t} &= -\frac{\kappa_e}{J_m}\sin(\pi_{(1)1}(f)) - \frac{b}{J_m}\pi_{(1)2}(f) + \frac{c}{J_m}C(f) \\ 0 &= y(\pi_{(1)}(f)) - y_r(\pi_{(1)}(f)) \end{aligned} \tag{6.18}$$

where $C_{(1)}(f) = \frac{J_m}{c}\left(\frac{\kappa_e}{J_m}\sin(\pi_{(1)1}(f)) + \frac{b}{J_m}\pi_{(1)2}(f) - \pi_{(1)1}(f)\right)$. We can design the fault tolerant regulation law for fault mode 1 as

$$u_{(1)} = \alpha_{(1)}(x,f) = C_{(1)}(f) + K(x) - K(\pi_{(1)}(f)) \tag{6.19}$$

It is clear that controller (6.19) solves the FTRP.

Similarly, we choose two mappings $\pi_{(2)}(f) = \begin{bmatrix} 2f_1 \\ 2f_2 \end{bmatrix}$, $\pi_{(3)}(f) = \begin{bmatrix} 4f_1 \\ 4f_2 \end{bmatrix}$, design

$$C_{(2)}(f) = \frac{J_m}{c}\left(\frac{\kappa_e}{J_m}\sin(\pi_{(2)1}(f)) + \frac{b}{J_m}\pi_{(2)2}(f) - \pi_{(2)1}(f)\right)$$

$$C_{(3)}(f) = \frac{J_m}{c}\left(\frac{\kappa_e}{J_m}\sin(\pi_{(3)1}(f)) + \frac{b}{J_m}\pi_{(3)2}(f) - \pi_{(3)1}(f)\right)$$

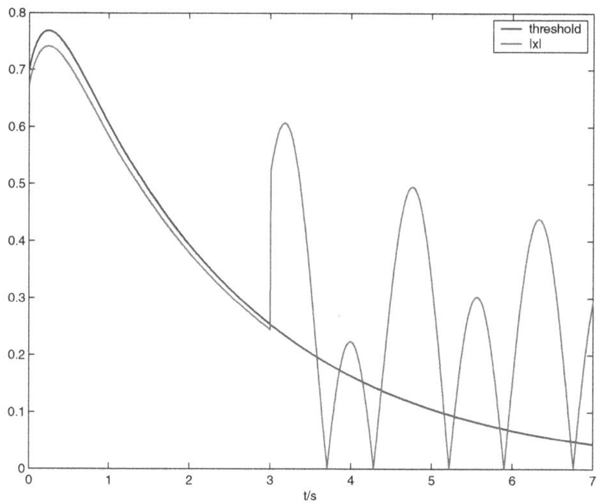

Fig. 6.2 Fault detection

The FTC law can be provided as

$$u_{(2)} = \alpha_{(2)}(x,f) = C_{(2)}(f) + K(x) - K(\pi_{(2)}(f)) \tag{6.20}$$
$$u_{(3)} = \alpha_{(3)}(x,f) = C_{(3)}(f) + K(x) - K(\pi_{(3)}(f)) \tag{6.21}$$

Controllers (6.20) and (6.21) solve the FTRP for fault modes 2 and 3 respectively. Assumption 6.1 is verified.

In the simulation, the parameters are $J_m = 0.935\ kgm^2$, $b = 1.17\ Nms/rad$, $\kappa_e = 0.311\ Nm/rad$, $c = 20.196\ Nm/V$. Assume that $f_{(1)}$ occurs at $t = 3s$, and Algorithm 6.1 is applied. Fig. 6.2 shows that the fault $f_{(1)}$ is detected at nearly $t = 3s$ using threshold (6.6). We choose $\beta = 1.5$. The non-repeated switching sequence obtained from (6.38) and (6.40) is $u_{(2)} \rightarrow u_{(1)}$. The dwell period of $u_{(1)}$ is $0.245s$; Once the fault is detected at $t = 3s$, $u_{(2)}$ is applied, then switch to $u_{(1)}$ at $t = 3.245s$. $f_{(2)}$ is assumed to occur at $t = 8s$. Fig. 6.3 shows the trajectories of the states and the regulated error, which means that the FTRP is solved.

6.1.2 FTC via Global Dissipativity

In this section, we extend the global passivity concept developed Section 2.4 and apply it to the supervisory FTC design.

Definition 6.2. *[14] A system (6.1)-(6.2) with $f \equiv 0$ is* dissipative *if there exists a nonnegative function $V : X \rightarrow \Re$, which satisfies $V(0) = 0$, called the* storage function, *and a* supply rate $W(y,u)$, *such that for all initial states $x(0) \in X$, $u \in U$, $y \in Y$ and $t \geq 0$*

$$\underbrace{V(x(t)) - V(x(0))}_{stored\ energy} \leq \underbrace{\int_0^t W(y(s),u(s))ds}_{supplied\ energy} \tag{6.22}$$

where $x(t)$ are the states at time t.

Definition 6.2 is more general than Definition 2.6. Similarly to assumptions 6.1 and 6.2, the following assumption ensures the recoverability of each fault mode and discernability of all modes.

Assumption 6.3. *There exist a family of functions $V_i(x) \in \mathscr{C}^1(\Re^n; R_{\geq 0})$ and functions $\alpha_1^i, \alpha_2^i \in \mathscr{K}_\infty$, $\phi_1^i < 0$, and $\phi_2^i \geq 0$ such that*

$$\forall i \in Q : \alpha_1^i(|x|) \leq V_i(x) \leq \alpha_2^i(|x|) \tag{6.23}$$

$$u = u_i(t) \Longrightarrow \begin{cases} f \in \mathscr{F}_i : \ V_i(x(t)) - V_i(x(t_{ik})) \leq \int_{t_{ik}}^t \phi_1^i(s)ds & (a) \\ f \in \mathscr{F}_j, j \neq i : \ V_i(x(t)) - V_i(x(t_{ik})) \leq \int_{t_{ik}}^t \phi_2^i(s)ds & (b) \end{cases} \tag{6.24}$$

Remark 6.3. *Assumption 6.3 explicitly addresses the behavior of the plant under the correct controller ($u = u_i(t)$ when $f \in \mathscr{F}_i$) or incorrect ones ($u = u_i(t)$ when $f \in \mathscr{F}_j, j \neq i$). For faults $f \in \mathscr{F}_i$, the controller $u_i(t)$ makes the plant still dissipative, as it can be seen from (6.24)(a), which means that all fault modes are recoverable. For faults $f \notin \mathscr{F}_i$, the function V_i may increase due to more stored energy. This implies that x may escape to a large region or infinity [60].*

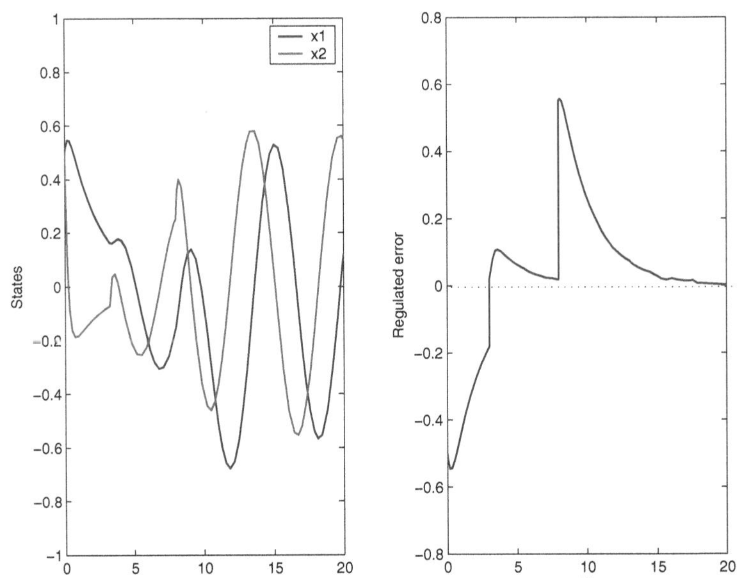

Fig. 6.3 Trajectories of x and e

We first address fault detection issue. Consider a time window where the control law and the fault mode are in adequacy, therefore (6.24)(a) holds. Once a fault occurs, the constraint (6.24)(a) may be violated. Similarly to the diagnosis idea in Section 2.4.1, we have

$$V_i(x(t)) - V_i(x(t_{ik})) \leq \int_{t_{ik}}^{t} \phi_1^i(s)ds + \underbrace{\int_0^t \left[\frac{\partial V_i}{\partial x}(x)\right]^\top f(x(s), u_i(s))ds}_{\text{"fault" energy } E_f} \tag{6.25}$$

As indicated in (6.25), the energy dissipativity property changes due to the fault. A fault detection law is given as

$$V_i(x(t)) - V_i(x(t_{ik})) > \int_{t_{ik}}^{t} \phi_1^i(s)ds \Longrightarrow \text{detection} \tag{6.26}$$

so that t_{fd} is the first time at which inequality (6.24)(a) is violated. Note that the faults with $E_f < 0$ are not necessary to be detected since they do not change the energy dissipativity.

Define $\sigma(t) : [0, \infty) \to Q$ and $t_0, t_1, t_2, ...$ as in Section 6.1.1. The following theorem provides a supervisory FTC scheme.

Theorem 6.2. *Consider a system (6.1)-(6.2) and a family of controllers satisfying (6.23)-(6.24) and assumption 6.3. Suppose that a fault $f \in \mathscr{F}_\iota$, $\iota \in Q$ occurs at $t = t_f$ and is detected at $t = t_{fd}$ via the threshold (6.26), then there exists a control switching scheme such that the origin of the system is stable for all $t \geq t_f$.*

Proof: Choose a constant $\beta > 0$. The switching law is designed as:

Algorithm 6.2. *Switching law of the controllers*

1. *Denote $t_0 = t_{fd}$; Let $s = 0$; Define $Q^\star \triangleq Q - \{\sigma(t_f)\}$; Set $\sigma(t_0) = i^\star$ where*

$$i^\star = \arg\min_{i \in Q^\star} V_i(x(t_0^-)) \tag{6.27}$$

2. *Choose t_{1+s} such that*

$$V_{\sigma(t_{1+s}^-)}(x(t_{1+s}^-)) - V_{\sigma(t_s)}(x(t_s)) \leq \frac{\beta}{M-1} \tag{6.28}$$

   If $V_{\sigma(t_{1+s}^-)}(x(t_{1+s}^-)) - V_{\sigma(t_s)}(x(t_s)) \leq \int_{t_s}^{t_{1+s}^-} \phi_1^{\sigma(t_s)}(s)ds$
   then apply the controller $u_{\sigma(t_s)}(t)$ $\forall t \geq t_{1+s}$; Stop the switching.
   else, go to 3.
3. *Let $Q^\star = Q^\star - \{\sigma(t_s)\}$; Set $\sigma(t_{1+s}) = i^\star$ where*

$$i^\star = \arg\min_{i \in Q^\star} V_i(x(t_{1+s}^-)) \tag{6.29}$$

   Apply the controller $u_{\sigma(t_{1+s})}(t)$ at $t = t_{1+s}$; Let $s = s + 1$; Go to 2. ■

We shall prove that Algorithm 6.2 implies the stability.

Note that at most $M-1$ switchings occur before the controller $u_\iota(t)$ related to $f \in \mathscr{F}_\iota$ is applied. We consider the worst case that $\sigma(t_{M-2}) = \iota$. The results for other cases are obtained straightly.

Since $\beta > 0$, and control mode $\sigma(t_0)$ is non-dissipative, it follows from (6.24)(b) that we can choose $t_1 > t_0$ such that $V_{\sigma(t_1^-)}(x(t_1^-)) - V_{\sigma(t_0)}(x(t_0)) \leq \frac{\beta}{M-1}$. We further have from (6.23) that

$$|x(t_1^-)| \leq (\alpha_1^{\sigma(t_0)})^{-1} \circ \left(\frac{\beta}{M-1} + \alpha_2^{\sigma(t_0)}(|x(t_0)|)\right) \tag{6.30}$$

Note that the fault detection law ensures that $x(t_0)$ is bounded, thus $x(t_1^-)$ is also bounded since the increasing stored energy is bounded during $[t_0, t_1)$.

By induction, it can be obtained at $t = t_{M-1}$ that

$$V_{\sigma(t_{M-1})}(x(t_{M-1})) - V_{\sigma(t_0)}(x(t_0)) - E_{tr}(x(t_0)) \leq \beta \tag{6.31}$$

where $E_{tr} = \sum_{k=1}^{N_{\sigma(t_{M-1})}} \left[V_{\sigma(t_k)} - V_{\sigma(t_k^-)}\right]$. The correct controller u_ι is selected and applied $\forall t \geq t_{M-1}$, we have

$$V_\iota(x(t)) - V_\iota(x(t_{M-1})) \leq \int_{t_{M-1}}^{t} \phi_1^\iota(s)ds, \quad t \geq t_{M-1} \tag{6.32}$$

for $\phi_1^\iota < 0$.

Combining (6.31) with (6.32) leads to

$$V_\iota(x(t)) - V_{\sigma(t_0)}(x(t_0)) - E_{tr}(x(t_0)) \leq \int_{t_{M-1}}^{t} \phi_1^\iota(s)ds + \beta \tag{6.33}$$

Since $\phi_1^\iota(s) < 0$, $\beta > 0$ is a bounded constant, there exists a time instant $t > t_{M-1}$ such that the inequality (6.33) satisfies the condition (2.114) in Definition 2.7 (global passivity). This means that the system is periodically fault tolerant dissipative during $[t_0, t)$. On the other hand, it follows from (6.26) that during $[t_f, t_0)$, the energy is still dissipative. The stability result follows from Theorem 2.7.

For the general case where $u_\iota(t)$ is selected before $M-2$ switchings occur, we can verify (6.33) with β^* instead of β where $\beta^* < \beta$. □

6.1.3 FTC via Gain Technique

The gain technique proposed in Section 2.5 is utilized to supervisory FTC design. Consider the system (6.2), the following assumption ensures the recoverability of each fault mode and discernability of all modes.

Assumption 6.4. *There exists a family of continuous non-negative functions $V_i(x)$: $\Re^n \to \Re_{\geq 0}$, and functions $\alpha_1^i, \alpha_2^i, \gamma \in \mathscr{K}_\infty$, $\phi_1^i \in \mathscr{KL}$, and $\phi_2^i \in \mathscr{GKL}$ such that*

$$\forall i \in Q : \alpha_1^i(|x|) \leq V_i(x) \leq \alpha_2^i(|x|) \tag{6.34}$$

$$u = u_i(t) \Longrightarrow \begin{cases} f \in \mathscr{F}_i : V_i(x(t)) \leq \phi_1^i(V_i(x(t_{ik})), t - t_{ik}) & (a) \\ f \in \mathscr{F}_j, j \neq i : V_i(x(t)) \leq \phi_2^i(V_i(x(t_{ik})), t - t_{ik}) & (b) \end{cases} \tag{6.35}$$

Let us consider a time window where the control law and the fault mode are in adequacy, therefore (6.35)(a) holds, and a simple fault detection law is given by

$$V_i(x(t)) > \phi_1^i(V_i(x(t_{ik})), t - t_{ik}) \Longrightarrow \text{detection} \tag{6.36}$$

so that t_{fd} is the first time at which inequality (6.35)(a) is violated.

Assumption 6.5. *There exists a known constant $\chi \geq 1$ such that*

$$\chi = \max_{j \in Q, k=1,2...} \frac{\phi_j(V_j(x(t_{jk})), 0)}{V_j(x(t_{jk}))} \tag{6.37}$$

Assumption 6.5 is similar to Assumption 2.11, which is required for the switching control design as shown in the following theorem. Note that $\phi_{\sigma(t_i)}$ will be taken instead of $\phi_2^{\sigma(t_i)}$ in (6.35)(b) only.

Theorem 6.3. *Consider a system (6.1) and a family of controllers satisfying (6.34)-(6.35) and assumptions 6.4-6.5. Suppose that a fault $f \in \mathscr{F}_\iota$, $\iota \in Q$ occurs at $t = t_f$ and is detected at $t = t_{fd}$ via the threshold (6.36), then there exists a control switching scheme such that x is bounded for all $t \geq t_f$.*

Proof: Choose a constant $\beta > \max[(M-2)(1+\chi)\chi^{M-2}, (M-2)(M-3)\chi^{M-3}]$, where χ is defined in (6.37). The switching law is designed as:

Algorithm 6.3. *Switching law of the controllers*

1. *Denote $t_0 = t_{fd}$; Let $s = 0$; Define $Q^\star \triangleq Q - \{\sigma(t_f)\}$; Set $\sigma(t_0) = i^\star$ where*

$$i^\star = \arg\min_{i \in Q^\star} V_i(x(t_0)) \tag{6.38}$$

2. *Choose t_{1+s} such that*

$$\sum_{k=0}^{s} \left(\prod_{j=k}^{s} \frac{\phi_{\sigma(t_j)}^{t_{j+1}-t_j}}{V_{\sigma(t_j)}^{t_j}} \right) \leq \frac{\beta}{(M-2-s)\chi^{M-2-s}} - 1 \tag{6.39}$$

If $V_{\sigma(t_s)}(x(t_{1+s})) \leq \phi_1^{\sigma(t_s)}(V_{\sigma(t_s)}(x(t_s)), t - t_s)$
then apply the controller $u_{\sigma(t_s)}(t)$ $\forall t \geq t_{1+s}$; Stop the switching.
else, go to 3.

3. *Let $Q^\star = Q^\star - \{\sigma(t_s)\}$; Set $\sigma(t_{1+s}) = i^\star$ where*

$$i^\star = \arg\min_{i \in Q^\star} V_i(x(t_{1+s})) \tag{6.40}$$

Apply the controller $u_{\sigma(t_{1+s})}(t)$ at $t = t_{1+s}$; Let $s = s + 1$; Go to 2. ■

Algorithm 6.3 implies the stability. The proof is quite similar to that of Theorem 2.12, and is omitted. □

It can be seen from Algorithm 6.3 that switching among a large number of controllers may result in a large β. In the following, the transient performance is improved by reducing the number of switchings.

Assumption 6.6. *There exists a family of continuous non-negative functions $\tilde{V}_i(x): \Re^n \to \Re_{\geq 0}$, $\forall i \in \mathcal{M}$ and $\tilde{\gamma} \in \mathcal{K}_\infty$, $\xi_i \in \mathcal{GKL}$ such that*

$$\tilde{V}_i(x(t)) \leq \xi_i(\tilde{V}_i(x(t_{jk})), t - t_{jk}) \qquad (6.41)$$
$$\forall f \in \mathcal{F}_i, u = u_j(x), j \neq i,\ t \geq t_{jk},\ k = 1, 2, \ldots$$

The following table shows the difference between Assumption 6.6 and Assumption 6.4. Assumption 6.4 assumes the existence of functions such that (6.35)(a) and (6.35)(b) are satisfied (rows 1 and 3 in the table) while Assumption 6.6 adds the existence of functions that satisfy also the conditions on row 2.

Table 6.1 Comparing Assumptions 6.4 and 6.6

	Fault	Control	Assumption
1	$f \in \mathcal{F}_i$	$u = u_i(x)$	$V_i(x(t)) \leq \phi_1^i\left(V_i\left(x\left(t_{ik}\right)\right), t - t_{ik}\right)$
2	$f \in \mathcal{F}_i$	$u = u_j(x), j \neq i$	$\tilde{V}_i(x(t)) \leq \xi_i\left(\tilde{V}_i\left(x\left(t_{jk}\right)\right), t - t_{jk}\right)$
3	$f \in \mathcal{F}_j, j \neq i$	$u = u_i(x)$	$V_i(x(t)) \leq \phi_2^i\left(V_i\left(x\left(t_{ik}\right)\right), t - t_{ik}\right)$

Note that the inequality (6.41) may still hold for $f \notin \mathcal{F}_i$. However, the converse is not true, i.e., if (6.41) is violated, it must hold that $f \notin \mathcal{F}_i$. Inequality (6.41) can be obtained *a priori* when a family of candidate FTC laws are designed.

Algorithm 6.4. *Accelerating switching law of the controllers*

1. *Denote $t_0 = t_{fd}$; Let $s = 0$; Define $\mathcal{M}^\star \triangleq Q - \{\sigma(t_f)\}$; Set $\sigma(t_0) = i^\star$ where*

$$i^\star = \arg\min_{i \in Q^\star} V_i(x(t_0))$$

2. *Choose t_{1+s} such that*

$$\sum_{k=0}^{s}\left(\prod_{j=k}^{s}\frac{\phi_{\sigma(t_j)}^{t_{j+1}-t_j}}{V_{\sigma(t_j)}^{t_j}}\right) \leq \frac{\beta}{(M-2-s)\chi^{M-2-s}} - 1 \qquad (6.42)$$

If $V_{\sigma(t_s)}(x(t_{1+s})) \leq \phi_1^{\sigma(t_s)}(V_{\sigma(t_s)}(x(t_s)), t - t_s)$
then apply the controller $u_{\sigma(t_s)}(x)$ $\forall t \geq t_{1+s}$; Stop the switching.
else, let $Q^\star = Q^\star - \{\sigma(t_s)\}$; Go to 3.

3. *Set $\sigma(t_{1+s}) = i^\star$ where*

$$i^\star = \arg\min_{i \in \mathcal{M}^\star} V_i(x(t_{1+s}))$$

If $\tilde{V}_{\sigma(t_{i^\star})}(x(t_{1+s})) > \xi_{\sigma(t_{i^\star})}(\tilde{V}_\sigma(t_{i^\star})(x(t_s)), t - t_s)$
then let $Q^\star = Q^\star - \{\sigma(t_{i^\star})\}$; *Go to 3.*
else, apply $u_{\sigma(t_{1+s})}(x)$ *at* $t = t_{1+s}$; *Let* $s = s + 1$; *Go to 2.* ■

The main idea behind Algorithm 6.4 is that at each switching instant, we check whether the fault mode satisfies (6.41), and remove incorrect candidate controllers from the switching sequence.

We shall prove that Algorithm 6.4 improves the transient behavior w.r.t. Algorithm 6.3. Denote $x(t_{A1})$ $\sigma(t_{A1})$ and $t_{A1|s}$ (respectively $x(t_{A2})$ $\sigma(t_{A2})$ and $t_{A2|s}$) the state trajectory, switching function and the sth switching time under Algorithm 6.3 (respectively Algorithm 6.4). We have the following result.

Corollary 6.1. *Consider a nonlinear system (3.33) and a family of controllers satisfying (4.41)-(6.35) and assumptions 6.4-6.6. Supposed that a fault* $f \in \mathscr{F}_\iota$, $\iota \in \mathscr{M}$ *occurs at* $t = t_f$ *and is detected at* $t = t_{fd}$ *via the threshold (6.6), then*

1) Algorithm 6.4 guarantees that x *is bounded for all* $t \geq t_f$ *and the system is ISS w.r.t.* $\bar{d}$ *after the correct controller* $u_\iota(t)$ *is applied.*
2) If $\sigma_{A2}(t_{A2|s}) = \sigma_{A1}(t_{A1|r}) = \iota$, *then* $|x_{A2}(t_{A2|s})| \leq |x_{A1}(t_{A1|r})|$.

Proof: 1) can be obtained following the same line as for Theorem 6.3.

2). Since the correct controller is selected after $s+1$ number of switchings under Algorithm 6.4, it can be concluded that $s \leq r \leq M-2$. Let us consider the worst case that $r = M-2$.

Choose $t_{A2|s}$ as (6.42), we obtain

$$\sum_{k=0}^{s-1}\left(\prod_{j=k}^{s-1}\frac{\phi_{\sigma(t_{A2|j})}^{t_{A2|j+1}-t_{A2|j}}}{V_{\sigma(t_{A2|j})}^{t_{A2|j}}}\right) \leq \frac{\beta}{(M-1-s)\chi^{M-1-s}} - 1 \tag{6.43}$$

Since $s \leq M-2$, we verify condition (2.129) with β^* instead of β where $\beta^* = \frac{\beta}{M-1-s} \leq \beta$ at $t = t_{A2|s}$.

It follows that

$$|x(t_{A2|s})| \leq (\alpha_1^{\sigma(t_{A2|s})})^{-1} \circ \beta^* \bar{\alpha}(|x(t_0)|) \tag{6.44}$$

where $\bar{\alpha}$ is defined in (2.143). Comparing (6.44) with (2.143) in Theorem 2.9 leads to the result. For the general case where $r < M-2$, the result can be obtained following above procedure. □

Example 6.2: Consider a one-link manipulator, whose revolution joint is actuated by a DC motor. The joint elasticity is modeled by a linear torsional spring [57]. The states are the angular positions and velocities of the motor and of the link $x = [\theta_m,\ \omega_m, \theta_l, \omega_l]^\top$. The control u is the torque delivered by the motor. The state-space model is

$$\begin{aligned}\dot{\theta}_m &= \omega_m \\ \dot{\omega}_m &= -\frac{\kappa}{J_m}(\theta_l - \theta_m) - \frac{b}{J_m}\omega_m + \frac{c}{J_m}u\end{aligned}$$

$$\begin{aligned} \dot{\theta}_1 &= \omega_1 \\ \dot{\omega}_1 &= -\frac{\kappa}{J_1}(\theta_1 - \theta_m) - \frac{mgh}{J_1}\sin(\theta_1) \end{aligned} \tag{6.45}$$

where J_m and J_1 denote respectively the inertia of the motor and of the link. κ is the elasticity constant, b denotes the related viscous friction coefficient, and c is the amplifier gain. The numerical values of the parameters given in [57] are: $J_m = 0.935\ kgm^2$, $J_1 = 23.303\ kgm^2$, $\kappa = 45.440 Nm/rad$, $b = 1.169\ Nms/rad$, $c = 20.196\ Nm/V$. $mgh = 7.760 Nm/rad$.

Table 6.2 Faulty cases

	Fault mode	Reason
Case 1	b is changed within [10m, 15m]	an increase in the friction of the motor
Case 2	κ reduces to 25% $\sim$ 50%	an unexpected change on elasticity condition
Case 3	κ reduces to 50% $\sim$ 75%	an unexpected change on elasticity condition
Case 4	c is changed within $[30Nm/V, 40Nm/V]$	amplifier malfunction

Table 6.2 describes four considered faulty cases, where cases 1-3 are concerned with process faults, and Case 4 is related to actuator faults. Consequently, we divide $\mathscr{F}$ into five parts as $\mathscr{F} \subset \bigcup_{i \in Q=\{1,2,\ldots,5\}} \mathscr{F}_i$, where $\mathscr{F}_i$ is related to the fault values in Case i. $\mathscr{F}_5$ denotes the fault-free situation. According to the FTC design procedure described in [57], we can design a nominal controller $u_5(x)$ for the healthy plant and four candidate controllers $u_i(x), i = 1,2,3,4$ for cases 1-4 respectively. The details are omitted here. Moreover, for each controller u_i, we can obtain $V_i(x) = x^\top H_i x$ where H_i is a positive definite matrix. V_5 denotes the function of the healthy plant.

In the simulation, suppose that Case 1 happens, $b = 11.69m$, we further have

$$\begin{aligned} V_1(x(t)) &\le e^{-1.1840t} V_1(x(0)), \ \forall f \in \mathscr{F}_1, u = u_1(x), \ t \ge 0 \\ V_2(x(t)) &\le e^{6.2893t} V_2(x(0)), \ \forall f \in \mathscr{F}_1, u = u_2(x), \ t \ge 0 \\ V_3(x(t)) &\le e^{18.8439t} V_3(x(0)), \ \forall f \in \mathscr{F}_1, u = u_3(x), \ t \ge 0 \\ V_4(x(t)) &\le e^{1.4031t} V_4(x(0)), \ \forall f \in \mathscr{F}_1, u = u_4(x), \ t \ge 0 \end{aligned}$$

It can be seen that Assumption 6.2 is satisfied. In fact, the system with the fault mode 1 is stabilized only by controller $u_1(x)$. Suppose that the initial states are $[1\ 0.4\ 0.5\ 0.1]^\top$. Case 1 occurs at $t = 1.5s$, Fig. 6.4 shows that the fault is detected at $t = 2.343s$ using threshold (6.36).

Now we apply Algorithm 6.3 to achieve the FTC goal. It can be obtained from (6.37) that $\chi = 1$, this satisfies Assumption 6.5. Since there are three unstabilizing controllers that may be activated, $M - 2 = 3$. We choose $\beta = 6.5 > 3 \times 2$. The non-repeated switching sequence obtained from (6.38) and (6.40) is $u_2 \to u_3 \to u_4 \to u_1$. Simple calculation based on (6.39) of Algorithm 6.3 leads to the dwell periods of three controllers: $0.0245s$ for $u_2(x)$; $0.0020s$ for $u_3(x)$; $0.3750s$ for $u_4(x)$. These dwell periods can be determined without checking the value $V^t_{\sigma(t)}$. Once the fault is detected, $u_2(x)$ is selected from (6.38) and applied at $t = 2.343s$, then switch to

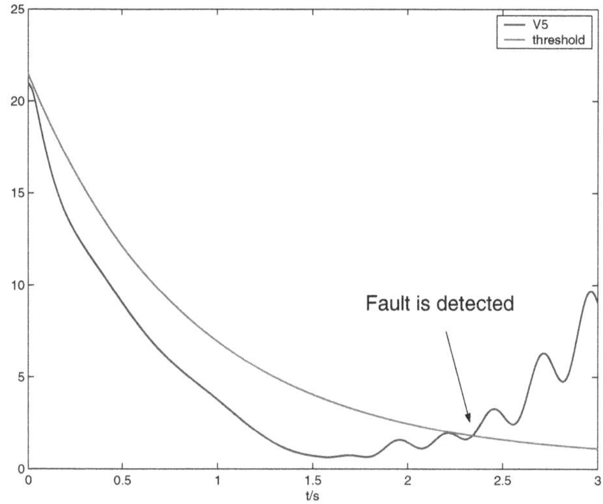

Fig. 6.4 Fault detection

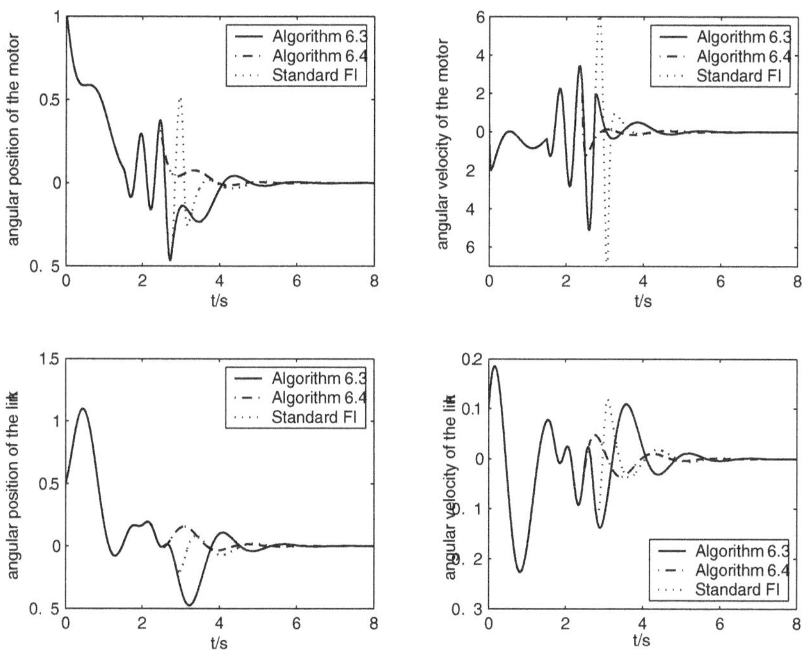

Fig. 6.5 State trajectories

$u_3(x)$ at $t = 2.3675s$, and switch to $u_4(x)$ at $2.3695s$. At $t = 2.7445s$, the fault is identified to be Case 1, the correct controller $u_1(x)$ is applied for $t \geq 2.7445s$. The solid lines in Fig.6.5 show the state trajectories, it can be seen that the FTC goal is achieved and during the delay $[1.5s, 2.7445s)$, the states are always bounded.

Now we illustrate Algorithm 6.4. It follows from the control design procedure in [57] that

$$\begin{aligned}
\tilde{V}_2(x(t)) &\leq e^{12.4639t}\tilde{V}_2(x(0)), \ \ \forall f \in \mathscr{F}_2, u \neq u_2(x), \ t \geq 0 \\
\tilde{V}_3(x(t)) &\leq e^{24.7468t}\tilde{V}_3(x(0)), \ \ \forall f \in \mathscr{F}_3, u \neq u_3(x), \ t \geq 0 \\
\tilde{V}_4(x(t)) &\leq e^{4.3206t}\tilde{V}_4(x(0)), \ \ \forall f \in \mathscr{F}_4, u \neq u_4(x), \ t \geq 0
\end{aligned}$$

The obtained switching sequence is the same as that using Algorithm 6.3. However, at the second switching instant, the controller $u_3(x)$ is removed from the sequence using the Step 3 of Algorithm 6.4. It follows from (6.42) that the dwell period of controller $u_4(x)$ becomes $0.0269s$. Thus we first apply $u_2(x)$ at $t = 2.343s$, then switch to $u_4(x)$ at $t = 2.3675s$. At $t = 2.3944s$, the fault is identified to be Case 1, the correct controller $u_1(x)$ is applied for $t \geq 2.3944s$. The dashed lines in Fig.6.5 show the state trajectories under Algorithm 6.4, it can be seen that during the delay $[1.5s, 2.7445s)$, the states are also bounded, and the transient performance is better than that under Algorithm 6.3.

6.2 Hybrid Control Based FTC for Automated Vehicles

In the final section of this chapter, we investigate the path tracking problem for four-wheel-steering and four-wheel-driving (4WS4WD) electric vehicles with input constraints, actuator faults and the external resistance. A hybrid FTC approach, which combines the linear quadratic control method in [118] and the control Lyapunov function technique in Section 3.2 is proposed. It not only maintains the vehicle's tracking performance in spite of faults, input constraints and the external resistance, but also reduces the cost of the fault tolerant process. A prototype vehicle in LAGIS laboratory is particularly focused on to illustrate the proposed approach.

6.2.1 *Background*

Electric vehicles (EV) are attracting a great deal of interest as a powerful solution to environmental and energy problems [16]. The four-wheel steering and four-wheel driving (4WS4WD) EV does not only take the advantage of a 4WD vehicle where the individual torque of each drive wheel can be controlled independently [106], but also benefits from the 4WS structure where both the steering positions of front wheels and rear wheels can be controlled [87]. Such structure significantly improves EV's lateral dynamics, especially in the situation of path tracking [96], [86, 100].

Faults may lead to vehicle's abnormal behaviors. The faults that mainly degrade the vehicle's performance include faults of sensors that provide important physical

Fig. 6.6 The Robucar^{TM} in LAGIS

characteristics (e.g., the vehicle speed, the sideslip angle) and actuator faults such as the malfunction of the steering systems and wheel torque controllers. The fault detection and isolation (FDI) techniques of vehicles have been investigated intensively by Isermann's group [33, 120], Ding's group [35], the PATH project [99], and also our LAGIS laboratory [26]. FTC approaches of vehicles have also been developed in order to guarantee the safety of the vehicle [11, 27].

However, few contribution has been made for the fault tolerant path tracking control of EV, e.g. [137], [146]. Path tracking of vehicles is one of the key issues in an intelligent transportation system. The tracking performance must be maintained in spite of faults, otherwise, traffic accidents may occur, which may lead to the vehicle destruction. Moreover, most of related FTC works do not address the issues of optimality, input constraints and the external resistance.

- Optimality means to reduce a cost function of the states and inputs of the vehicle systems as much as possible that is needed for FTC.
- Input constraints are involved to prevent the vehicles from skidding or spinning when FTC is activated.
- External resistance includes the air resistance, wind effects, the deformation of the wheels, and the internal friction of the vehicle. These factors always affect the vehicle.

In this section, we focus on the optimal fault tolerant path tracking control for a 4WS4WD EV in LAGIS as shown in Fig. 6.6. This prototype vehicle, named RobuCar^{TM}, is built by the Robosoft Company [157]. Several important types of actuator faults are considered as in [137]. A hybrid control approach is proposed, which combines the linear quadratic (LQ) based progressive accommodation (PA) method [118] and the control Lyapunov function (CLF) technique in Section 3.2. The motivation of developing such control structure is to maintain the vehicle's tracking performance in spite of faults, input constraints and the external resistance, and meanwhile, reduce the cost function of states and inputs that results from the FTC algorithm. This work focuses on the FTC design and we do not consider the

FDI technique of vehicles. The readers interested by fruitful results on such FDI techniques are referred to [33, 26]. The sensor faults are also not involved, some related work can be seen in [124].

6.2.2 Vehicle Model and Fault Setting

The features of the RobuCar^{TM} dynamics are described in Fig. 6.7. Our system comprises a 4WS4WD vehicle body, four wheels, and a reference path for tracking. The distance between the center of gravity (CG) and the front axle (resp. rear axle) is l_f (resp. l_r), l_d is one half of the tread. $r_{ei} (i = 1,2,3,4)$ denotes the effective radius of the wheel i.

The state variables are the speed of CG v, the sideslip angle β, the yaw rate γ, the perpendicular distance y_c between the vehicle and the reference path, the angle ϕ between the vehicle and the tangent to the path curvature ρ_{ref}. The traction forces f_{xi} and f_{yi} are transmitted from the road surface via the wheels to the vehicle chassis. The input variables to be applied are the steering angle δ_i and the torque T_i. Denote $\delta_f \triangleq \delta_1 = \delta_2$ and $\delta_r \triangleq \delta_3 = \delta_4$ as the steering angles of front wheels and rear wheels respectively.

The detailed dynamical equations of the vehicle body, wheel, and path tracking can be seen in [96], [1] and [26], thus are omitted here. Around the free-rolling equilibrium point: $v = v_0$, $\beta = 0$, $\gamma = 0$, $y_c = 0$, $\phi = 0$, and $\rho_{ref} = 0$, a linearized vehicle model can be obtained as

$$\dot{x} = Ax + BKu_s + R \tag{6.46}$$

where $x = [(v - v_0)\ \beta\ \gamma\ y_c\ \phi]^\top$ are measurable states, $u_s = [T_1^c + T_1^r \quad T_2^c + T_2^r \quad T_3^c + T_3^r \quad T_4^c + T_4^r \quad \delta_f^c \quad \delta_r^c]^\top$ are torques and steering controllers' output vector, T_j^c is used for the path tracking control design, while T_j^r is applied to overcome the

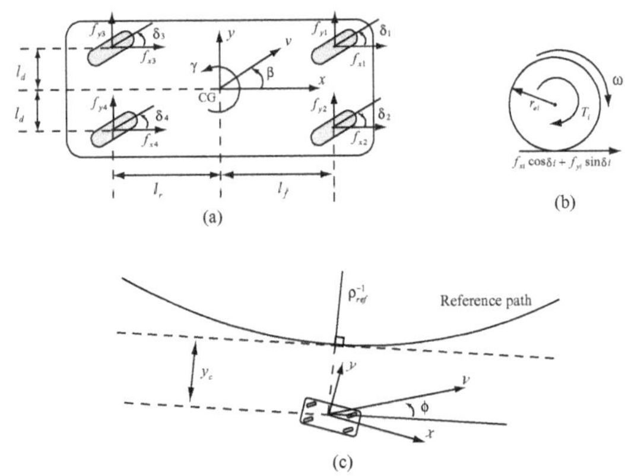

Fig. 6.7 The vehicle system

external resistance, denoted as $R = [\bar{R}\ 0\ 0\ 0\ 0]^\top$. $\bar{R}$ is assumed to be a known constant around the free-rolling equilibrium point. Denote the plant input vector as $\bar{u} = [T_1\ T_2\ T_3\ T_4\ \delta_f\ \delta_r]^\top$ as shown in Fig.6.7. $\bar{u} = Ku_s$, with K defined as the actuator gain matrix, and $K = diag[\eta_1, \eta_2, \ldots, \eta_6]$, $\eta_i = 1$ in the healthy situation, and will be defined later for the faulty cases. Moreover

$$A = \begin{bmatrix} 0 & 0 & 0 & 0 & 0 \\ 0 & -\frac{C_f+C_r}{mv_0} & -\frac{l_fC_f-l_rC_r}{mv_0^2}-1 & 0 & 0 \\ 0 & -\frac{l_fC_f-l_rC_r}{J_z} & -\frac{l_f^2C_f+l_r^2C_r}{J_zv_0} & 0 & 0 \\ 0 & 0 & 0 & 0 & -v_0 \\ 0 & -\frac{C_f+C_r}{mv_0} & -\frac{l_fC_f-l_rC_r}{mv_0^2} & 0 & 0 \end{bmatrix},$$

$$B = \begin{bmatrix} \frac{1}{mr_{e_1}} & \frac{1}{mr_{e_2}} & \frac{1}{mr_{e_3}} & \frac{1}{mr_{e_4}} & 0 & 0 \\ 0 & 0 & 0 & 0 & \frac{C_f}{mv_0} & \frac{C_r}{mv_0} \\ \frac{-l_d}{J_zr_{e_1}} & \frac{l_d}{J_zr_{e_2}} & \frac{-l_d}{J_zr_{e_3}} & \frac{l_d}{J_zr_{e_4}} & \frac{l_fC_f}{J_z} & -\frac{l_rC_r}{J_z} \\ 0 & 0 & 0 & 0 & 0 & 0 \\ 0 & 0 & 0 & 0 & \frac{C_f}{mv_0} & \frac{C_r}{mv_0} \end{bmatrix}$$

where m denotes the mass of the vehicle. J_z is the moment of inertia. The constant coefficients C_f and C_r are cornering stiffness of the front and rear wheels. $C_f = C_{f1} + C_{f2}$, $C_r = C_{r1} + C_{r2}$. The pair (A, B) is controllable. Note that the developed model is more general than the usual vehicle's lateral model as in [106], [86] and [100] where only δ_f and δ_r are applied as the inputs, and the resistance factors are not considered.

The structure of B ensures the existence of constant torques T_i^r, $i = 1,2,3,4$ such that

$$\sum_{i=1}^{4}\left(\frac{\eta_i T_i^r}{mr_{e_i}}\right) = -\bar{R} \tag{6.47}$$

$$\sum_{i=1}^{4}\left(\frac{(-1)^i l_d \eta_i T_i^r}{J_z r_{e_i}}\right) = 0 \tag{6.48}$$

This implies that 4 constant torques T_i^r can be applied to overcome the external resistance.

The input constraints have to be considered for the saturation property of the wheel slip which is related to the road condition. The relation between input (T_i, δ_i) to the wheel slip S_i at the free-rolling equilibrium point can be given as in [96] and [1]

$$S_i = \begin{bmatrix} \frac{T_i k_i}{r_{ei}C_{fi}} \\ -\beta - \frac{l_f}{v_0}\gamma + \delta_i \end{bmatrix}, \quad i = 1,2. \tag{6.49}$$

$$S_i = \begin{bmatrix} \frac{T_i k_i}{r_{ei} C_{ri}} \\ -\beta - \frac{l_r}{v_0}\gamma + \delta_i \end{bmatrix}, \quad i = 3,4. \tag{6.50}$$

where k_i represents the tire-tread-profile attenuation factor. From (6.49)-(6.50), it can be seen that T_i and δ_i need to be constrained to ensure the magnitude of S_i below the prescribed value c, i.e., $|S_i| \leq c$, with $|\cdot|$ the Euclidean norm. More precisely, since $\eta_i T_i^r$ is a constant, a constant bound can be imposed on $\eta_i T_i^c$, while a state dependent bound should be imposed on $\eta_i \delta_i^c$.

The control objective in the healthy situation is to let the vehicle track the reference path, i.e. to make the origin of the system (6.46) asymptotical stable, and meanwhile, restrict the magnitude of S_i into the prescribed region to prevent the vehicle from skidding or spinning.

Once an actuator fault occurs at $t = t_f$, the system (6.46) can be represented as

$$\dot{x} = Ax + B_f u_s + R \tag{6.51}$$

where $B_f \triangleq BK$ denotes the fault input distribution matrix. It is assumed that (A, B_f) is still controllable. In this work, both faults of steering systems and wheel torque control systems are considered. Fig. 6.8 shows the schematic diagram of the RobuCar^{TM}. Four faulty cases are investigated:

- **(F1)** The failure of one steering controller (front or rear), which may result from the broken wires, the malfunction power amplifier or the steeling motor breakdown. In this case, the steering actuator float with zero moment and does not contribute to the control authority. Consequently, $\eta_5 = 0$ or $\eta_6 = 0$, which is consistent with $B_{f5} = \mathbf{0}$ or $B_{f6} = \mathbf{0}$, where B_{fi} denotes the ith column of B_f.
- **(F2)** The loss of control effectiveness of steering controllers, which does not destroy the steering controller, but influences its control gain. In this case, η_5 and η_6 represent the loss of effectiveness factors and are such that $0 < \eta_5 < 1, 0 < \eta_6 < 1$. If $\eta_i = 0$, this faulty case is consistent with F1.

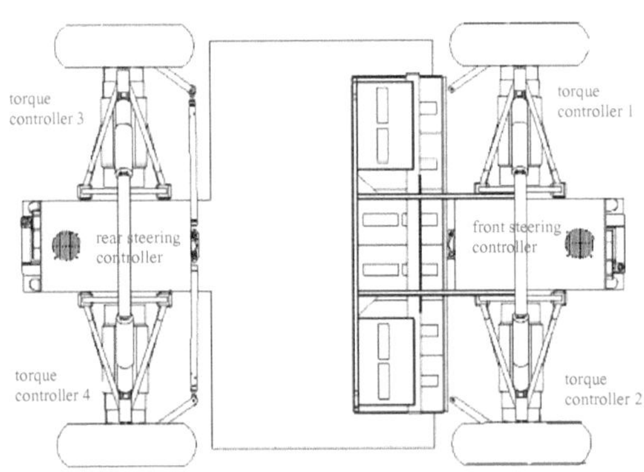

Fig. 6.8 The schematic diagram of the RobuCar^{TM}

- **(F3)** The failure of wheel torque controllers, which may result from the inverter failure, the brake system failure or the wheel motor failure, such that no torque input is generated. In this case, $\eta_i = 0$, or $B_{fi} = \mathbf{0}$, $i \in \{1,2,3,4\}$.
- **(F4)** The loss of control effectiveness of wheel torque controllers, which does not destroy the torque controller, but influences its control gain. Consequently, $0 < \eta_i < 1$, $i \in \{1,2,3,4\}$.

The FTC objective in this work is *to let the vehicle track the reference path in spite of input constraints, the external resistance and actuator faults F1-F4.*

In the sequel, we consider that the torque inputs T_i^r are chosen to overcome the resistance term R, i.e. equations (6.47)-(6.48) are solvable. This implies that at least 2 wheel torque controllers are available. Consequently, equations (6.46) and (6.51) are rewritten as

$$\dot{x} = Ax + Bu \tag{6.52}$$

$$\dot{x} = Ax + B_f u \tag{6.53}$$

where $u = [T_1^c \; T_2^c \; T_3^c \; T_4^c \; \delta_f^c \; \delta_r^c]^\top$.

We will first recall the progressive accommodation (PA) strategy proposed in [118], and analyze its availability in the presence of input constraints, then combine such optimal FTC approach with the CLF based bounded controller. The resulting hybrid control approach takes both advantages of the optimal control and the bounded control.

6.2.3 Hybrid FTC Scheme

The LQ optimal control objective is to transfer the system state from the initial value $x(0) = x_0$ to some final value $x(\infty)$, while minimizing the cost function

$$J(u,x_0) = \int_0^\infty (u^\top R u + x^\top Q x)dt$$

where Q and R are symmetric matrices. From the classical theory, the solution is given by $u = -R^{-1}B^\top P_n x \triangleq -F_n x$ where P_n is the unique positive definite solution of the algebraic Riccati equation $P_n A + A^\top P_n + Q - P_n B R^{-1} B^\top P_n = 0$.

In the practical faulty situations, three time instants, namely t_f, t_{fdi}, t_{ftc} have to be considered, leading to four time windows :

$[0, t_f[$	Nominal system, (A,B) is controlled by $u = -F_n x$
$[t_f, t_{fdi}[$	Diagnostic delay, (A,B_f) is controlled by $u = -F_n x$
$[t_{fdi}, t_{ftc}[$	FTC delay, (A,B_f) is controlled by $u = -F_n x$
$[t_{ftc}, \infty)$	Fault is accommodated, (A,B_f) is controlled by the FTC law

The pair (A,B) is changed into (A,B_f) at time t_f due to the actuator faults. Once B_f has been identified at $t = t_{fdi}$ under some FD schemes, the classic FTC law can be designed as $u = -R^{-1}B_f^\top P_f x$ and applied at $t = t_{ftc}$, where P_f is the unique positive definite solution of

$$P_fA + A^\top P_f + Q - P_fB_fR^{-1}B_f^\top P_f = 0 \tag{6.54}$$

The delay $t_{ftc} - t_{fdi}$ is mainly due to the computation of the Riccati equation (6.54).

The PA strategy aims at minimizing the cost in $t_{ftc} - t_{fdi}$. Such strategy is based on the following Newton-Raphson scheme:

Let P_i be the unique solution of the Lyapunov equation

$$P_i(A_f - B_fF_{i-1}) + (A_f - B_fF_{i-1})^\top P_i = -Q - F_{i-1}^\top RF_{i-1} \tag{6.55}$$

where $F_i = R^{-1}B_f^\top P_i$ for all $i = 1, 2, \cdots$ and the initial F_0 is given.

The Newton-Raphson scheme is one of the effective solutions for (6.54). The computation of (6.55) is much faster than (6.54). The PA strategy is to apply $u_i = -F_ix$ as soon as it is obtained. The system behavior after the fault occurrence is therefore

$$\begin{aligned}
\dot{x} &= (A - B_fF_n)x, \quad t \in [t_f, t_0[\\
\dot{x} &= (A - B_fF_0)x, \quad t \in [t_0, t_1[\\
\dot{x} &= (A - B_fF_i)x, \quad t \in [t_i, t_{i+1}[, \quad i = 1, 2, \ldots
\end{aligned}$$

where $t_0 > t_{fdi}$ and F_0 define the algorithm initialization. It has been proven in [118] that $lim_{i \to \infty} P_i = P_f$, and the PA strategy significantly reduces the loss of cost that results from the classic FTC law in the time delay $t_{ftc} - t_{fdi}$.

Now we consider the input constraints. In the fault-free situation, $u = -F_nx$ is applied. We can find a region

$$\Psi = \{x \in \Re^n | x^\top P_nx \leq r\} \tag{6.56}$$

where r is small enough such that $\forall x \in \Psi$, the ith input $|u_i| < u_i^{\max}(x)$, $\forall i = \{1, \ldots, 6\}$. $u_i^{\max}(x) > 0$ is a constant or a state dependent bound of the ith input from (6.49)-(6.50). It follows that if the initial state $x(0)$ is chosen within Ψ, then $u = -F_nx$ is always available.

In the faulty situation during $[t_i, t_{i+1}[$, the PA control law $u = -F_ix$ is applied. Note that after the fault occurs, T_i^r is adjusted to overcome R, denote

$$\Delta\eta_iT_i^r \triangleq \eta_iT_i^r - \eta_{i(n)}T_i^r{}_{(n)}$$

where $\eta_{i(n)}T_i^r{}_{(n)}$ is related to the normal situation, and $\eta_iT_i^r$ corresponds to the new controller in the faulty case. If all $\eta_i \neq 0$ $i = 1, 2, 3, 4$, then each $\eta_iT_i^r$ keeps a unique constant throughout the process, i.e. $\Delta\eta_iT_i^r = 0$, and does not affect the bound of the T_i^c. Similarly, define the region

$$\bar{\Psi}_i = \{x \in \Re^n | x^\top P_ix \leq \varepsilon_i\} \tag{6.57}$$

where ε_i is small enough such that $\forall x \in \bar{\Psi}_i$, $|u_i| < u_i^{*\max}(x) \triangleq \frac{u_i^{\max}(x) - \Delta\eta_iT_i^r}{\eta_i}$, for $\eta_i \neq 0$, and $u_i^{*\max}(x) = 0$, for $\eta_i = 0$, where $u_i^{\max}(x) - \Delta\eta_iT_i^r$ $(i = 1, 2, 3, 4)$ is assumed to be positive, and T_5^r, T_6^r do not exist. If $\eta_i = 0$, it follows that $u_i = 0$ from the LQ control method. Note that if $x(t_i) \in \bar{\Psi}_i$, then $u = -F_i x$ is available throughout the interval $[t_i, t_{i+1}[$. We also obtain the following property

Proposition 6.2. *If $x(t_i) \in \bar{\Psi}_i$ such that*

$$|(-R^{-1}B_f^\top)_j| \cdot |P_i| \cdot |x| \leq \frac{u_i^{*\max}(x)}{\eta_i}, \forall j = \{1, ..., 6\}$$

then the PA strategy is available throughout the interval $[t_i, \infty)$.

Proof: The result follows the fact that the iterating algorithm (6.55) leads to $P_f \leq \cdots \leq P_{i+1} \leq P_i \leq \cdots \leq P_1$ [118]. Since $x(t_i) \in \bar{\Psi}_i$, then under the controller $-F_i x$, $\bar{\Psi}_i$ is an invariant set for x, i.e., $x(t) \in \bar{\Psi}_i, \forall t \in [t_i, t_{i+1}[$. $|(-R^{-1}B_f^\top)_j| \cdot |P_1| \cdot |x| \leq \frac{u_i^{*\max}(x)}{\eta_i}$ implies that $|(-R^{-1}B_f^\top)_j| \cdot |P_{i+1}| \cdot |x| \leq \frac{u_i^{*\max}(x)}{\eta_i}$, thus $-F_{i+1}x$ is available throughout the interval $[t_{i+1}, t_{i+2}[$, $\bar{\Psi}_i$ is still an invariant set for x. Finally, it can be concluded that the optimal FTC strategy is available for $t \in [t_i, \infty)$. □

Such property is useful to reduce the computation level. If we have checked at $t = t_i$ that $|(-R^{-1}B_f^\top)_j| \cdot |P_1| \cdot |x| \leq \frac{u_i^{*\max}(x)}{\eta_i}$, then we do not have to check at every following instants t_κ, for $\kappa \geq i$.

However, we can not always guarantee the availability of the PA strategy. If $x(t_i) \notin \bar{\Psi}_i$, such strategy would lead to the input saturation and the system's performance will be degraded.

To avoid the input saturation, a CLF based bounded FTC method in Chapter 3.2 is developed, which will be combined with the PA strategy.

Reformulate the faulty system (6.53) as

$$\dot{x} = Ax + Bu + Bf \tag{6.58}$$

where the fault is represented as an additive term Bf. $f = (K - I)u$, I is the unit matrix, and $K = diag[\eta_1, ..., \eta_6]$ with $0 \leq \eta_i \leq 1$ defined in Section 6.2.2. Since the system inputs are bounded, it is reasonable to assume that actuator faults are bounded, i.e., $|f| \leq \bar{f}$, where $\bar{f} > 0$. It is also assumed that $|\Delta\eta_i T_i^r| \leq \bar{\Delta}_i$ for $\bar{\Delta}_i > 0$ and $u_i^{\max}(x) - \bar{\Delta}_i$ $(i = 1, 2, 3, 4)$ is positive.

Consider a Lyapunov function $V = x^\top Px$ for the system (6.58), where P is a positive definite symmetric matrix that satisfies the Riccati equation $A^\top P + PA - PBB^\top P = -W$ for a positive definite matrix W.

V can be regarded as a *control Lyapunov function* for system (6.58). The continuous bounded FTC law can be designed as

$$u_i = -\Upsilon_i(V)(L_{B_i}V)^\top(x) \triangleq b_i(x), \quad i = 1, ..., 6 \tag{6.59}$$

with

$$\Upsilon_i(V) = \begin{cases} \dfrac{\vartheta(V) + \sqrt{\vartheta(V)^2 + (u_i^{*\max}(x)|(L_{B_i}V)^\top|)^4}}{|(L_{B_i}V)^\top|^2\left[1 + \sqrt{1 + (u_i^{*\max}(x)|(L_{B_i}V)^\top|)^2}\right]}, & (L_{B_i}V)^\top \neq 0 \\ 0, & (L_{B_i}V)^\top = 0 \end{cases}$$

where $\vartheta(V) \triangleq \frac{1}{6}(L_{Ax}V + \rho V + |L_I V|\bar{f})$, L denotes the Lie derivative, i.e., $L_{Ax}V = x^\top(A^\top P + PA)x$, $(L_{B_i}V)^\top = 2B_i^\top Px$, and $\rho > 0$. $u_i^{\star\max} \triangleq u_i^{\max}(x) - \bar{\Delta}_i$.

For all initial states, the stability region of system (6.58) is defined by the set

$$\Omega = \{x \in \Re^n | V(x) \leq c^{\max}\} \tag{6.60}$$

where $c^{\max}$ is small enough such that $\vartheta(V) < \min_{i\in\{1,2,\ldots,6\}} u_i^{\star\max}|(L_{B_i}V)^\top|$ for all $x \in \Omega$.

Proposition 6.3. *For the initial state $x(0) \in \Omega$, the bounded controller $u = b(x)$ with $b(x) \triangleq [b_1(x)\ldots b_6(x)]^\top$ in (6.59) makes the origin of the system (6.58) asymptotically stable in spite of faults.*

Proof: The result can be straightly obtained from Lemma 3.1. □

Proposition 6.3 provides a result for the multiple state dependent input constraint form, i.e. $|u_i| < u_i^{\star\max}(x)$, $i = 1,\ldots,6$. It can be seen that for any $x(0) \in \Omega$, the controller $u = b(x)$ can always be applied and does not need to be modified in the presence of faults.

Based on above analysis, a hybrid control method can be provided as

$$u = \begin{cases} -F_n x, & \text{for } x \in \Psi \cap \Omega,\ t \in [0, t_{fdi}[, \text{ with } x(0) \in \Psi \cap \Omega \\ b(x), & \text{for } x \in \Psi \cap \Omega,\ t \in [t_{fdi}, t_1[\\ -F_i x, & \text{for } x \in \bar{\Psi}_i \cap \Omega,\ t \in [t_i, t_{i+1}[\\ b(x), & \text{for } x \notin \bar{\Psi}_i \cap \Omega,\ t \in [t_i, t_{i+1}[,\ i = 1,2,\ldots \end{cases} \tag{6.61}$$

where Ψ, $\bar{\Psi}_i$ and Ω are defined respectively in (6.56), (6.57) and (6.60). Fig. 6.9 shows the block diagram of the control system.

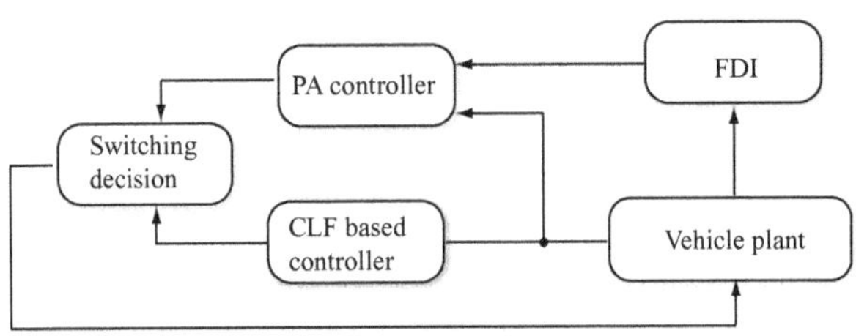

Fig. 6.9 The block diagram of the FTC system

Discussion

1. Compared with the convex conjugacy technique [43] that requires $x \in \Psi$, the bounded hybrid controller (6.61) restricts x into a relative small region $\Psi \cap \Omega$, which, however, leads to a low computation level. Since $b(x)$ can be designed off-line, we do not have to solve the backward Hamiltonian system every time when x reaches the bound of $\Psi \cap \Omega$ as in [43].

2. Since the initial state $x(0) \in \Psi \cap \Omega$, the controller $-F_n x$ ensures that $x(t_f) \in \Psi \cap \Omega$. Nothing can be said about the state trajectory during the diagnosis delay $t_{fdi} - t_f$. Effective diagnosis approaches can significantly shorten this delay. Assuming $x(t_{fdi}) \in \Psi \cap \Omega$ is quite acceptable in the practical application.
3. Applying $b(x)$ at the beginning of the FTC process $[t_{fdi}, t_1[$ shortens the initial time of PA strategy in [118]. Moreover, in each interval $[t_i, t_{i+1}[$, once $x \notin \bar{\Psi}_i \cap \Omega$, the controller $b(x)$ can always make x return to $\bar{\Psi}_i \cap \Omega$ as in Proposition 6.2, such that the controller $-F_i x$ is available.
4. The region Ω in (6.60) is based on a fixed norm bound of faults $\bar{f}$. This region could be zoomed in since faults impossibly exist all the time. The reader can see the related work in Section 3.2.
5. In this work, a straight reference path is considered, i.e. the curvature $\rho_{ref} \approx 0$. For the curving path with $\rho_{ref} \neq 0$, an additional term $[0\ \ 0\ \ 0\ v_0 \rho_{ref}]^\top$ should be added in the system equation (6.46). The robust design of LQ control [119] and CLF based control can be applied.
6. If three wheel torque controllers are faulty (F3), the remaining wheel torque can not overcome the resistance factor. This also leads to the robust problem as in D5.

6.2.4 Simulation Results

The proposed FTC method is now applied to the path tracking of RobuCar^{TM} system (6.46). The parameters are given in Table 6.3. The vehicle starts the path tracking with the initial values $v(0) = 5$ m/s, $\beta(0) = 0$ rad, $\gamma(0) = 0$ rad/s, $y_c(0) = 0.2$ m, and $\phi(0) = 0$ rad. In accordance with the road condition and vehicle data stated in Table 6.1, the resistance factor $\bar{R}$ is assumed to be -0.5 m/s^2. We set the wheel slip constraint as $c = 0.3$, the attenuation factor in (6.49)-(6.50) is $k_i = 0.2$. The input constraints are imposed as -0.0004 Nm $\leq T_i^c \leq 0.0004$ Nm, $(-0.18 + 0.08\gamma + \beta)$ rad $\leq \delta_i^c \leq (0.18 + 0.08\gamma + \beta)$ rad, for $i = 1, ..., 4$. Since the vehicle's speed is around a constant v_0, small torques T_i^c are required. The objective is to obtain the tracking behavior as fast as possible (under no fault and faulty conditions) while maintaining the input constraints. We will consider in the following the four faulty cases F1-F4 described in Section 6.2.2 and will illustrate the tracking performance.

We first consider the fault of wheel torque control system. Suppose that the inverters of the two front wheels broke down at $t = 0.3$ s. These failures make the motor torques of the two wheels become zero, i.e., η_1 and η_2 abruptly change from 1 to 0 after 0.3 s, $B_{f1} = B_{f2} = \mathbf{0}$. The consequence is a big yaw moment and the unstable vehicle motion. In addition, there is a 75 percent loss of control effectiveness of rear right wheels after $t = 0.3$ s, i.e., $\eta_4 = 0.25$.

Both Q and R are chosen as the unit matrices. The classic FTC law $u_f = -F_f x$ can be obtained after 2 iterations of (6.55), i.e., $F_f = F_2$. Assume it takes 0.1 s for fault diagnosis, 0.1 s for the initialization of PA strategy, and 0.1 s for each iteration of PA. The classic FTC approach would apply $-F_n x$ until $t = 0.5$ s and then $-F_2 x$, while the PA strategy applies $-F_n x$ until 0.4 s and then applies $b(x)$ at 0.4 s, and the sequence $-F_1 x$ and $-F_2 x$ at respective times 0.5 and 0.6 s. Fig. 6.10 shows the input

Table 6.3 Parameters of RobuCar^{TM} and the reference path

Parameter	Value
m (kg)	350
l_f (m)	0.401
l_r (m)	0.802
l_d (m)	0.605
r_{ei} (m)	0.350
C_f (N/rad)	2000
C_r (N/rad)	2000
J_z (kgm^2)	82
v_0 (m/s)	5
ρ_{ref} (m^{-1})	0

trajectories. Due to the complete failures of two wheels' inverters, the inputs T_1, T_2 are not provided any more after 0.3 s. The tracking performance is maintained by the tradeoff among T_3, T_4, δ_f and δ_r. Although T_i^r $(i = 1,2,3,4)$ are adjusted abruptly after the fault occurs. The original input constraints imposed on δ_i^c and T_i^c are still available.

Fig.6.10 shows the trajectories of the PA controller output vector. It can be seen that both δ_f^c and δ_r^c are adjusted to compensate for the big yaw moment due to faults. All the inputs are within the constraints, the PA controller is always available. Fig.6.11 shows the torques T_j^r for the resistance rejection, T_1^r and T_2^r are not provided any more after 0.3 s. Once the fault is diagnosed at 0.4 s, both T_3^r and T_4^r are adjusted to overcome the resistance.

Fig.6.12 illustrates the vehicle motion behavior, the tracking goal is achieved at nearly $t = 2$ s. After a very short overshoot at the beginning, v is always maintained at 5 m/s, this validates the linearized model (6.46). The input trajectories and vehicle motion behavior under the classic FTC law are similar as that in Fig. 6.10, thus

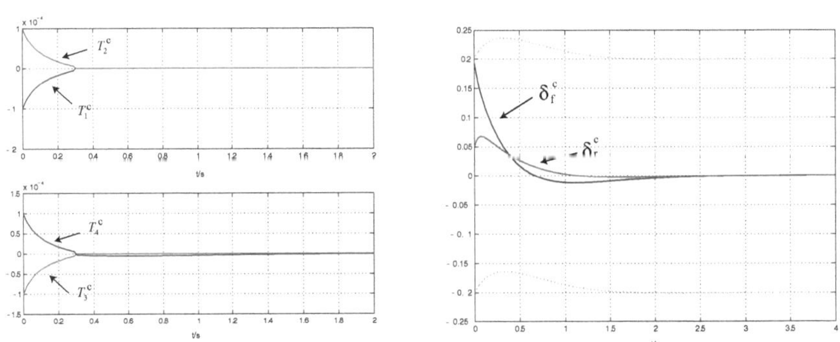

(a) Torque inputs (b) Steering angle inputs

Fig. 6.10 Input trajectories

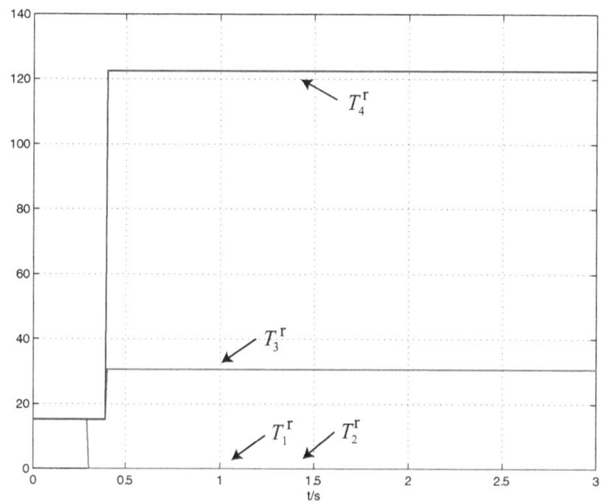

Fig. 6.11 Torque inputs for resistance rejection

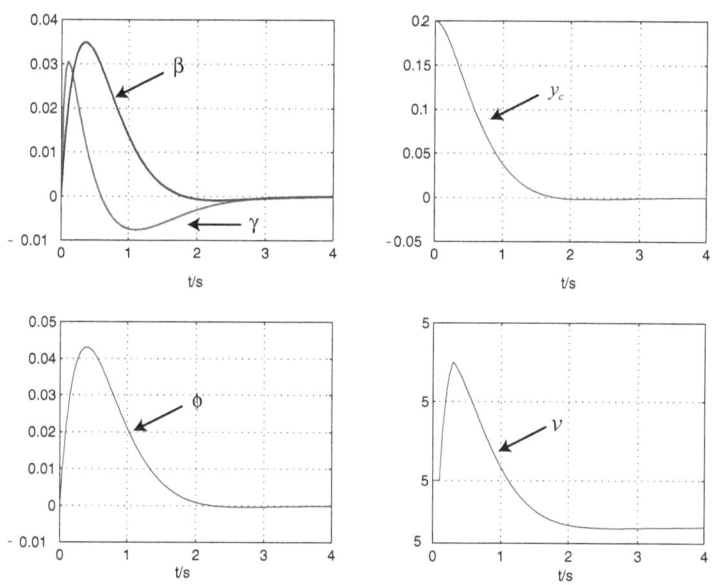

Fig. 6.12 Vehicle motion behavior

are not presented here. Fig.6.13 gives the evolution of the system cost with the classic and PA methods. It is seen that the PA approach widely improves system performance during the fault accommodation transient.

Now we address the fault of steering system. Suppose that the front steering system is broken at $t = 2$ s that leads to $\eta_1 = 0$. Such failure is also consistently

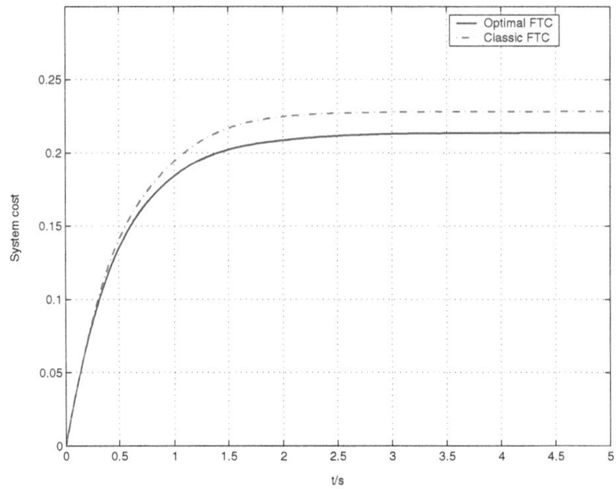

Fig. 6.13 Cost comparison

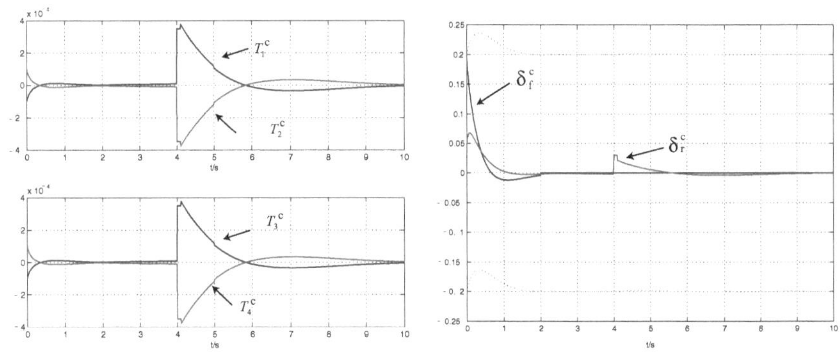

(a) Torque inputs

(b) Steering angle inputs

Fig. 6.14 Input trajectories

represented by $B_{f5} = \mathbf{0}$. In addition, there is a 90 percent loss of control effectiveness in the power amplifier of the rear steering actuator after $t = 2$ s, i.e., $\eta_6 = 0.1$. In this case, the tracking performance is maintained only by applying T_1, T_2, T_3, T_4 and δ_r.

The classic FTC law $u_f = -F_f x$ can be obtained after 3 iterations of (6.55), i.e., $F_f = F_3$. To illustrate our approach, assume it takes 2 s for FD, 0.1 s for the initialization of PA strategy, and 0.9 s for each iteration of PA. However, $-F_1 x$ exceeds the input bound of T_i at $t = 4.1$ s, thus the CLF based controller is applied until 4.2 s, and then the PA controller is activated. $-F_2 x$ satisfies the property of Proposition 6.3 at $t = 5$ s, which implies that the PA control is always available after 5 s.

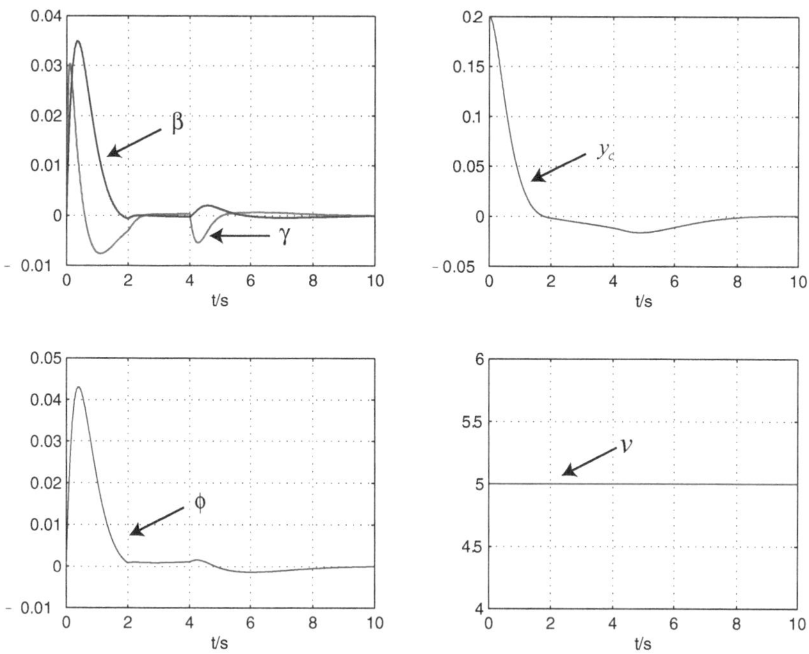

Fig. 6.15 Vehicle motion behavior

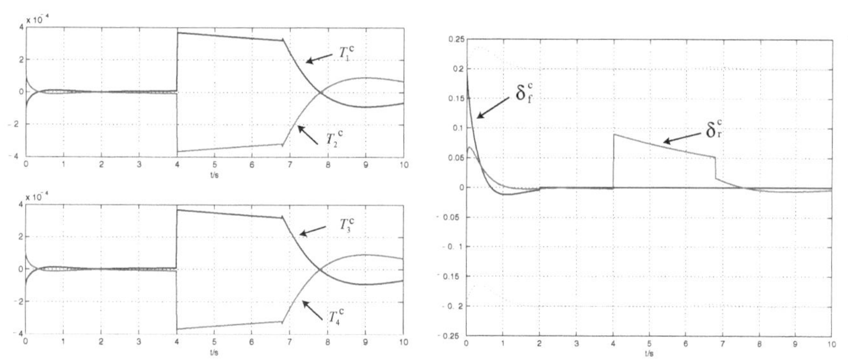

(a) Torque inputs (b) Steering angle inputs

Fig. 6.16 Input trajectories

Fig. 6.14 shows the trajectories of the hybrid controller. It can be seen that all T_i^c are adjusted abruptly to compensate for the faults. All the torques $T_i^r = 15.315$ Nm $i = 1,2,3,4$, which do not change since no fault occurs at torque control system. Fig. 6.16 illustrates the vehicle motion behavior. After $t = 2$ s, the trajectory of the vehicle deviates from the reference path, while the tracking goal is achieved at nearly $t = 8$ s, v is also maintained at 5 m/s. Fig. 6.16 illustrates the trajectories of

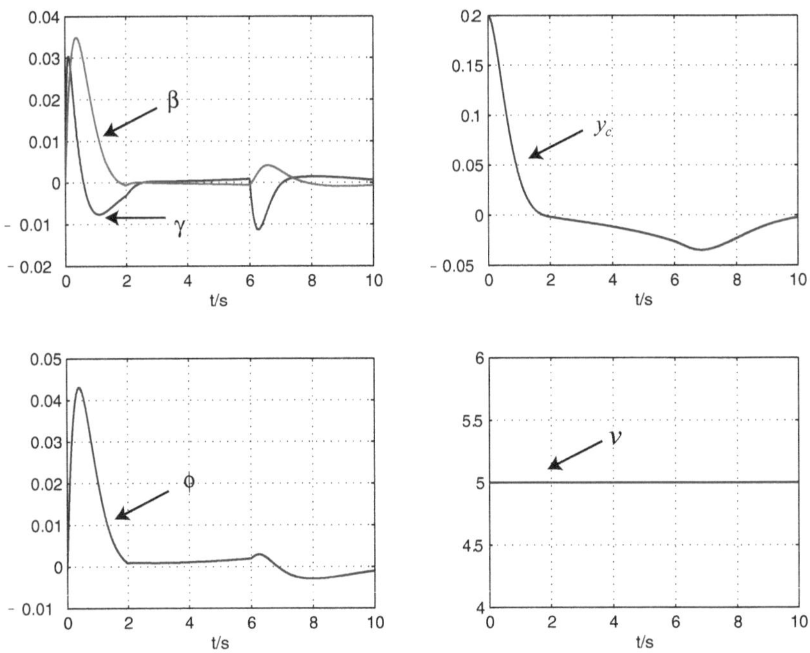

Fig. 6.17 Vehicle motion behavior

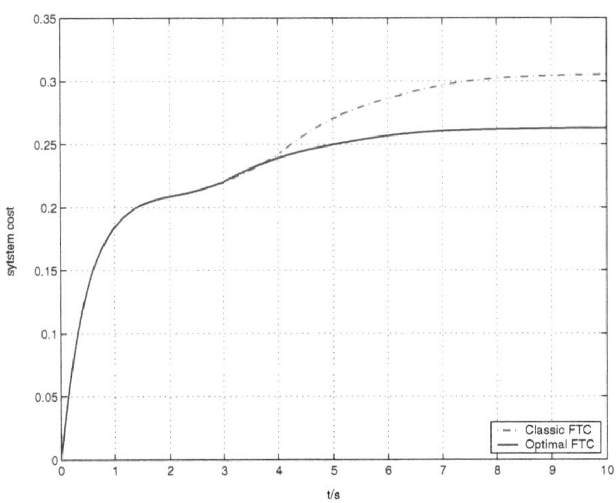

Fig. 6.18 Cost comparison

the controller that combines the classic LQ method and CLF technique. The classic LQ controller exceeds the constraints at 5.9 s and is not applied until 6.8 s. It can be seen that much more control effort has to be made than the hybrid control one.

Fig.6.17 illustrates the vehicle motion behavior. The vehicle tracks the path again at nearly $t = 10$ s. Fig. 6.18 gives the evolution of the system cost with the classic and PA methods, which also implies the good system performance during the fault accommodation transient under the proposed hybrid approach.

6.3 Conclusion

This chapter has discussed the supervisory FTC problem using hybrid system approaches. Three novel switching control based FDI/FTC schemes have been proposed for general nonlinear systems. The good feature of these three switching schemes is that no additional model or filter is needed to compare with the plant. However, how to improve the transient performance deserves further investigations.

This chapter has also proposed an optimal hybrid FTC approach with application to the path tracking control problem for 4WS4WD RobuCar^{TM} vehicle in LAGIS. Several important types of actuator faults are addressed. More directions would be associated with the robust fault tolerant path tracking control design of 4WS4WD vehicles.

Chapter 7
Conclusion and Future Research Directions

FTC of HS is a hot research topic that intersects two communities of fault diagnosis/tolerance and HS. This book has presented several interesting theories and applications on FTC for HS. It has been shown that both the continuous system theories and DES theories can be applied. This conclusion seems natural since HS consists of continuous and discrete dynamics. However, it deserves to point out that the utilizations of these two main theories in HS field are quite different from that in their own fields.

Due to the special structures and properties of HS, many non-hybrid system FTC methods are unavailable directly for HS. Continuous system theories for non-hybrid systems have to be modified and the switching properties must be taken into account, the difficulty of such work are reflected in Chapters 2-4. DES theories also can not be applied directly. Compared with pure DES, the continuous dynamics of HS have to be considered as indicated in Chapter 5.

There are still many open problems to be further investigated. We shall conclude this book by providing some future research directions, which we hope could be a helpful guide to interested readers when exploring FTC for HS.

1. To consider optimality as a FTC goal besides the continuous stability and the discrete specification. The optimality is very important for the modern systems with considerations for the environment and energy problems. Optimal FTC goal not only requires the stability of the faulty systems but also needs it to be as optimal as possible in spite of faults. Such goal could be potentially achieved by combining the optimal theories of HS [6, 97] and the proposed FTC methods in this book.
2. To relax the constraints about the structure of HS, e.g., consider the stability at non-zero equilibriums. Many HS that are widely used in process control have non-zero equilibriums [83]. On the other hand, the time-variant continuous vector fields as described in [65] also deserve further investigations.

H. Yang et al.: Fault Tolerant Control Design for Hybrid Systems, LNCIS 397, pp. 181–182.
springerlink.com

3. To combine continuous system theories with DES ones such that an integrated fault tolerance framework can be provided with application to real systems. In many real situations, a complex system may have various faults (both continuous and discrete ones) occurring simultaneously. The nondeterministic finite automata model developed in [77] maybe a good tool to address this issue.

References

1. Ackermann, J.: Robust Control. Springer, London (1993)
2. Alur, R., Henzinger, T.A., Lafferriere, G., Pappas, G.J.: Discrete abstraction of hybrid systems. Proc. of IEEE 88, 971–984 (2000)
3. Asarin, E., Schneider, G., Yovine, S.: Algorithmic analysis of polygonal hybrid systems, part I: Reachability. Theoretical Computer Science 379, 1-2(12), 231–265 (2007)
4. Basile, F., Carbone, C., Chiacchio, P., Boel, R.K., Avram, C.C.: A hybrid model for urban traffic control. In: Proc. of 2004 IEEE International Conference on Systems, Man and Cybemetics, The Hague, Netherlands, pp. 1795–1800 (2004)
5. Batt, G., Ropers, D., de Jong, H., Geiselmann, J., Page, M., Schneider, D.: Qualitative analysis and verification of hybrid models of genetic regulatory networks. In: Morari, M., Thiele, L. (eds.) HSCC 2005. LNCS, vol. 3414, pp. 134–150. Springer, Heidelberg (2005)
6. Bemporad, A., Giorgetti, N.: Logic-based methods for optimal control of hybrid systems. IEEE Trans. Automatic Control 51(6), 963–976 (2006)
7. Benveniste, A., Fabre, E., Haar, S., Jard, C.: Diagnosis of asynchronous discrete-event systems: a net unfolding approach. IEEE Trans. on Automatic Control 48(5), 714–727 (2003)
8. Bernard, O.: Global qualitative description of a class of nonlinear dynamical systems. Artificial Intelligence 136(1), 29–59 (2002)
9. Bhat, S., Bernstein, D.: Finite-time stability of continiuous automonous systems. SIAM Journal of Control and Optimization 38(3), 751–766 (2000)
10. Blanke, M., Kinnaert, M., Lunze, J., Staroswiecki, M.: Diagnosis and Fault-Tolerant Control, 2nd edn. Springer, Heidelberg (2006)
11. Blanke, M., Thomsen, J.S.: Electrical steering of vehicles: fault tolerant analysis and design. Microelectronics and Reliability 46(9-11), 1421–1432 (2006)
12. Bošković, J.D., Mehra, R.K.: Stable multiple model adaptive flight control for accommodation of a large class of control effector failures. In: Proc. of the 1999 American Control Conference, pp. 1920–1924 (1999)
13. Branicky, M.S.: Multiple Lyapunov functions and other analysis tools for switched and hybrid systems. IEEE Transactions on Automatic Control 43(1), 475–482 (1998)
14. Byrnes, C.I., Isidori, A., Willems, J.C.: Passivity, feedback equivalence, and the global stabilization of minimum phase nonlinear systems. IEEE Transactions on Automatic Control 36(11), 1228–1240 (1991)
15. Cao, Y., Ying, M.: Similarity-based supervisory control of discrete-event systems. IEEE Trans. Automatic Control 51(2), 325–330 (2006)

16. Chan, C.C.: The state of the art of electric, hybrid, and fuel cell vehicles. Proc. of the IEEE 95(4), 704–718 (2007)
17. Chatterjee, D., Liberzon, D.: On stability of randomly switched nonlinear systems. IEEE Trans. on Automatic Control 52(12), 2390–2394 (2007)
18. Chen, J., Patton, R.J.: Robust Model-based Fault diagnosis for Dynamics Systems. Kluwer Academic Publishers, Boston (1999)
19. Chiasson, J.N.: Nonlinear differential-geometric techniques for control of a series dc moto. IEEE Trans. Control System Technology 2(1), 35–42 (1994)
20. Cocquempot, V., El Mezyani, T., Staroswiecki, M.: Fault detection and isolation for hybrid systems using structured parity residuals. In: Proc. of 5th Asian Control Conference, pp. 1204–1212 (2004)
21. David, R., Alla, H.: On hybrid Petri nets. J. Discrete Event Dynamic Systems:Theory and Applications 11(1-2), 9–40 (2001)
22. Decarlo, R.A., Branicky, M.S., Pettersson, S., Lennartson, B.: Perspectives and results on the stability and stabilizability of hybrid systems. Proceedings of the IEEE 88(7), 1069–1082 (2000)
23. Ding, S.X.: Model-based Fault Diagnosis Techniques: Design Schemes, Algorithms, and Tools. Springer, Heidelberg (2008)
24. Dotoli, M., Fanti, M.P., Mangini, A.M.: Real time identification of discrete event systems by Petri nets. In: Proc. of 1st IFAC Workshop on Dependable Control of Discrete Systems, ENS Cachan, France (2007)
25. Dotoli, M., Fanti, M.P.: An Urban Traffic Network Model via Coloured Timed Petri Nets. Control Engineering Practice 14(10), 1213–1229 (2006)
26. Dumont, P.E., Aïtouche, A., Bayart, M.: Fault detection of actuator faults for electric vehicle. In: Proc. of 16th IEEE International Conference on Control Applications, Mumbai, India, pp. 1067–1072 (2007)
27. Dumont, P.E., Aïtouche, A., Merzouki, R., Bayart, M.: Fault tolerant control on an electric vehicle. In: Proc. of International Conference on Industrial Technology, Singapore, pp. 2450–2455 (2006)
28. El-Farra, N.H., Mhaskar, P., Christofides, P.D.: Output feedback control of switched nonlinear systems using multiple lyapunov functions. Systems and Control Letters 54(12), 1163–1182 (2005)
29. Ezzine, J., Haddad, A.H.: Controllability and observability of hybrid systems. Int. J. Control 49(6), 2045–2055 (1989)
30. Febbraro, A.D., Giglio, D., Sacco, N.: Urban traffic control structure based on hybrid Petri nets. IEEE Trans. on Intelligent Transportation Systems 5(4), 224–237 (2004)
31. Febbraro, A.D., Sacco, N.: On modelling urban transportation networks via hybrid Petri nets. Control Engineering Practice 12(10), 1225–1239 (2004)
32. Feng, W., Zhang, J.F.: Stability analysis and stabilization control of multi-variable switched stochastic systems. Automatica 42(1), 169–176 (2006)
33. Fischer, D., Börner, M., Schmitt, J., Isermann, R.: Fault detection for lateral and vertical vehicle dynamics. Control Engineering Practice 15(3), 315–324 (2007)
34. Floquet, T., Barbot, J.P., Perruquetti, W., Djemai, M.: On the robust fault detection via a sliding mode disturbance observer. Int. J. Control 77(7), 622–629 (2004)
35. Gao, Z., Ding, S.X., Ma, Y.: Robust fault estimation for vehicle lateral dynamic systems. In: Proc. of IFAC Safeprocess 2006, Beijing, China, pp. 1039–1043 (2006)
36. Gertler, J.J.: Fault Detection and Diagnosis in Engineering Systems. Marcel Dekker, New York (1998)
37. Ghosh, M.K., Arapostathis, A., Marcus, S.: Ergodic control of switching diffusions. SIAM J. Control and Optimization 35(6), 1952–1988 (1997)

38. Girault, A.: A hybrid controller for autonomous vehicles driving on automated highways. Transportation Research Part C: Emerging Technologies 12(6), 421–452 (2004)
39. Giua, A., Seatzu, C.: Observability of place/transition nets. IEEE Trans. on Automatic Control 47(9), 1424–1437 (2002)
40. Giua, A., Seatzu, C., Basile, F.: Observer-based state-feedback control of timed Petri nets with deadlock recovery. IEEE Trans. on Automatic Control 49(1), 17–29 (2004)
41. Giua, A., Seatzu, C.: Fault detection for discrete event systems using Petri nets with unobservable transitions. In: Proc. of the joint 44th IEEE Conference on Decision and Control, European Control Conference, pp. 6323–6328 (2005)
42. Goebel, R., Sanfelice, R., Teel, A.R.: Hybrid dynamical systems. IEEE Control Systems Magazine 29(2), 28–93 (2009)
43. Goebel, R., Subbotin, M.: Continuous time linear quadratic regulator with control constraints via convex duality. IEEE Trans. on Automatic Control 52(5), 886–892 (2007)
44. Grizzle, J.W., Abba, G., Plestan, F.: Asymptotically Stable Walking for Biped Robots: Analysis via Systems with Impulse Effects. IEEE Trans. on Automatic Control 46(1), 51–64 (2001)
45. Guan, Z., Hill, D.J., Shen, X.: On hybrid impulsive and switching systems and application to nonlinear control. IEEE Trans. on Automatic Control 50(7), 1058–1062 (2005)
46. Guéguen, H., Zaytoon, J.: On the formal verification of hybrid systems. Control Engineering Practice 12(10), 1253–1267 (2004)
47. Hespanha, J.P.: Uniform stability of switched linear systems: extensions of LaSalle's invariance principle. IEEE Trans. on Automatic Control 49(4), 470–482 (2004)
48. Hespanha, J.P., Liberzon, D., Angeli, D., Sontag, E.D.: Nonlinear norm-observability notions and stability of switched systems. IEEE Trans. on Automatic Control 50(2), 154–168 (2005)
49. Hespanha, J.P., Morse, A.S.: Stability of switched systems with average dwell time. In: Proceedings of the 38th IEEE Conference on Decision and Control, Phoenix, USA, pp. 2655–2660 (1999)
50. Holloway, L.E., Krogh, B.H.: Synthesis of feedback logic for a class of controlled Petri nets. IEEE Trans. on Automatic Control 35(5), 514–523 (1990)
51. Hsieh, F.-S.: Fault-tolerant deadlock avoidance algorithm for assembly processes. IEEE Trans. on Systems, Man, and Cybernetics-Part A: Systems and Humans 34(1), 65–79 (2004)
52. Huang, J.: Nonlinear Output Regulation: Theory and Applications. SIAM, Philadelphia (2004)
53. Iordache, M.V., Antsaklis, P.J.: Supervision based on place invariants: a survey. J. Discrete Event Dynamic Systems:Theory and Applications 16(4), 451–492 (2006)
54. Isermann, R.: Fault-Diagnosis Systems: An Introduction from Fault Detection to Fault Tolerance. Springer, Heidelberg (2006)
55. Isidori, A.: Nonlinear Control Systems, 3rd edn. Springer, London (1995)
56. Jiang, B., Staroswiecki, M., Cocquempot, V.: Fault diagnosis based on adaptive observer for a class of nonlinear systems with unknown parameters. Int. J. Control 77(4), 415–426 (2004)
57. Jiang, B., Staroswiecki, M., Cocquempot, V.: Fault accommodation for nonlinear dynamic systems. IEEE Trans. on Automatic Control 51(9), 1578–1583 (2006)
58. Jiang, B., Wang, J.L., Soh, Y.C.: An adaptive technique for robust diagnosis of faults with independent effect on system output. Int. J. Control 75(11), 792–802 (2002)
59. Jiang, B., Chowdhury, F.N.: Parameter fault detection and estimation of a class of nonlinear systems using observers. Journal of the Franklin Institute 342, 725–736 (2005)

60. Jiang, B., Yang, H., Shi, P.: Switching fault tolerant control design via global dissipativity. Int. J. Systems Science (in press, 2009)
61. Kailath, T.: Linear Systems. Prentice-Hall, Englewood Cliffs (1980)
62. Khalil, H.K.: Nonlinear Systems, 3rd edn. Prentice-Hall, Upper Saddle River (2002)
63. Krstic, M., Kanellakopoulos, I., Kokotovic, P.: Nonlinear and Adaptive Control Design. Wiley, New York (1995)
64. Kuipers, B.J.: Qualitative Reasoning: Modeling and Simulation with Incomplete Knowledge. MIT Press, Cambridge (1994)
65. Lee, T.C., Jiang, Z.P.: Uniform asymptotic stability of nonlinear switched systems with an application to mobile robots. IEEE Trans. Automatic Control 53(5), 1235–1252 (2008)
66. Lee, S., Park, M.: State observer for MIMO nonlinear systems. IEE Proceedings: Control Theory and Applications 150(4), 421–426 (2003)
67. Lefebvre, D., Delherm, C.: Diagnosis of DES with Petri net models. IEEE Trans. on Automation Science and Engineering 4(1), 114–118 (2007)
68. Li, Z.G., Wen, C.Y., Soh, Y.C.: Observer-based stabilization of switching linear systems. Automatica 39(3), 517–524 (2003)
69. Li, Z.G., Soh, Y.C., Wen, C.Y.: Switched and Impulsive Systems: Analysis, Design and Application. Springer, Berlin (2005)
70. Liberzon, D., Morse, A.S., Sontag, E.D.: Output-input stability and minimum-phase nonlinear systems. IEEE Trans. Automatic Control 47(4), 422–436 (2002)
71. Liberzon, D.: Output-input stability implies feedback stabilization. Systems and Control Letters 53(4), 237–248 (2004)
72. Liberzon, D.: Switching in Systems and Control. Birkhauser, Boston (2003)
73. Lin, Y., Sontag, E.D.: A universal formula for stabilization with bounded controls. Systems and Control Letters 16, 393–397 (1991)
74. Liu, B., Marquez, H.J.: Quasi-exponential input-to-state stability for discrete-time impulsive systems. Int. J. Control 80(4), 540–554 (2007)
75. Liu, S.J., Zhang, J.F., Jiang, Z.P.: Decentralized adaptive output-feedback stabilization for large-scale stochastic nonlinear systems. Automatica 43(2), 238–251 (2007)
76. Liu, Y., Zhao, J.: Output regulation of a class of switched linear systems with disturbances. In: Proc. of the American Control Conference, Arlington, VA, USA, pp. 25–27 (2001)
77. Lunze, J., Nixdorf, B.: Discrete reachability of hybrid systems. International Journal of Control 76(14), 1453–1468 (2003)
78. Lunze, J., Nixdorf, B., Richter, H.: Process supervision by means of a hybrid model. Journal of Process Control 11(1), 89–104 (2001)
79. Lunze, J.: Fault diagnosis of discretely controlled continuous systems by means of discrete-event models. J. Discrete Event Dyn. Syst. 18(2), 181–210 (2007)
80. Lygeros, J., Johansson, K.H., Simić, S.N., Zhang, J., Sastry, S.: Dynamical properties of hybrid automata. IEEE Trans. Automatic Control 48(1), 2–17 (2003)
81. Mahmoud, M., Jiang, J., Zhang, Y.M.: Stochastic stability analysis of fault-tolerant control systems in the presence of noise. IEEE Trans. on Automatic Control 46(11), 1810–1815 (2001)
82. Marino, R., Tomei, P.: Nonlinear Control Design–Geometric, Adaptive and Robust. Prentice Hall, London (1995)
83. Mhaskar, P., El-Farra, N.H., Christofides, P.D.: Predictive control of switched nonlinear systems with schedules mode transitions. IEEE Trans. on Automatic Control 50(11), 1670–1680 (2005)

84. Miller, R.K., Michel, A.N.: Ordinary Differential Equations. Academic Press, New York (1982)
85. Moody, J.O., Antsaklis, P.J.: Petri net supervisors for DES with uncontrollable and unobservable transitions. IEEE Trans. on Automatic Control 45(3), 462–476 (2000)
86. Moriwaki, K.: Autonomous steering control for electric vehicles using nonlinear state feedback H_∞ control. Nonlinear Analysis 63(5-7), e2257–e2268 (2005)
87. Mostavi, M.R., Shariatpanahi, M., Kazemi, R.: A novel optimal four wheel steering control. In: Proc. of IEEE International Conference on Industrial Technology, Tunisia, pp. 1596–1601 (2004)
88. Munoz de la Pena, D., Christofides, P.: Stability of nonlinear asynchronous systems. In: Proceedingsof 46th IEEE Conference on Decision and Control, New Orleans, LA, USA, pp. 4576–4583 (2007)
89. Murata, T.: Petri nets: properties, analysis and applications. Proc. of IEEE 77(4), 541–580 (1989)
90. Narasimhan, S., Biswas, G.: Model-based diagnosis of hybrid systems. IEEE Trans. on Systems, Man, and Cybernetics-Part A: Systems and Humans 37(3), 348–361 (2007)
91. Ortega, R., Van Der Schaft, A.J., Mareels, I., Maschke, B.: Putting energy back in control. IEEE Control Systems Magazine 21(2), 18–33 (2001)
92. Paoli, A., Lafortune, S.: Safe diagnosability for fault-tolerant supervision of discrete-event systems. Automatica 41(8), 1335–1347 (2005)
93. Parisini, T., Sacone, S.: Fault diagnosis and controller re-configuration: an hybrid approach. In: Proc. of IEEE ISIC/CIRA/ISAS Joint Conference, pp. 163–168 (1998)
94. Parisini, T., Sacone, S.: Stable hybrid control based on discrete-event automata and receding-horizon neural regulators. Automatica 37, 1279–1292 (2001)
95. Patton, R.J., Frank, P.M., Clark, R.N.: Issues of fault diagnosis for dynamic systems. Spring, London (2000)
96. Peng, S.T.: On one approach to constraining the combined wheel slip in the autonomous control of a 4WS4WD vehicle. IEEE Trans. on Control Systems Technology 15(1), 168–175 (2007)
97. Pepyne, D.L., Cassandras, C.G.: Optimal control of hybrid systems in manufacturing. Proc. of the IEEE 88(7), 1108–1123 (2000)
98. Pola, G., Bujorianu, M.L., Lygeros, J., Benedetto, M.D.D.: Stochastic hybrid models: an overview. In: Proc. of the IFAC Conference on Analysis and Design of Hybrid System (2003)
99. Rajamani, R., Howell, A.S., Chen, C., Hedrick, J.K., Tomizuka, M.: A complete fault diagnostic system for automated vehicles operating in a platoon. IEEE Trans. on Control System Technology 9(4), 553–564 (2001)
100. Raksincharoensak, P., Nagai, M., Mouri, H.: Investigation of automatic path tracking control using four-wheel steering vehicle. In: Proc. of IEEE International Vehicle Electronics Conference, Noordwijk, Netherlands, pp. 73–77 (2001)
101. Ramadge, P.J., Wonham, W.M.: The control of discrete event systems. Proc. of IEEE 77(1), 81–98 (1989)
102. Ramírez-Treviño, A., Ruiz-Beltrán, E., Rivera-Rangel, I., López-Mellado, E.: Online fault diagnosis of discrete event systems: a Petri net-based approach. IEEE Trans. on Automation Science and Engineering 4(1), 31–39 (2007)
103. Qu, Z., Ihlefeld, C.M., Jin, Y., Saengdeejing, A.: Robust fault-tolerant self-recovering control of nonlinear uncertain systems. Automatica 39(10), 1763–1771 (2003)
104. Rajamani, R., Cho, Y.M.: Existence and design of observers for nonlinear systems: relation to distance to unobservability. Int. J. Control 69(5), 717–731 (1998)

105. Saboori, A., Zad, S.H.: Robust nonblocking supervisory control of discrete-event systems under partial observation. Systems and Control Letters 55(10), 839–848 (2006)
106. Sakai, S., Sado, H., Hori, Y.: Motion Control in an electric vehicle with four independently driven in-wheel motors. IEEE/ASME Trans. on Mechatronics 4(1), 9–16 (1999)
107. Sampath, M., Lafortune, S., Teneketzis, D.: Active diagnosis of discrete event systems. IEEE Trans. on Automatic Control 43(7), 908–929 (1998)
108. Sánchez, A.M., Montoya, F.J.: Safe supervisory control under observability failure. J. Descrete Event Dynamics Systems 16(4), 493–525 (2006)
109. Sava, A.T., Alla, H.: Combining hybrid Petri nets and hybrid automata. IEEE Trans. on Robotics and Automation 17(5), 670–678 (2001)
110. Seatzua, C., Gromovb, D., Raischb, J., Corona, D., Giua, A.: Optimal control of discrete-time hybrid automata under safety and liveness constraints. Nonlinear Analysis 65(6), 1188–1210 (2006)
111. Silva, M., Recalde, L.: On fluidification of Petri nets: from discrete to hybrid and continuous models. Annual Reviews in Control 28(2), 253–266 (2004)
112. Skorohod, A.V.: Asymptotic Methods in the Theory of Stochastic Differential Equations. American Mathematical Society, Providence (1989)
113. Sontag, E., Wang, Y.: New characterizations of input-to-state stability. IEEE Trans. on Automatic Control 41(9), 1283–1294 (1996)
114. Sontag, E., Wang, Y.: On characterizations of the input-to-state stability property. Systems and Control Letters 24(5), 351–359 (1995)
115. Sontag, E.D.: Smooth stabilization implies coprime factorization. IEEE Trans. on Automatic Control 34(4), 435–443 (1989)
116. Spong, M.I., Marino, R., Peresada, S.M., Taylor, D.G.: Feedback linearizing control of switched reluctance motors. IEEE Trans. on Automatic Control 32(5), 371–379 (1987)
117. Staroswiecki, M., Gehin, A.-L.: From control to supervision. Annual Reviews in Control 25, 1–11 (2001)
118. Staroswiecki, M., Yang, H., Jiang, B.: Progressive accommodation of parametric faults in linear quadratic control. Automatica 43(12), 2070–2076 (2007)
119. Staroswiecki, M.: Robust fault tolerant linear quadratic control based on admissible model matching. In: Proc. of 45th IEEE Conference on Decision and Control, pp. 3506–3511 (2006)
120. Straky, H., Kochem, M., Schmitt, J., Isermann, R.: Influences of braking system faults on vehicle dynamics. Control Engineering Practice 11(3), 337–343 (2003)
121. Tiwari, A., Khanna, G.: Series abstractions for hybrid automata. In: Tomlin, C.J., Greenstreet, M.R. (eds.) HSCC 2002. LNCS, vol. 2289, pp. 465–478. Springer, Heidelberg (2002)
122. Tomlin, C.J., Mitchell, I., Bayen, A.M., Oishi, M.: Computational techniques for the verification of hybrid systems. Proceedings of the IEEE 91(7), 986–1001 (2003)
123. Trontis, A., Spathopoulos, M.P.: Hybrid control synthesis for eventuality specifications using level set methods. Int. J. Control 76(16), 1599–1627 (2003)
124. Unger, I., Isermann, R.: Fault tolerant sensors for vehicle dynamics control. In: Proc. of 2006 American Control Conference, Minnesota, USA, pp. 3948–3953 (2007)
125. Vu, L., Chatterjee, D., Liberzon, D.: Input-to-state stability of switched systems and switching adaptive control. Automatica 43(4), 639–646 (2007)
126. Wang, W., Zhou, D.H., Li, Z.: Robust state estimation and fault diagnosis for uncertain hybrid systems. Nonlinear Analysis 65(12), 2193–2215 (2006)
127. Willems, J.C.: Dissipative Dynamical Systems. European Journal of Control 13, 134–151 (2007)

128. Wu, Y., Hadjicostis, C.N.: Algebraic approaches for fault identification in discrete-event systems. IEEE Trans. on Automatic Control 50(12), 2048–2053 (2005)
129. Xie, W.X., Wen, C.Y., Li, Z.G.: Input-to-state stabilization of switched nonlinear systems. IEEE Trans. on Automatic Control 46(7), 1111–1116 (2001)
130. Xu, A., Zhang, Q.: Nonlinear system fault diagnosis based on adaptive estimation. Automatica 40(7), 1181–1193 (2004)
131. Yang, H., Jiang, B., Staroswiecki, M.: Supervisory fault tolerant control for a class of uncertain nonlinear systems. Automatica 45(10), 2319–2324 (2009)
132. Yang, H., Jiang, B., Cocquempot, V.: A fault tolerant control framework for periodic switched nonlinear systems. Int. J. of Control 82(1), 117–129 (2009)
133. Yang, H., Jiang, B., Cocquempot, V.: Fault tolerance analysis for stochastic systems using switching diffusion processes. Int. J. of Control 82(8), 1516–1525 (2009)
134. Yang, H., Cocquempot, V., Jiang, B.: Robust fault tolerant tracking control with applications to hybrid nonlinear systems. IET Control Theory and Applications 3(2), 211–224 (2009)
135. Yang, H., Cocquempot, V., Jiang, B.: On stabilization of switched nonlinear systems with unstable modes. Systems and Control Letters 58(10-11), 703–708 (2009)
136. Yang, H., Jiang, B., Cocquempot, V.: Observer-based fault-tolerant control for a class of hybrid impulsive systems. Int. J. of Robust and Nonlinear Control (in press, 2009)
137. Yang, H., Cocquempot, V., Jiang, B.: Optimal fault tolerant path tracking control for 4WS4WD electric vehicles. IEEE Trans. on Intelligent Transportation Systems (in press, 2009)
138. Yang, H., Cocquempot, V., Jiang, B.: Fault tolerant control for a class of hybrid systems with uncontrollable switching. Int. J. of Systems Science 40(10), 1063–1075 (2009)
139. Yang, H., Jiang, B., Cocquempot, V., Shi, P.: Fault tolerant control design via hybrid Petri nets. Asian J. of Control (in press, 2009)
140. Yang, H., Jiang, B., Cocquempot, V.: Qualitative fault tolerant analysis for a class of hybrid systems. Nonlinear Analysis: Hybrid Systems 2(3), 846–861 (2008)
141. Yang, H., Cocquempot, V., Jiang, B.: Fault tolerance analysis for switched systems via global passivity. IEEE Trans. on Circuits and Systems II 55(12), 1279–1283 (2008)
142. Yang, H., Jiang, B., Staroswiecki, M.: Observer based fault tolerant control for a class of switched nonlinear systems. IET Control Theory and Applications 1(5), 1523–1532 (2007)
143. Yang, H., Jiang, B., Cocquempot, V.: Observer-based fault tolerant control for constraint switched systems. Int. J. of Control, Automation and Systems 5(6), 707–711 (2007)
144. Yang, H., Jiang, B., Cocquempot, V., Staroswiecki, M.: Adaptive fault tolerant strategy for a class of hybrid systems with faults independently effecting on outputs. In: Proc. of 6th IFAC Symposium on Fault Detection, Supervision and Safety of Technical Processes, Beijing, China, pp. 1021–1026 (2006)
145. Yang, H., Jiang, B., Cocquempot, V.: Fault accommodation for hybrid systems with continuous and discrete faults. In: Proc. of 10th International Conference on Hybrid Systems: Computation and Control, Pisa, Italy (2007)
146. Yang, H., Cocquempot, V., Jiang, B.: Fault tolerant strategy for hybrid longitudinal control system of automated vehicles. In: Proc. of 46th IEEE Conference on Decision and Control, New Orleans, LA, USA, pp. 3176–3181 (2007)
147. Yang, Z.: An algebraic approach towards the controllability of controlled switching linear hybrid systems. Automatica 38(7), 1221–1228 (2002)
148. Ye, H., Michel, A.N., Hou, L.: Stability theory for hybrid dynamical systems. IEEE Transactions on Automatic Control 43(4), 461–474 (1998)

149. Yoon, T.-W., Kim, J.-S., Morse, A.S.: Supervisory control using a new control-relevant switching. Automatica 43, 1791–1798 (2007)
150. Yu, W., Moreno, M.A., Li, X.: Observer-based neuro identifier. IEE Proceedings: Control Theory and Applications 147(2), 145–152 (2000)
151. Zefran, M., Bullo, F., Stein, M.: A notion of passivity for hybrid systems. In: Proceedings of the 40th IEEE Conference on Decision and Control, Orlando, Florida, pp. 768–773 (2001)
152. Zhang, X., Parisini, T., Polycarpou, M.M.: Adaptive fault-tolerant control of nonlinear uncertain systems: an information-based diagnostic approach. IEEE Trans. Automatic Control 49(8), 1259–1274 (2004)
153. Zhang, Y.M., Jiang, J.: Integrated active fault-tolerant control using IMM approach. IEEE Trans. on Aerospace and Electronic Systems 37(4), 1221–1235 (2001)
154. Zhang, Y.M., Jiang, J.: Bibliographical review on reconfigurable fault-tolerant control systems. Annual Reviews in Control 32(2), 229–252 (2008)
155. Zhao, F., Koutsoukos, X., Haussecker, H., Reich, J., Cheung, P.: Monitoring and fault diagnosis of hybrid systems. IEEE Trans. on Systems, Man, and Cybernetics-Part B: Cybernetics 99(14), 1225–1240 (2005)
156. Zhao, J., Hill, D.J.: Passivity and stability of switched systems: A multiple storage function method. System and Control Letters 57(2), 158–164 (2008)
157. `http://www.robosoft.fr/eng/`

Index

Lecture Notes in Control and Information Sciences

Edited by M. Thoma, F. Allgöwer, M. Morari

Further volumes of this series can be found on our homepage:
springer.com

Vol. 397: Yang, H.; Jiang, B.; Cocquempot, V.:
Fault Tolerant Control Design for
Hybrid Systems
191 p. 2010 [978-3-642-10680-4]

Vol. 396: Kozlowski, K. (Ed.):
Robot Motion and Control 2009
475 p. 2009 [978-1-84882-984-8]

Vol. 395: Talebi, H.A.:
Neural Network-Based State
Estimation of Nonlinear Systems
appro. 200 p. 2010 [978-1-4419-1437-8]

Vol. 394: Pipeleers, G.; Demeulenaere, B.;
Swevers, J.:
Optimal Linear Controller Design for
Periodic Inputs
177 p. 2009 [978-1-84882-974-9]

Vol. 393: Ghosh, B.K.; Martin, C.F.;
Zhou, Y.:
Emergent Problems in Nonlinear
Systems and Control
285 p. 2009 [978-3-642-03626-2]

Vol. 392: Bandyopadhyay, B.; Deepak, F.;
Kim, K.-S.:
Sliding Mode Control Using Novel Sliding
Surfaces
137 p. 2009 [978-3-642-03447-3]

Vol. 391: Khaki-Sedigh, A.; Moaveni, B.:
Control Configuration Selection for
Multivariable Plants
232 p. 2009 [978-3-642-03192-2]

Vol. 390: Chesi, G.; Garulli, A.;
Tesi, A.; Vicino, A.:
Homogeneous Polynomial Forms for
Robustness Analysis of Uncertain Systems
197 p. 2009 [978-1-84882-780-6]

Vol. 389: Bru, R.; Romero-Vivó, S. (Eds.):
Positive Systems
398 p. 2009 [978-3-642-02893-9]

Vol. 388: Jacques Loiseau, J.; Michiels, W.;
Niculescu, S-I.; Sipahi, R. (Eds.):
Topics in Time Delay Systems
418 p. 2009 [978-3-642-02896-0]

Vol. 387: Xia, Y.;
Fu, M.; Shi, P.:
Analysis and Synthesis of
Dynamical Systems with Time-Delays
283 p. 2009 [978-3-642-02695-9]

Vol. 386: Huang, D.;
Nguang, S.K.:
Robust Control for Uncertain Networked
Control Systems with Random Delays
159 p. 2009 [978-1-84882-677-9]

Vol. 385: Jungers, R.:
The Joint Spectral Radius
144 p. 2009 [978-3-540-95979-3]

Vol. 384: Magni, L.; Raimondo, D.M.;
Allgöwer, F. (Eds.):
Nonlinear Model Predictive Control
572 p. 2009 [978-3-642-01093-4]

Vol. 383: Sobhani-Tehrani E.:
Khorasani K.;
Fault Diagnosis of Nonlinear Systems
Using a Hybrid Approach
360 p. 2009 [978-0-387-92906-4]

Vol. 382: Bartoszewicz A.;
Nowacka-Leverton A.:
Time-Varying Sliding Modes for Second
and Third Order Systems
192 p. 2009 [978-3-540-92216-2]

Vol. 381: Hirsch M.J.; Commander C.W.;
Pardalos P.M.; Murphey R. (Eds.):
Optimization and Cooperative
Control Strategies: Proceedings of the 8th
International Conference on Cooperative
Control and Optimization
459 p. 2009 [978-3-540-88062-2]

Vol. 380: Basin M.:
New Trends in Optimal Filtering and Control for Polynomial and Time-Delay Systems
206 p. 2008 [978-3-540-70802-5]

Vol. 379: Mellodge P.; Kachroo P.:
Model Abstraction in Dynamical Systems: Application to Mobile Robot Control
116 p. 2008 [978-3-540-70792-9]

Vol. 378: Femat R.; Solis-Perales G.:
Robust Synchronization of Chaotic Systems Via Feedback
199 p. 2008 [978-3-540-69306-2]

Vol. 377: Patan K.:
Artificial Neural Networks for the Modelling and Fault Diagnosis of Technical Processes
206 p. 2008 [978-3-540-79871-2]

Vol. 376: Hasegawa Y.:
Approximate and Noisy Realization of Discrete-Time Dynamical Systems
245 p. 2008 [978-3-540-79433-2]

Vol. 375: Bartolini G.; Fridman L.; Pisano A.; Usai E. (Eds.):
Modern Sliding Mode Control Theory
465 p. 2008 [978-3-540-79015-0]

Vol. 374: Huang B.; Kadali R.:
Dynamic Modeling, Predictive Control and Performance Monitoring
240 p. 2008 [978-1-84800-232-6]

Vol. 373: Wang Q.-G.; Ye Z.; Cai W.-J.; Hang C.-C.:
PID Control for Multivariable Processes
264 p. 2008 [978-3-540-78481-4]

Vol. 372: Zhou J.; Wen C.:
Adaptive Backstepping Control of Uncertain Systems
241 p. 2008 [978-3-540-77806-6]

Vol. 371: Blondel V.D.; Boyd S.P.; Kimura H. (Eds.):
Recent Advances in Learning and Control
279 p. 2008 [978-1-84800-154-1]

Vol. 370: Lee S.; Suh I.H.; Kim M.S. (Eds.):
Recent Progress in Robotics: Viable Robotic Service to Human
410 p. 2008 [978-3-540-76728-2]

Vol. 369: Hirsch M.J.; Pardalos P.M.; Murphey R.; Grundel D.:
Advances in Cooperative Control and Optimization
423 p. 2007 [978-3-540-74354-5]

Vol. 368: Chee F.; Fernando T.
Closed-Loop Control of Blood Glucose
157 p. 2007 [978-3-540-74030-8]

Vol. 367: Turner M.C.; Bates D.G. (Eds.):
Mathematical Methods for Robust and Nonlinear Control
444 p. 2007 [978-1-84800-024-7]

Vol. 366: Bullo F.; Fujimoto K. (Eds.):
Lagrangian and Hamiltonian Methods for Nonlinear Control 2006
398 p. 2007 [978-3-540-73889-3]

Vol. 365: Bates D.; Hagström M. (Eds.):
Nonlinear Analysis and Synthesis Techniques for Aircraft Control
360 p. 2007 [978-3-540-73718-6]

Vol. 364: Chiuso A.; Ferrante A.; Pinzoni S. (Eds.):
Modeling, Estimation and Control
356 p. 2007 [978-3-540-73569-4]

Vol. 363: Besançon G. (Ed.):
Nonlinear Observers and Applications
224 p. 2007 [978-3-540-73502-1]

Vol. 362: Tarn T.-J.; Chen S.-B.; Zhou C. (Eds.):
Robotic Welding, Intelligence and Automation
562 p. 2007 [978-3-540-73373-7]

Vol. 361: Méndez-Acosta H.O.; Femat R.; González-Álvarez V. (Eds.):
Selected Topics in Dynamics and Control of Chemical and Biological Processes
320 p. 2007 [978-3-540-73187-0]

Vol. 360: Kozlowski K. (Ed.):
Robot Motion and Control 2007
452 p. 2007 [978-1-84628-973-6]

Vol. 359: Christophersen F.J.:
Optimal Control of Constrained Piecewise Affine Systems
190 p. 2007 [978-3-540-72700-2]

Vol. 358: Findeisen R.; Allgöwer F.; Biegler L.T. (Eds.): Assessment and Future Directions of Nonlinear Model Predictive Control
642 p. 2007 [978-3-540-72698-2]

Vol. 357: Queinnec I.; Tarbouriech S.; Garcia G.; Niculescu S.-I. (Eds.):
Biology and Control Theory: Current Challenges
589 p. 2007 [978-3-540-71987-8]

Kolibri 360 GmbH
Ferdinand-Jühlke-Straße 7
99095 Erfurt, Deutschland
produktsicherheit@kolibri360.de